Trust in Risk Management

Earthscan Risk and Society Series
Series editor: Ragnar E. Löfstedt

Facility Siting
Risk, Power and Identity in Land-Use Planning
Edited by Åsa Boholm and Ragnar E. Löfstedt

Hazards, Vulnerability and Environmental Justice
Susan L. Cutter

The Perception of Risk
Paul Slovic

Risk, Media and Stigma
Understanding Public Challenges to Modern Science and Technology
Edited by James Flynn, Paul Slovic and Howard Kunreuther

Risk, Uncertainty and Rational Action
Carlo C. Jaeger, Ortwin Renn, Eugene A. Rosa and Thomas Webler

The Social Contours of Risk Volume 1
The Social Contours of Risk Volume 2
Jeanne X. Kasperson and Roger E. Kasperson

Social Trust and the Management of Risk
Edited by George Cvetkovich and Ragnar E. Löfstedt

The Tolerability of Risk
A New Framework for Risk Management
Edited by Frédéric Bouder, David Slavin and Ragnar E. Löfstedt

Transboundary Risk Management
Edited by Joanne Linnerooth-Bayer, Ragnar E. Löfstedt
and Gunnar Sjöstedt

Trust in Risk Management
Uncertainty and Scepticism in the Public Mind
Edited by Michael Siegrist, Timothy C. Earle and Heinz Gutscher

Trust in Risk Management
Uncertainty and Scepticism in the Public Mind

Edited by
Michael Siegrist, Timothy C. Earle and Heinz Gutscher

publishing for a sustainable future

London • Washington, DC

First published by Earthscan in the UK and USA in 2007
Published in paperback 2010

Copyright © Michael Siegrist, Timothy C. Earle and Heinz Gutscher, 2007

All rights reserved

ISBN-13: 978-1-84971-106-7

Typeset by MapSet Ltd, Gateshead, UK
Printed and bound in the UK by TJ International
Cover design by Yvonne Booth

For a full list of publications please contact:

Earthscan
Dunstan House
14a St Cross Street
London, EC1N 8XA, UK
Tel: +44 (0)20 7841 1930
Fax: +44 (0)20 7242 1474
Email: earthinfo@earthscan.co.uk
Web: www.earthscan.co.uk

Earthscan publishes in association with the International Institute for Environment and Development

A catalogue record for this book is available from the British Library

Library of Congress Cataloging-in-Publication Data

Trust in risk management : uncertainty and scepticism in the public mind / edited by Michael Siegrist, Timothy C. Earle and Heinz Gutscher.
    p. cm.
 ISBN-13: 978-1-84407-424-2 (hardback)
 ISBN-10: 1-84407-424-2 (hardback)
 1. Risk management. 2. Trust. I. Siegrist, Michael. II. Earle, Timothy C. III. Gutscher, Heinz.
 HD61.T78 2007
 658.15'5—dc22

2006100480

The paper used for this book is FSC-certified and totally chlorine-free. FSC (the Forest Stewardship Council) is an international network to promote responsible management of the world's forests.

Contents

List of Figures and Tables   vii
List of Contributors   ix
Preface   xi
Acknowledgements   xv
List of Acronyms and Abbreviations   xvii

1. Trust, Risk Perception and the TCC Model of Cooperation   1
   Timothy C. Earle, Michael Siegrist and Heinz Gutscher

2. Social Identity and the Group Context of Trust: Managing Risk and Building Trust through Belonging   51
   Michael A. Hogg

3. Trust and Risk: Implications for Management   73
   Eric M. Uslaner

4. A Social Judgement Analysis of Trust: People as Intuitive Detection Theorists   95
   Mathew P. White and J. Richard Eiser

5. Scepticism, Reliance and Risk Managing Institutions: Towards a Conceptual Model of 'Critical Trust'   117
   Nick Pidgeon, Wouter Poortinga and John Walls

6. Societal Trust in Risk Analysis: Implications for the Interface of Risk Assessment and Risk Management   143
   Lynn Frewer and Brian Salter

7. Rebuilding Consumer Trust in the Context of a Food Crisis   159
   Lucia Savadori, Michele Graffeo, Nicolao Bonini, Luigi Lombardi, Katya Tentori and Rino Rumiati

8	Trust and Risk in Smallpox Vaccination *Ann Bostrom and Emily Atkinson*	173
9	The What, How and When of Social Reliance and Cooperative Risk Management *George Cvetkovich and Patricia L. Winter*	187
10	Getting Out of the Swamp: Towards Understanding Sources of Local Officials' Trust in Wetlands Management *Branden B. Johnson*	211
11	Antecedents of System Trust: Cues and Process Feedback *Peter de Vries, Cees Midden and Anneloes Meijnders*	241
12	Trust and Confidence in Crisis Communication: Three Case Studies *Michael Siegrist, Heinz Gutscher and Carmen Keller*	267

List of Figures and Tables

FIGURES

1.1	The trust, confidence and cooperation (TCC) model	8
1.2	Model of the relationship between risk perception and trust	27
4.1	The aggregate relationship between trust and perceived discrimination ability	107
4.2	The aggregate relationship between trust and perceived response bias	108
4.3	Trust as a function of source type and perceived risk of mobile phones	110
5.1	Critical trust	128
5.2	Trust in various information sources and agreement about involvement in decision-making about genetically modified (GM) food	135
7.1	Final path model (salmon) with standardized regression weights	166
7.2	Final path model (chicken) with standardized regression weights	167
8.1	Beliefs regarding the likelihood of side effects from smallpox vaccination in the Blendon et al (2003) study compared to this study	177
8.2	Search for information about smallpox, and beliefs regarding the incidence of smallpox in the US and in the world during the past five years	178
8.3	Hypothetical vaccination decisions before and immediately after mental models questions	181
9.1	Model of salient value similarities, trust and evaluations	194
11.1	Route planner interface	253
11.2	Effects of consensus and process feedback in before and after interaction measurements	254
11.3	Effects of consensus and process feedback on staked credits	255
11.4	Effects of consensus and process feedback in before and after interaction measurements	258

11.5 Effects of consensus and process feedback on staked credits 260
12.1 The core parts of the dual-mode model of trust and confidence 269

Tables

1.1	Empirical studies of the relation between risk perception and trust	36
3.1	Dimensions of trust	76
4.1	The four possible outcomes of a simple binary detection task	99
4.2	Perceived discrimination ability, response bias and trust in various sources	105
5.1	Average trust ratings of focus group participants	124
5.2	Items and factor loadings after varimax rotation	129
5.3	'To what extent do you trust the following organizations and people to tell the truth about GM food?'	131
5.4	Factor loadings after varimax rotation	132
5.5	'How much do you agree or disagree that the following should be involved in making decisions about genetically modified food?'	133
8.1	Beliefs about the effectiveness of smallpox vaccine	177
8.2	Hypothetical questions to those in favour of vaccination	182
8.3	Hypothetical questions to those against vaccination	183
9.1	Characteristics of implicit and explicit modes of information processing	189
9.2	Four identified patterns of trust	196
9.3	Regression analyses of evaluations (effectiveness and approval) of management practices for total samples, states and genders	196
10.1	Trust criteria ratings of importance	223
10.2	Principal axis factoring	227
10.3	Multiple regression analyses of trust judgements on criteria and knowledge variables (betas)	230

List of Contributors

Emily Atkinson Georgia Institute of Technology, US
Nicolao Bonini University of Trento, Italy
Ann Bostrom Georgia Institute of Technology, US
George Cvetkovich Western Washington University, US
Peter de Vries Universiteit van Twente, The Netherlands
Timothy C. Earle Western Washington University, US
J. Richard Eiser University of Sheffield, UK
Lynn Frewer Wageningen University, The Netherlands
Michele Graffeo University of Trento, Italy
Heinz Gutscher University of Zurich, Switzerland
Michael A. Hogg Claremont Graduate University, US
Branden B. Johnson New Jersey Department of Environmental Protection, US
Carmen Keller University of Zurich, Switzerland
Luigi Lombardo University of Trento, Italy
Anneloes Meijnders Eindhoven University of Technology, The Netherlands
Cees Midden Eindhoven University of Technology, The Netherlands
Nick Pidgeon Cardiff University, UK
Wouter Poortinga University of East Anglia, UK
Rino Rumiati University of Trento, Italy
Brian Salter University of East Anglia, UK
Lucia Savadori University of Trento, Italy
Michael Siegrist University of Zurich, Switzerland

Katya Tentori University of Trento, Italy
Eric M. Uslaner University of Maryland, US
John Walls University of East Anglia, UK
Mathew P. White University of Plymouth, UK
Patricia L. Winter US Forest Service, US

Preface

Most researchers agree that trust is an important factor in risk management. It affects judgements of risk and benefit, and, directly or indirectly, it affects technology acceptance and other forms of cooperation. There is little agreement among researchers, however, on how trust in risk management should be studied. Progress in this area is dependent upon trust researchers exploring their differences and identifying the common ground that they share. The results of such an exploration would contribute significantly to the development of the tools that risk managers need to communicate information about performance effectively.

In order to facilitate a dialogue among researchers studying trust within the domain of risk management, we organized the Zurich Conference on Trust and Risk Management. This conference stands in the tradition established by the Bellingham International Social Trust conference. Papers presented at prior meetings of the Bellingham International Social Trust group were published in book form in *Social Trust and the Management of Risk* by George Cvetkovich and Ragnar E. Löfstedt (Earthscan, 1999).

Leading researchers in the field of trust and risk management were invited to participate in the Zurich trust meeting. The conference was structured as an active workshop. Each participant was requested to prepare a paper on trust and its implications for risk management, and to distribute the paper to the other participants prior to the meeting. At the workshop, each paper was briefly introduced by a discussant. Open discussion by all participants followed. Upon completion of the workshop, participants revised their papers based on the comments and suggestions they received. This book, which consists of the revised versions of the workshop papers, is one major result of the Zurich meeting.

Chapter 1 by Earle, Siegrist and Gutscher is an introduction to the great variety of trust studies, including those within the field of risk management. The authors describe their dual-mode model of cooperation based on trust and confidence (the TCC model). This model integrates most of the existing literature of trust and trust-related concepts. Earle and colleagues argue that the distinction between trust (based on morality information) and confidence (based on performance information) is crucial to a better understanding of the antecedents of cooperation. With regard to risk perception, the TCC model shows how trust dominates knowledge of performance. This model has clear consequences for risk communication efforts: without a solid foundation of trust, communicating information indicating good past performance may be of little value.

Chapters 2 and 3 discuss the foundations of trust. Both Hogg and Uslaner demonstrate that researchers and practitioners in risk management can benefit from trust research developed in other disciplines. Social identity and the group-related dimension of trust are the key issues discussed in Chapter 2. The author sketches out a social identity analysis of the social psychology of group and inter-group trust and risk management. This chapter provides an important basis for a better understanding of trust. In Chapter 3, Uslaner presents data that suggest that a more trusting environment produces less conflict in the firm and between firms. Trust also promotes diversity and better relations among different groups. Trust makes it easier to work in a globalized economy, and countries with greater levels of trust have higher rates of economic growth. Based on a wide variety of trust-related literature and survey evidence, the author elaborates two fundamental types of trust: moralistic trust and strategic trust. The moral dimension is based upon values, while the strategic one depends upon experience. Moralistic trust rests on an optimistic view of the world and one's ability to control it; strategic trust reflects expectations about how people will behave. While moralistic trust is quite stable over time, strategic trust is fragile. The author analyses the influence of these two dimensions on, among other issues, business life, corruption, neighbourhood safety and the legal system.

In Chapter 4, White and Eiser introduce a new approach to understanding trust in the context of risk management by extending a theory of decision-making under conditions of uncertainty (the Signal Detection Theory). It is claimed that members of the public act as intuitive detection theorists when deciding whether to trust a specific source of risk-related information. Support for the approach is provided by the findings of a survey investigating issues of risk perception and trust in relation to the potential effects of mobile phone technology. Chapter 4 provides interesting aspects of the dimensions of trust and proposes new insights into trust processes.

In Chapter 5, Pidgeon, Poortinga and Walls propose a model of critical trust. Critical trust can be conceptualized as a practical form of reliance on a person or institution combined with a degree of scepticism. This concept is based on the assumption that what is frequently called 'trust' or 'distrust' exists along a continuum, ranging from uncritical emotional acceptance to downright rejection. In order to illustrate the conceptual theme, the authors draw upon three separate empirical studies and discuss the implications of their findings for theories of trust, and for risk communication and management practice, more generally.

Frewer and Salter examine the historical context surrounding risk analysis and public trust in Chapter 6, as well as the importance of citizen/consumer trust. The authors discuss whether the functional separation of the different components of risk analysis (that is, risk assessment and risk management) results in distrust. They demonstrate that there is evidence that the continued separation is likely to increase public distrust in risk analysis in the future. Finally, it is argued that a more integrated risk analysis framework may be needed if citizen/consumer confidence in the risk analysis process is to be developed and maintained.

When a food accident occurs, the best way of re-establishing product demand to its original level is to restore consumer trust. How to rebuild consumer trust in the context of a food crisis is the subject of the contribution by Savadori and colleagues in Chapter 7. In a first step, the authors discuss how consumers respond to a food crisis. Next, they empirically examine the relative influence of trust and attitude on consumption intentions in the context of a hypothetical dioxin food scare. Results show that shared values were the best predictors of consumption in the event of a scare – even more important than having positive attitudes.

Universal vaccination is widely considered to be one of the top public health successes of the last century. Some observers, however, fear that the public is becoming increasingly averse to the risks of vaccine. Recent focus on smallpox as a potential weapon for biotcrrorism has increased the importance of understanding how people think about vaccination. Chapter 8 by Bostrom and Atkinson analyses the role of trust and risk perception in smallpox vaccination. In an empirical study, the authors examine trust in sources of information about smallpox vaccine or disease, behavioural intentions, and mental models of both smallpox disease and vaccination. The results support recent findings on the importance for trust of prior attitudes, as well as the importance of trust when knowledge is lacking, and the tendency of those who respect expertise to trust experts more.

In Chapter 9, Cvetkovich and Winter examined people's perceptions of the cooperative risk management of US national forests. The authors offer some substantiated suggestions on how to overcome a recognized lack of consensus on definitions of key concepts regarding social reliance and trust. After having defined the key terms, the authors discuss the nature of trust and its underlying social psychological processes. Finally, the circumstances determining the importance of trust to judgements about cooperative risk management are identified.

Risk analysts have increasingly focused on sources of trust in institutional risk management because trust seems critical to how people perceive hazards, and risk managers want their messages about risk magnitudes and risk management actions to be credible. The contribution by Johnson in Chapter 10 illustrates some conceptual and methodological issues on sources of trust with data from a survey of local officials on wetland management. These data provoke questions, worth more systematic investigation, about the role of 'familiarity' and various trust-target attributes in trust judgements, and about how these attributes might vary across types of judges and trust targets.

Trust plays an important role, not only in interactions with other persons, but also in relations with computer-based systems. Confidence is the subject of Chapter 11 by de Vries, Midden and Meijnders. When interacting with systems, such as decision aids, people may have similar experiences, as when interacting with human counterparts. Users, too, may lack information concerning a system's behaviour and the outcomes that it provides. As with interpersonal trust, the authors point out, meaningful interaction requires sufficient levels of trust to enable reductions of uncertainty regarding a particular system and its capabilities. Two experiments examining the effects of recommendations and process feedback are described, with results showing that system trust does not necessarily rely on objectified information, or solely on past behaviour, but may also be based on simple cues and, possibly, on inferred agency, intentions or values, analogous to trust in a social context.

Chapter 12 by Siegrist, Gutscher and Keller describes three case studies focused on trust and confidence in crisis communication. In a crisis, the authors argue, most people do not have the knowledge they need to make informed decisions. People need trust in order to reduce the complexity they are faced with. The disposal of an oil platform, a food crisis in Europe, and the successful handling of a listeriosis crisis in the US are described and analysed within the TCC framework introduced in Chapter 1. The TCC model can be used to explain why some crisis management strategies fail and others succeed.

Michael Siegrist, University of Zurich
Heinz Gutscher, University of Zurich
Timothy C. Earle, Western Washington University, US
October 2006

Acknowledgements

The Zurich Conference on Trust and Risk Management and this book would not have been possible without the financial support and intellectual input of many. The meeting was hosted by the Swiss Re Centre for Global Dialogue, Rüschlikon, Switzerland. We thank Fritz Gutbrodt and his team for making the meeting an unforgettable experience for all participants. The Swiss National Science Foundation (10CO11-101545) and the University of Zurich provided further financial resources for organizing the conference. Hans Kastenholz provided important editorial guidance for the entire book. The following people participated in the conference and provided valuable contributions towards the improvement of the chapters in this book: Ann Bostrom; Wernher Brucks; George Cvetkovich; J. Richard Eiser; Lynn Frewer; Hans Geiger; Michael A. Hogg; Branden B. Johnson; Roger E. Kasperson; Carmen Keller; Cees Midden; Robert O'Connor; Nick Pidgeon; Wouter Poortinga; Lucia Savadori; Eric Uslaner; and Mathew P. White.

List of Acronyms and Abbreviations

AMA	American Medical Association
ANES	American National Election Studies
ANOVA	analysis of variance
BPEO	best practicable environmental option
BSE	bovine spongiform encephalopathy
CDC	Centers for Disease Control and Prevention
CEO	chief executive officer
CFA	confirmatory factor analyses
CO_2	carbon dioxide
COO	chief operating officer
COS	carbonyl sulphide
DEFRA	UK Department for Environment, Food and Rural Affairs
DEP	Department of Environmental Protection
DOE	US Department of Energy
EC	environmental commission
EFSA	European Food Safety Authority
EL	elected
EMF	electromagnetic field
EPA	US Environmental Protection Agency
ESRC	Economic and Social Research Council
EU	European Union
FAO	UN Food and Agriculture Organization
FDA	US Food and Drug Administration
FSA	UK Food Standards Agency
GM	genetically modified
GT	general trust
HSE	UK Health and Safety Executive
HSM	heuristic systematic model
ICPSR	Inter-University Consortium for Political and Social Research
IDT	intuitive detection theorist
IEGMP	Independent Expert Group on Mobile Phones (UK)
IOM	Institute of Medicine

kg	kilogram
MORI	Market and Opinion Research International
MSI	mass sociogenic illness
MSI	Medical School Inspection
n	total population sample size
NJDEP	New Jersey Department of Environmental Protection
NRC	Nuclear Regulatory Commission
NZZ	*Neue Zürcher Zeitung*
PB	planning board
PCA	principal components analysis
PCB	polychlorinated biphenyl
SARF	Social Amplification of Risk Framework
SBMWA	Stony Brook–Millstone Watershed Association
SD	standard deviation
SDT	signal detection theorist/theory
STAR	Science to Achieve Results programme
SVS	salient values similarity
TCC	trust, confidence and cooperation
UEA	University of East Anglia
UK	United Kingdom
UN	United Nations
UNEP	United Nations Environment Programme
US	United States
USDA	US Department of Agriculture
VIS	vaccine information statement
WHO	World Health Organization
Y2K	year 2000
ZB	zoning board

1 Trust, Risk Perception and the TCC Model of Cooperation[1]

Timothy C. Earle, Michael Siegrist and Heinz Gutscher

Within the broad field of environmental and technological risk management, one of the first researchers to examine the nature of trust and the significance of the relation between trust and risk perception was the founder of risk perception research, Paul Slovic. Drawing on work by others (for example, Bella et al, 1988; Pijawka and Mushkatel, 1991/1992; Renn and Levine, 1991; Kasperson et al, 1992) and by himself and his collaborators (Slovic et al, 1991; Flynn et al, 1992), Slovic pointed out that high public concern about a risk issue (for example, nuclear power) is associated with distrust of the managers responsible for that issue; low public concern (as in the case, for example, of medical uses of radiation) is associated with trust in risk managers (Slovic, 1993). In general, trust in risk management is negatively related to risk perception. This is an important observation because it opens a possible pathway to affecting public risk perception and improving risk management: if we understood trust, and if we could affect levels of trust, then we might also be able to affect levels of risk perception and, ultimately, risk acceptance/rejection.

Developing some means of affecting public risk perception and risk acceptance – means that would be compatible with our participatory form of democracy – became important to risk managers when early risk perception research showed that public thinking about risks differed from, and was often unaffected by, assessments of risk by technical experts (Slovic, 2000). The field of risk communication research was developed during the 1980s to devise ways of bridging the public–expert risk judgement gap. In the 1990s, Slovic argued that risk communication had not yet lived up to its promise (Slovic, 1993). The primary reason cited by Slovic for this failure was lack of attention to the key role of trust in risk communication. Given a context of trust, he observed, risk communication seemed easy. But, lacking trust, risk communication seemed impossible. Slovic concluded that 'trust is more fundamental to conflict resolution than is risk communication' (Slovic, 1993, p677). Today, more than a decade later, the wisdom of Slovic's judgement is evident, particularly in the case of issues that are of high personal or moral importance to the individuals or groups involved.

In addition to arguing for the significance of the relation between trust and risk perception, Slovic also examined the nature of trust. If trust is fundamental to improved risk communication and risk management, then a useful understanding of it must be developed. Slovic centred his discussion of trust on what was then – and which stubbornly remains – the widespread belief that trust is fragile: 'It is typically created rather slowly, but it can be destroyed in an instant – by a single mishap or mistake' (Slovic, 1993, p677). Later in this chapter, we will present our argument against this belief in 'trust asymmetry'; but here we point only to one apparent flaw in Slovic's case for it. Slovic attempts to distinguish two types of information:

1   negative (trust-destroying) events; and
2   positive (trust-building) events.

In doing so, Slovic treats these events as though their meanings were self-evident to all persons: what is negative or positive for one is negative or positive for all. Later in his article, however, Slovic points out that 'initial trust or distrust colours our interpretation of events, thus reinforcing our prior beliefs'. If this is so – and a good deal of evidence suggests it is (see, for example, Plous, 1991) – then negative/positive may be more a matter of subjective judgement than of objective fact, particularly when morally important issues are at stake. It is the distinction between information that has moral implications versus information that has only performance implications, we will argue throughout this chapter, which is central to understanding trust.

Although Slovic may turn out not to have been fully correct on the nature of trust, he certainly was right in insisting on its significance to risk management theory and practice. As indicated by articles published in *Risk Analysis*, 'the dominant journal for scholarly literature on risk communication' (Gurabardhi et al, 2004) and the journal in which Slovic published his article on trust and perceived risk (Slovic, 1993), researchers have worked steadily over the past decade to increase our understanding of trust, publishing from five to eight articles per year on the topic. Fourteen articles dealing with trust were published in *Risk Analysis* in 2003, seven articles in 2004. Given this record of sustained interest and enquiry, the task of this chapter is to attempt to assess what has been learned about trust and the relation between trust and risk perception.

Two inescapable facts of trust research prevent us from approaching this task in the traditional manner of a straightforward literature review within the field of risk management (by which we mean studies dealing with the management of environmental and technological risks or hazards). First, as in most other fields in which it now plays an increasingly important role, there is, as yet, little agreement within risk management on the nature of trust or on how it should be studied. This lack of order among studies of trust in risk management requires us to create a structured environment within which we can understand, connect and differentiate between the confusing variety facing us. Second, just as lack of agreement characterizes studies of trust within fields of enquiry, there is great diversity in concepts and methods – at least judging from surface appearances – across fields. As a consequence, we cannot adopt an approach that places trust in risk management within a well-ordered context of trust studies from other fields and use the structure of the latter to interpret the meaning of the variety in the former.

In sum, we need not only to create a means of understanding trust in risk management, we need to provide a means of comprehending the connections among studies of trust in diverse fields, and of understanding what is basic and common across fields. We need to provide a general conceptual model that accommodates and gives meaning to the great variety of trust studies, including those within the field of risk management. Our approach, then, is to proceed from the most general to the most specific: from the nature of trust, to trust studies in a variety of fields, to trust in risk management, and, finally, to the relation between trust and risk perception.

CONCEPTUAL BACKGROUND

For many years, trust research was tightly circumscribed, studied primarily within the framework of the 'prisoner's dilemma' (Deutsch, 1958, 1973) or as a trait variable (Rotter, 1971, 1980). Recently, however, in various fields of the social sciences, such as psychology (Kramer, 1999), management (Rousseau et al, 1998), marketing (Geyskens et al, 1998, 1999), political science (Ostrom, 1998; Sullivan and Transue, 1999; Levi and Stoker, 2000) and risk management (Slovic, 1993; Cvetkovich and Löfstedt, 1999; Siegrist, 2000), empirical research devoted to trust has begun to flourish. Results of this research have demonstrated the importance of trust for cooperation. However, this research has also spawned a confusing variety of measures, constructs and theories that has frustrated progress towards a more useful understanding of trust. In this chapter, we attempt to identify the common ground underlying this diverse body of research and to provide a unifying framework that can serve as a base for future progress. Our overall goal is to evaluate the hypothesis that there are two principal pathways to cooperation: trust and confidence. We will argue throughout that this distinction is crucial to a useful understanding of the antecedents of cooperation. In support of this claim we will show that confidence depends upon and presupposes trust, and that attempts to achieve cooperation that do not take this dependency into account (implicitly or explicitly) are likely to fail.

First, we define trust and confidence and explicate the key differences between these two closely related concepts. We then propose a dual-mode model of cooperation based on trust and confidence (the trust, confidence and cooperation, or TCC, model) as a means of understanding and integrating much of the existing literature on trust and trust-related concepts. This is followed by a series of brief reviews of the empirical research on trust and confidence in a number of contexts and disciplines, concluding with risk management. The results of these reviews are used to evaluate our proposed model. Finally, we identify connections between our TCC model and other dual-mode models, and we discuss the implications of our model for future studies of trust and confidence in risk management. Throughout this chapter, it will be clear that the authors' shared disciplinary background is social psychology. We have attempted, however, to mitigate the narrowness of our perspective, to some degree, by drawing upon as wide and diverse a sampling of trust research as the confines of this chapter would permit.

Trust and confidence

We define trust as the willingness, in expectation of beneficial outcomes, to make oneself vulnerable to another based on a judgement of similarity of intentions or values. This definition is close to that of Rousseau et al (1998); but here we want to emphasize that trust is based on social relations, group membership and shared values. Confidence is defined as the belief, based on experience or evidence, that certain future events will occur as expected. Both trust and confidence support cooperation. But whereas confidence has a specific performance criterion, trust is placed in the freedom of the other. In the case of trust, the other is free to act in ways that indicate shared values, regardless of whether specific acts are expected or not. In the case of confidence, the other must act in specifically expected ways.

The distinction between trust and confidence derives from the work of Luhmann (1979, 1988); recently, Seligman (1997, 1998, 2001) has extended and clarified a number of Luhmann's ideas (see also Earle and Cvetkovich, 1995). These and other theorists also distinguish two basic types of trust: one within groups, and one across groups (Giddens, 1990; Becker, 1996). Within-group trust includes what we call social trust (trust at a distance, based on limited information) and interpersonal trust (trust close at hand, based on repeated interaction). Across-group trust is called general trust (GT) or trust among strangers. Any judgement of confidence presupposes a relation of trust. That is, the descriptions on which confidence is based are justified and accepted only within the group or community of trust that generated them. Also, any judgement of trust presupposes judgements of confidence; communities must have predictions to judge. In short, predictions cannot be separated from communities; each is required by the other.

The crucial point that is generally overlooked is the dependence of confidence on trust (for another argument along this line, see O'Neill, 2004). We all describe within communities; we can't do otherwise. However, since our descriptions are linked to our communities, and accepted as justified only within them, we normally are not made aware of the dependence of the one upon the other – and the potential rejection of our descriptions within other communities. To take a very simple example, one might claim that one's confidence that the Earth will circle the sun is not based on a relation of trust. But one only has the idea of 'the Earth circling the sun' as a consequence of one's membership in a particular community. There is nothing given about that or any other description. This, of course, can become a serious matter when the descriptions one makes provoke more variable and contested effects on others than a description of the Earth's relation to the sun.

Summary of conceptual background

Several additional important aspects of trust and confidence – suggested areas for future enquiries rather than established features – follow from our discussion:

- Confidence, being less demanding than trust, is the normal mode of operation. It has, therefore, the characteristics of a background variable, and it is experienced as a positive mood. Trust has the characteristic of a figure, and it is experienced as an affect or emotion. In the flow of experience, confidence is a much more common phenomenon than trust. Loss of confidence makes trust necessary for establishing a new record of experience or past performance, allowing the replacement of trust by confidence.
- Trust requires a choice among alternatives. The options include trusting other entities and not trusting at all. An individual chooses to trust another entity after considering other courses of action. Confidence does not require a choice among alternatives. Instead, confidence involves exposure to an external system. An individual expresses confidence in an entity by relying on it to properly control his/her external exposure.
- Trust entails risk. The risk of trust is in the freedom of the other. Individuals who trust take the risk that the other will act contrary to their interests. Confidence is occasioned by danger (following Luhmann, 1993, the distinction between risk and danger centres on the former being the product of agency and the latter being the product of objectivity). When individuals express confidence in an entity by relying on it, they expose themselves to the dangers that they expect the relied-on entity to control.

- Trust, in its general form, can be forward looking and creative. Since it is based on similarity (and with sufficient imagination and effort, similarity can be created as needed), trust can be a force for change, a powerful problem-solving tool. Confidence can only look back. Since it is based on familiarity, and we can only be familiar with what has passed, confidence is a force for stasis, for the perpetuation of things as they are.
- Because it is based on the freedom and unpredictability of the other, trust, in its general form, is a very demanding type of sociality to maintain over time. General trust cannot supply the foundation for daily life. Instead, it serves a transitory function as a bridge between two stable states, each of which is characterized by confidence. Because it is based on familiarity, confidence is an undemanding form of sociality to maintain over time. As long as the social system is stable, shared confidence in the system can supply the necessary foundation for daily life. When the system becomes unstable, however, confidence is lost, and trust is required to provide a transition to – to give birth to – a new, stable state.

With this conceptual background in place, we can now describe our dual-mode model of cooperation based on trust and confidence.

THE TRUST, CONFIDENCE AND COOPERATION (TCC) MODEL

The TCC model of cooperation is designed to serve several useful purposes. The first is unification. It provides a framework within which all expressions of trust and confidence can be interpreted and related to one another. The second is specification. To a greater extent than available alternatives, it identifies the basic psychological processes involved in judgements of trust and confidence. The third is clarification. At the centre of the TCC model is an explicit account of the interaction between trust and confidence, a major source of confusion in other approaches. The final purpose is generation of new insights. By unifying and bringing more specificity and clarity to the understanding of trust and confidence, the TCC model points to potentially fruitful connections with other areas of social psychological and applied research, and suggests novel research hypotheses.

The TCC model of cooperation postulates that trust is based on social relations, on shared values. Shared values can be measured in many different ways. In empirical studies, as we will show in our reviews, trust can be indicated variously by measures of in-group membership, morality, benevolence, integrity, inferred traits and intentions, fairness and caring. All of these, we will argue, can be taken to mean good intentions relative to those of the trusting person – shared values. In addition, judgements of shared values can be affected by general trust or a disposition to trust strangers.

In our conceptual framework, the basis for confidence is past performance or institutions designed to constrain future performance. As in the case of shared values, past performance and performance-controlling institutions can be measured in many different ways. Our reviews of empirical studies will show that confidence can be indicated by measures of familiarity; evidence; regulations; rules/procedures; contracts; record-keeping/accounting; social roles; ability; experience; control; competence; and standards. In parallel to general trust, judgements of past performance can be affected by general confidence, a disposition to be confident about the future. Both trust and confidence, in a range of combinations, can lead to various forms of cooperation.

The TCC model is shown in Figure 1.1, with social trust on the upper path and confidence on the lower. At the far left of the model, the information perceived by a person is divided into two types: that which is judged to be relevant to 'morality' and that which is judged relevant to 'performance' (note that here, and throughout the model, the elements represent subjective judgements). This division of information, although central in studies of impression formation (Skowronski and Carlston, 1989; Peeters and Czapinski, 1990, who call morality information 'other profitable' and performance information 'self profitable'), has been overlooked in most studies of trust and confidence. In our conceptual framework, morality information is equivalent to actions reflecting the values of an agent, and performance information is simply the behaviour of an object. The agent's values define a relationship of trust, and it is within that relationship that the performance information and the confidence to which it leads are judged.

In empirical social psychology, the importance of the morality/performance distinction is demonstrated, first, by studies that show that individuals tend to organize impressions of others along two dimensions – social desirability (morality) and intellectual desirability (performance) (Rosenberg et al, 1968) – and, second, by studies that show that morality information tends to dominate performance information (Wojciszke, 1994; Wojciszke et al, 1998; Van Lange and Sedikides, 1998; De Bruin and Van Lange, 1999a, 1999b, 2000). By 'dominate' we mean that, to an observer, morality information is more important and that it conditions the interpretation of performance information. For example, given positive morality information, negative performance is judged much less harshly than it would be if the morality information were negative. A direct demonstration of the biasing effects of shared values and trust on performance evaluation (though trust itself was not measured) is given in Lipshitz et al's (2001) study of the 'one-of-us effect'.

The elements of the TCC model are aligned in parallel pairs for trust and confidence. Note that the model identifies constituent (in-coming)

8 *Trust in Risk Management*

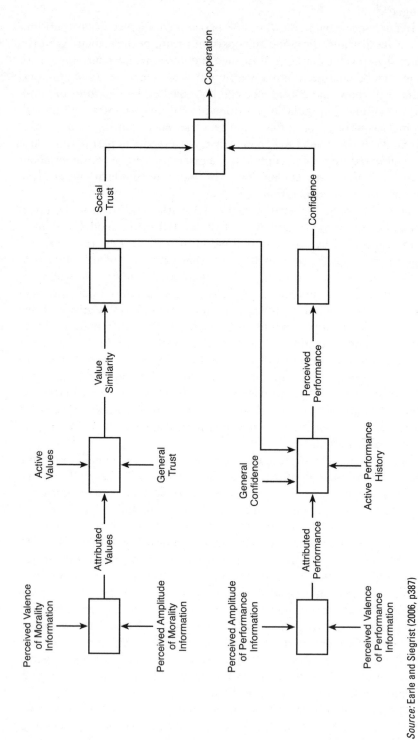

Figure 1.1 The trust, confidence and cooperation (TCC) model

Source: Earle and Siegrist (2006, p387)

and product (exiting) elements, but does not specify how the former are combined to produce the latter:

1 *Perceived amplitude of (morality/performance) information*: the judged degree to which the given information has (morality/performance) implications.
2 *Perceived valence of (morality/performance) information*: the judged degree of positivity/negativity of the given information (1 and 2 combine to form 3).
3 *Attributed (values/performance)*: the (values/performance) attributed by the observer to the other.
4 *Active values/active performance history*: in the case of values, these are the values that are currently active for the observer – which may be the product of existing social trust relations. In the case of performance, this is whatever history of relevant performance that is currently active for the observer.
5 *General trust/general confidence*: general trust is defined and discussed in previous sections. General confidence is the performance-based counterpart of the values-based general trust: the belief that things, in general, are under control, uncertainty is low and events will occur as expected (3, 4 and 5 combine to form 6).
6 *Value similarity/perceived performance*: value similarity is the judged similarity between the observer's currently active values and the values attributed to the other. Perceived performance is the observer's interpretation of the other's performance; note that this is a product not only of 3, 4 and 5, but also of social trust, element 7 (6 leads to 7).
7 *Social trust/confidence*: these elements are defined and discussed in previous sections (7 leads to 8).
8 *Cooperation*: any form of cooperative behaviour between a person and another person or group of persons, or between a person and an organization/institution.

The TCC model of cooperation unifies a wide variety of trust research by showing how the key forms of trust and confidence interact. In doing so, the model also suggests new research hypotheses. The key features of the model are as follows:

- It shows how social trust is based on morality-relevant information, while confidence is based on performance-relevant information.
- It shows how, in times of low social uncertainty, when morality information is not relevant, social trust does not play the dominant role in cooperation.
- It shows how social trust becomes more important in times of uncertainty, when morality information is relevant.

- It shows how social trust affects judgements of confidence both directly and via effects on perceived performance.
- As a complement to general trust, it identifies the previously overlooked concept of general confidence.

Two recent studies using structural equation techniques have confirmed the primary paths described in the TCC model of cooperation (Siegrist et al, 2003; Earle and Siegrist, 2006). Both of these studies were conducted within applied risk management contexts, the former dealing with electromagnetic field (EMF) risks, the latter with petroleum exploration and development in the Arctic and traffic congestion in Seattle. In contrast to those narrowly focused empirical studies, our goal here is to evaluate the model across a broad range of studies conducted within a diverse set of disciplines.

Evaluating the TCC model of cooperation

A striking characteristic of trust research, both conceptual and empirical, is its diversity. In the conceptual realm, some theorists identify four forms of trust (Kramer, 1999), others three (Misztal, 1996), and still others two (Ring, 1996). A clear distinction between trust and confidence is a feature of some discussions (Hardin, 2000); but confusion between the two – or, perhaps, indifference – appears to be more common. Since our main focus is on the empirical evaluation of a specific model, our review of conceptual approaches is brief and designed to demonstrate that, behind the diverse labels, there is widespread, if at times unacknowledged, recognition of the distinction between trust and confidence.

If the conceptual approaches to trust are varied, the empirical studies are kaleidoscopic. Due to pressures towards uniformity, some domains of research, such as trust in government, have adopted more or less standard measures and methods. In areas that lack such pressures, such as trust in organizations, measures and methods tend to vary from study to study. Variety within domains, however, tends to lie mostly on the surface, being manifested, for example, in the number and wording of the items used to measure trust. It is among the domains of research, which are defined by specific applied concerns, that the greatest variance is found. We therefore have structured our review by research domain. This enables us both to focus on commonalities within domains and to seek connections among them. The domains are general trust, experimental studies of trust and our domain of central interest: trust in risk management. These three groupings do not exhaust the universe of empirical studies of trust, of course, but they represent the main centres of interest relevant to the TCC model and to risk management (major domains not covered here include interpersonal trust, trust within organizations and trust in government). The most promising approach to improving our understanding of trust in risk management – in addition to a careful review of that domain –

is by attending to and learning from studies of trust in other related domains.

CONCEPTUAL DISCUSSIONS OF TRUST AND CONFIDENCE

The non-empirical literature on trust and related concepts is vast and complex; we therefore limit our brief review to a sampling of influential accounts from the social sciences. And our focus is on commonalities rather than distinctions. Specifically, we want to demonstrate – beneath the confusing clutter of labels – the ubiquity of two core concepts: trust based on shared values and confidence based on performance.

Most treatments of trust in the social sciences have either identified two or three varieties of trust or have contrasted trust with one or two related concepts. Granovetter (1985), for example, discusses three accounts of economic action: undersocialized (similar to confidence), oversocialized (similar to general trust – that is, trust in 'most persons') and socialized (trust based on social relations). Chiles and McMackin (1996) use a framework similar to Granovetter's in their discussion of transaction cost economics. Misztal (1996) identifies three forms of trust: habitus (confidence in persons), passion (trust based on social relations) and policy (confidence in institutions). Zucker's (1986) typology is similar to Misztal's: process trust (confidence in persons), characteristic trust (trust based on social relations) and institutional trust (confidence in institutions). The trio of concepts discussed by Lewicki and Bunker (1995) differs slightly from those of Misztal and Zucker: calculus-based trust (confidence based on cost-benefit analysis), knowledge-based trust (confidence based on predictability) and identification-based trust (trust based on social relations).

Sitkin and Roth (1993) focus their discussion on the contrast between trust (confidence based on performance) and distrust (trust based on shared values/distrust based on lack of shared values). This distinction allows these authors to argue that trust can persist in the face of negative outcomes. The differentiation that Das and Teng (1998) make between trust and confidence is similar to ours: trust is based on shared values, and confidence is based on performance. To these two concepts, Das and Teng add control, which is based on organizational culture and which is similar to the concept of institutional trust that others have discussed. According to Das and Teng, both trust and control contribute to confidence (in a more recent formulation, Das and Teng (2001) use the terms goodwill trust and competence trust to differentiate trust from confidence). For Jones and George (1998), trust is based on shared values and leads to cooperation; confidence is not clearly differentiated from trust. Focusing on relative capacity to withstand disappointment, Ring (1996) identifies two forms of trust: fragile (confidence) and resilient (trust).

Although his adoption of the term 'confidence' seems to be recent (Hardin, 2000, 2001a), the distinction between trust and confidence has long been central in Hardin's work (1991, 1993, 1995, 1999). Hardin applies trust to both persons and institutions – as most of us do in everyday language – and then argues that trust doesn't make sense in the case of institutions: 'In actual life we might not trust an organization, but might merely depend on its apparent predictability by induction from its past behaviour' (Hardin, 1998, p22). Institutions (organizations, groups, etc.) are different from persons (more complex, conflicted, etc.) and cannot be trusted because persons cannot know them sufficiently well. But a person can have confidence in an institution based on his/her knowledge of its past performance. Levi (1998) adopts an ordinary language approach to trust and confidence: trust implies risk; when the risk is very low, trust becomes confidence. Often, according to Levi, trust 'depends on confidence in institutions that back up the trustee' (Levi, 1998, p79). Following Luhmann (1979, 1988) and Seligman (1997, 1998, 2001), Tonkiss and Passey (1999) discuss the roles of trust (based on shared values) and confidence (based on institutional and contractual forms) in voluntary organizations (see also Smith, 2001). Finally, two recent reviews of trust research (Rousseau et al, 1998; Kramer, 1999) both identify two general factors distinguishing various approaches: forms of instrumental/calculative trust (confidence) and forms of social/relational trust (trust). Kramer (1999), for example, discusses four forms of trust: history based (interpersonal trust based on interaction, combining both trust and confidence); category based (social trust based on social relations); role based (confidence in persons); and rule based (confidence in institutions).

Summary

Our review of conceptual discussions of trust reveals a strong basis of common ground beneath a surface of disparate terminology. Trust (based on social relations and shared values) is different from confidence (based on experience or evidence). The object of trust is a person (or a person-like organization or institution – that is, an organization perceived to have person-like qualities, such as values). The object of confidence is an organization or institution (or a person who is perceived to have institutional qualities, such as a record of performance). Trust and confidence may interact, probably in ways that are context specific. Both trust and confidence can contribute to various forms of cooperative behaviour. In sum, there is broad agreement that trust should be distinguished from confidence primarily because the target of the former is (treated as) an agent, while the target of the latter is (treated as) an object.

Implications for the study of trust in risk management

As in many other applied contexts, the study of trust in risk management has suffered from a lack of conceptual clarity and consistency. Later in this chapter, our review of empirical studies of trust in risk management will show that individual research programmes have tended to rely either on simple single-item measures or on unique *ad hoc* definitions and measures. As an initial step towards conceptual coherence, and in accord with the general agreement found across a variety of disciplines, studies of trust in risk management should adopt an approach that distinguishes between agent-based trust and object-based confidence. A good example of the benefits of such an approach is that it would replace the mistaken notion of 'trust asymmetry' with the understanding that trust generally tends to be resilient, while confidence generally tends to be fragile.

EMPIRICAL STUDIES OF TRUST AND CONFIDENCE

General trust (GT)

The concept of GT (alternatively known as generalized interpersonal trust) was first developed as a personality variable (Rotter, 1980; for a review, see Jones et al, 1997); but its primary application has been in the study of groups. GT is measured with survey items that refer to 'most people' and which are included in such large-scale periodic surveys as the General Social Survey and the World Values Survey, as well as in smaller-scale independent studies. One of the General Social Survey items, for example, asks this question: 'Some people say that most people can be trusted. Others say you can't be too careful in your dealings with people. How do you feel about it?' Unlike other trust domains, confusion between trust and confidence has never been a problem in studies of general trust. In fact, the parallel concept of general confidence has been ignored.

Studies of GT are often based on large representative samples and are concerned with broad questions of democratic governance and social capital (for reviews, see Portes, 1998; Sullivan and Transue, 1999; Adler and Kwon, 2002). Other studies are designed to explore more modest hypotheses. In an example of the latter approach, Stolle (1998) directly compared two types of trust – GT and in-group trust – and found a negative relation between them. Stronger in-group trust was associated with weaker GT. In addition, Stolle (2001a, 2001b) concluded that GT was positively influenced by personal resources and experiences (that is, life satisfaction), but not by increased group memberships (that is, civic engagement). The relationship between civic engagement and GT, demonstrated by Brehm and Rahn (1997) and others, was due, Stolle argued, to self-selection.

Stolle's findings support the argument long made by Yamagishi and colleagues (Yamagishi, 1986, 1988, 2001; Yamagishi and Sato, 1986;

Yamagishi and Yamagishi, 1994; Hayashi and Yamagishi, 1998; Yamagishi et al, 1998) that GT is negatively related to strong ties in small groups (also see Earle and Cvetkovich, 1995, 1997; Uslaner, 2001). Yamagishi has proposed an 'emancipation theory of trust', based on game theoretic studies, in which 'general trust is considered to play the role of emancipating people from the confinement of existing relationships by providing booster power for launching them from the security of old relationships into the unknown world of opportunities' (Hayashi and Yamagishi, 1998, p287). People seek to reduce social uncertainty (Luhmann, 1979, 1988; Yamagishi and Yamagishi, 1994). One way of doing this is through assurance (that is, confidence) in existing relationships; this is associated with low levels of social uncertainty. Another route is through GT, which is associated with high levels of social uncertainty (Kollock, 1993, 1994; Oswald and Fuchs, 1998; Molm et al, 2000). Between the two poles of the uncertainty continuum lies trust, which, according to Yamagishi, 'is based on the inference of the interaction partner's personal traits and intentions' (Yamagishi et al, 1999, p156).

Summary

Reflecting its origins as a personality variable, GT is measured by items that refer to 'most people' rather than to specific persons. Most studies use items drawn either from those included in large periodic surveys, such as the General Social Survey, or from those developed by Yamagishi and colleagues (for example, Yamagishi, 1988). These studies provide strong evidence that the primary antecedent of GT is life satisfaction and the positive affect associated with it. There is also strong support for a negative relation between in-group trust and GT: strong trust within group boundaries is associated with weak trust across boundaries. A causal link between civic engagement and GT receives equivocal support, and the association may be due to self-selection. Regarding the consequences of GT, there is ample evidence that GT supports a wide variety of cooperative behaviour, particularly in relation to strangers (Uslaner, 2001).

Implications for trust in risk management

Since general trust is most strongly related to cooperation in situations characterized by lack of information, it may play an important role in managing risks associated with emerging technologies that do not have established groups of proponents and opponents. For well-known, established hazards – or for emerging technologies that have sparked moral debate – general trust is likely to be dominated by specific trust, particularly for individuals for whom those hazards are personally important. For individuals for whom those hazards are not personally important, however, general trust might contribute significantly to cooperation. At present, little is known about the role of general trust in risk management contexts.

Recently, Siegrist and colleagues (2005) made an initial attempt to assess the effects on risk perception of both general trust and general confidence (which, unlike general trust, had received little research attention of any kind previously). Results showed significant effects for both factors. To increase our understanding of when they may be important and how they might be used to facilitate cooperation, future studies of trust in risk management should include measures of both general trust and general confidence.

EXPERIMENTAL STUDIES

Studies of trust in applied settings, using surveys and questionnaires, and correlation-based techniques, attempt to examine trust as it exists in nature. Another approach is to explore trust experimentally, creating laboratory simulations based on what are believed to be the essential elements of trust. Most experimental studies of trust rely on the well-developed structure of social dilemma theory; exceptions to this rule employ less abstract and more context-specific simulations.

Social dilemma studies

The well-known social dilemma simulations (prisoner's dilemma, assurance, ultimatum, etc.) capture the essentials of situations in which behaviour that is beneficial to individuals is collectively damaging. A general conclusion of research in this area is that trust is a key to resolving most social dilemmas and to generating cooperation (Kollock, 1998a; Ostrom, 1998, 2003; Cook and Cooper, 2003). Specific results can be organized according to a scheme suggested by Messick (1999, 2000) Dawes and Messick (2000) and Weber et al (2004). Drawing on a theory by March (1994), Messick points to three key factors that affect choices in social dilemmas. The first is situation construal: 'What kind of situation is this?' The second is identity: 'What kind of person am I?' The third consists of rules: 'What does a person like me do in a situation like this?' Since the third factor is dependent upon the first two, trust and cooperation in social dilemmas critically depend upon signals for the construal of situation and identity (for a conceptual discussion of signals and trust, see Bacharach and Gambetta, 2001).

A fundamental aspect of situation construal is the perceived uncertainty of, or lack of control over, the behaviour of the other. One way of reducing uncertainty and increasing control is through a system of sanctions. Tenbrunsel and Messick (1999) showed that sanctions acted as a signal that the situation was business related and that cooperation should be based on an analytic calculation of costs and benefits. Lack of sanctions signalled an ethical situation that, for most participants, led directly to cooperation. Thus, sanctions were associated with building confidence, while lack of control was associated with trust. Similar situation-signalling effects have been demonstrated by Bohnet and colleagues (2001), Larrick and Blount (1997), and Pillutla and Chen (1999) (for general discussions of the relations between law and trust, see Kahan, 2001; Ribstein, 2001; Hall, 2002).

In a field application of the social dilemma framework, Morrison (1999) showed that identity construal (activation of social identity) affected perceptions of similarity, shared needs and values, and commitment to cooperate. This study demonstrated that, because people belong to multiple groups, identity construal – and cooperation – can vary across contexts. Kollock (1998b) also showed that trust and cooperation in social dilemmas is a within-group phenomenon (the link between social identity and trust is the subject of conceptual discussions by Kramer et al, 2001, and Messick and Kramer, 2001). Shared group membership is one way of signalling trustworthiness; other, more subtle, ways include facial resemblance (DeBruine, 2002) and emotional displays (Frank, 1988; Frank et al, 1993a; Eckel and Wilson, 2003). Frank argued that there are two pathways to perceived trustworthiness, one via reputation for moral behaviour, the other through cues to emotions. One reason that emotions would be effective cues to trustworthiness is that they would be difficult or costly to fake (for a discussion of costly signals and reassurance, see Kydd, 2000). In a comparison between economists and non-economists, Frank et al (1993b) also demonstrated that self-construal is a strong predictor of cooperation.

Non-dilemma studies

Addressing a central aspect of situation construal, Kollock (1994) showed that trust (measured by a single, bipolar untrustworthy/trustworthy item) was strongly associated with outcome uncertainty. Uncertainty and vulnerability signalled trust (see also Sniezek and Van Swol, 2001). A key distinction between two forms of exchange, reciprocal and negotiated, is that risk and uncertainty are greater in the former. Molm et al (2000) demonstrated that, compared with negotiated exchange, reciprocal exchange was associated with higher trust, greater commitment and more positive affect. Finally, in a series of field studies with experimental elements – not directly dealing with trust but focusing, instead, on volunteerism and social movement participation – Simon and colleagues (1998, 2000) identified two independent pathways to cooperation: one based on cost-benefit calculations, the other on activated social identity. Interpreted in our terms, one pathway, calculation based, was confidence related; the other, identity based, was trust related.

Summary

Experimental studies of trust have focused on identifying signals of trust, aspects of situations and individuals that are reliably related to trust. A number of studies have demonstrated that situations that are characterized by uncertainty and vulnerability are associated with trust. One factor that can introduce uncertainty into a situation is the simple presence of other people, and affective reactions to other persons have been shown to be linked to trust. There is abundant evidence, however, that activated group identity is strongly related to trust: individuals tend to trust others who are similar to them in ways that define their activated group identity. But trust is not the only pathway to cooperation: some situations signal confidence rather than trust. Studies have shown that sanctions, accountability, cost-benefit calculations and other forms of analytical control are linked to confidence.

Implications for trust in risk management

The signals identified by the experimental studies of trust – those that identify the trust pathway to cooperation and those that mark the confidence pathway – are simply specific manifestations of two general types of information: that which has value implications (signalling trust) and that which has performance implications (signalling confidence). This simple scheme can be used to improve our understanding of why people think and act the way they do in certain risk management contexts. Ultimately, with that improved understanding, the scheme could be used to design risk management contexts that support the two pathways to cooperation.

TRUST IN RISK MANAGEMENT

Leading authorities on the social and psychological aspects of risk management have long pointed to the central importance of trust (Renn and Levine, 1991; Kasperson et al, 1992). This interest in trust was originally driven by a desire to understand – and correct – risk communication failures. Following earlier psychological studies of communication and trust (for example, Hovland et al, 1953), risk communication researchers pointed out that success depended not only upon the content of the message, but also upon a relationship of trust between the sender and receiver. But what did trust do? According to these early accounts, information provided by a trusted person could be accepted as objective, competent and responsible. On the basis of such information, an individual could make confident judgements of the risks associated with a particular activity and decide whether to support or oppose it in some way. Thus, trust was said to affect both risk perception and cooperation.

During the past 15 years, a growing number of empirical studies have explored the nature of trust in risk management and the relations between trust, risk perception and cooperation. One general result of these studies is to call into question earlier conceptions of trust. Instead of signalling objectivity and competence, some researchers have argued – and demonstrated – that trust indicates a sharing of important values between the trusting person and the trustee (Earle and Cvetkovich, 1995, 1999). For example, in a study of the role played by trust in public opposition to the siting of a high-level nuclear waste repository, Pijawka and Mushkatel (1991/1992) identified three groups of trustees: federal institutions, federal agencies, and state and local institutions. These researchers found that trust in federal institutions and agencies was negatively related to several measures of risk perception, while trust in state and local institutions was positively related to them. These results suggest that trust was driven by policy agreement and that trust, in turn, affected judgements of risk. Respondents and trustees who favoured the repository (federal institutions and agencies) saw little risk; respondents and trustees in opposition (state and local institutions) saw high risk. A similar set of relations was demonstrated by Groothuis and Miller (1997) in their study of the siting of a hazardous waste disposal facility. Respondents who trusted the government and the news media (which favoured the project) had a lower perception of risks than respondents who trusted technical authorities (who opposed the project). More specifically, trust in the waste disposal business was related to lower risk perception, while trust in an environmental spokesperson was related to higher risk perception. These studies suggest that neither the factors on which trust is based, nor the relations between trust and risk perception are as simple and straightforward as some theorists have claimed.

Our aim in this section is to closely examine the relation between risk perception and trust. We will do this by reviewing relevant empirical studies and by interpreting the results of those studies within the framework of our TCC model. By relevant, we mean studies exploring public risk perception and trust within the context of the management of (mainly) technological hazards, such as nuclear power, gene technology and electromagnetic fields. The risks perceived by the public are seen to be by-products of the technology. The trustees trusted by the public are (mainly) those responsible for the management (either industrial or regulatory) of the technology. Thus, our approach does not include studies based on models, such as that proposed by Das and Teng (2004), in which risk and trust are associated with a single entity and are simply mirror images of each other.

Many studies of risk perception and trust are atheoretical, employing *ad hoc* definitions and measures. Others are the products of well-developed – or, at least, developing – theoretical positions. Prominent among the latter are studies by Slovic and colleagues (for example, Flynn et al, 1992) and Sjöberg (for example, Sjöberg, 1999). Interestingly, these two researchers have opposing views on the relation between risk perception and trust. Although often characterized by Sjöberg (2000b) and others as being based on a specific small set of factors (for example, new–old, dread and number of persons exposed), Slovic has explicitly stated that no particular factor is essential to his psychometric model: 'There is no universal set of characteristics for describing risk' (Slovic, 2000, p395). According to Slovic, perceived risk is multidimensional, with the causal factors varying across persons and contexts. And, in some circumstances, trust is an important determinant of perceived risk. Slovic's psychometric model can be characterized as a concrete approach to risk perception, designed to explain the judgements of particular persons in specific places.

In contrast to Slovic, Sjöberg has adopted an abstract approach to risk perception, with the aim of identifying the small set of universal factors that explains most of the variance in the judgements of most of the people most of the time. Sjöberg's prolific programme of research has identified a number of factors, such as 'tampering with nature' (Sjöberg, 2000a), that he believes strongly affect perceived risk, as well as a large number of factors that he believes have little effect. Prominent among the latter is trust (Sjöberg, 2001). Thus, Slovic's concrete, contextualized approach depicts trust as a strong determinant of perceived risk in some situations, whereas Sjöberg's abstract and universalist approach ascribes little importance to trust under any (or most) circumstances. Through our review of empirical studies, we hope to gain some insight into this dispute over the nature of the relation between risk perception and trust.

Most of the empirical studies of the relation between risk perception and trust have been conducted within the past decade by researchers from a wide variety of disciplinary backgrounds. In identifying studies for inclusion in our review, we have attempted to present a broad and fair representation of the field, while not aiming for total inclusivity. The latter goal is probably not practically achievable since there remains little agreement among researchers on such basic matters as the naming, defining and measuring of constructs – particularly in the case of trust. We identified studies through keyword database searches using not only 'risk perception' and 'trust', but also such variants as 'risk', 'perceived risk', 'risk judgement', 'confidence', 'credibility' and 'cooperation'. And, of course, we also took advantage of references by researchers to earlier studies. Again, due to lack of agreement among researchers, it was impossible to employ any single definition of trust as a criterion by which to judge whether a given study was a study of trust. If a study claimed to be a study of trust, we accepted that claim. As a result, our final set of 45 studies was characterized by a great deal of surface variability, with a wide variety of labels, definitions (or lack thereof) and measures. After identifying the studies, then, our next task was to develop a simple scheme for classifying and making some common sense of them.

One way to approach studies of risk perception and trust is by examining the tasks that researchers presented to their respondents. This method has the advantage of seeing studies from the neutral, cross-study point of view of respondents: what were respondents told? What were respondents asked to do? This sort of question, focusing on data collection methods, can be answered in straightforward theory-neutral ways. For example, what kind of judgements were respondents asked to make? A preliminary examination of our set of studies showed that respondents were asked to make two general types of judgement, which we call *attribute* and *rating*. In an attribute judgement, a respondent is asked to judge a target on one or several attributes said to be indicators of trust (or trustworthiness). Katsuya (2002), for example, asked respondents to indicate, on a seven-point scale, their agreement with the statement: 'The government openly provides the public with information about nuclear power.' The target of this item is the government, which is responsible for nuclear power risk management, and the attribute being judged is the government's openness with nuclear power information. Among the studies that employed attribute judgements, there was great variation in targets and attributes, with both varying over study contexts. In contrast to attribute judgements, ratings vary little, if at all, across studies that used them. Although the exact wording used in the questionnaire, including the rating scale, is not given (this, unfortunately, is not unusual), respondents in a study by Bassett et al (1996, p312) were asked to provide 'trust ratings for institutions associated with nuclear waste'. The institutions included the nuclear industry, national environmental groups, the US Department of Energy (DOE), the Nuclear Regulatory Commission (NRC) and the US Environmental Protection Agency (EPA). Studies that employed ratings, but not those that used attribute judgements, often included multiple targets.

In addition to *type of judgement*, our preliminary examination of the studies suggested three other simple classification criteria. The first is the *type of target*. A basic distinction that judges make with regard to targets, and the subject of recent psychological studies (Malle and Knobe 1997a, 1997b; Malle, 2001), is whether to treat the target as an agent or an object. A target treated as an agent is one to which behaviour-controlling beliefs, desires and intentions are attributed. The behaviour of a target treated as an object is described and predicted on bases other than beliefs, desires and intentions. Trust assumes agency, and one can trust any entity, person or non-person to whom one ascribes beliefs, desires and intentions. For the great majority of studies in our set, the target of judgement was an agent. For example, in asking their respondents to rate their level of trust for the developer of a hazardous waste facility and for the governmental institutions responsible for monitoring the safety of the facility, Ibitayo and Pijawka (1999) implied that the behaviour of both the developer and the institutions was controlled by their respective beliefs, desires and intentions. Because he wanted an 'affectively neutral' approach to the question of 'whether experience shows that the behaviours of specialized individuals and institutions can be counted on', Freudenburg (1993) provides a clear case of a target presented as an object. Freudenberg asked his respondents to judge their trust in 'current scientific and technical ability to build safe, efficient nuclear waste disposal sites', 'the ability of private enterprise to develop cost-effective, safe disposal sites in the United States', and the ability of 'national government agencies to safely administer a system of nuclear waste sites'. By focusing respondents' attention on the abilities of targets, Freudenburg implied that judgements of future behaviour should be based on knowledge of past behaviour.

Our third classification criterion is *tangibility of target*. A target can be presented as a recognizable concrete entity or as a generalized abstraction. Presented with a concrete entity, respondents can base their judgements on their knowledge of, and experience with, that entity. For example, Hine et al (1997) asked their respondents to make judgements on various attributes of the Atomic Energy Control Board of Canada. In this case, respondents may or may not have had much knowledge of the target; but they did know exactly what the target was. Presented with abstract targets, however, respondents must rely on their familiarity with stereotypes and other generalizations that may bear little resemblance to any specific things. Among the abstract targets included in a study by Priest (2001), for example, were 'media', 'industry', 'government', 'farmers', 'churches', 'scientists' and 'doctors'. Judgements of abstract targets present problems of interpretation that are absent from judgements of concrete targets.

The final classification criterion in our simple scheme, *referent of judgement*, distinguishes between two types of information about the target: morality information and performance information. *Referent of judgement* is linked to *target of judgement*, with morality information used to describe agents and performance information used to describe objects. The distinction between morality information (that is, information indicative of the target's important values) and performance information (information about particular behaviours that have no moral significance) has long been used in psychological studies of impression formation (see, for example, Rosenberg et al, 1968), but has been generally ignored in studies of trust in risk management. The oft-cited principle of 'trust asymmetry' (that trust is hard to gain and easy to lose; see Slovic, 1993), for example, distinguishes only between 'negative' and 'positive' information, with the former overpowering the latter. Morality and performance information are given equal weight. A number of studies have shown, however, that morality information tends to dominate, and control the interpretation of, performance information (Wojciszke, 1994; Van Lange and Sedikides, 1998; Wojciszke et al, 1998; De Bruin and Van Lange, 1999a, 1999b, 2000). Thus, contrary to the asymmetry principle, positive morality information may, in some cases, overpower negative performance information, and trust may turn out to be easy to gain and hard to lose. With regard to our set of studies, *referent of judgement* can be applied, of course, only to studies that employ attribute judgements and not to those that use ratings. Bord and O'Connor (1992), for example, used a morality information referent when they asked respondents about their 'perceptions of industry's concern for public health'. Williams and Hammitt (2001), on the other hand, employed a performance information referent when they asked respondents whether they believed that 'Pesticide residues on fruits and vegetables are safe if they meet government standards.'

Classifying the studies

Since we examined a very diverse set of studies of risk perception and trust, we focused on the data collection methods used by researchers: what respondents were told and what they were asked to do. This respondent-centred approach suggested four simple criteria by which the studies could be classified:

1. *target of judgement* (agent or object);
2. *tangibility of target* (concrete or abstract);
3. *type of judgement* (attribute or rating); and
4. *referent of judgement* (morality or performance).

Although the classification criteria were simple, some of the classification judgements were not. Two judges independently classified all of the studies on each of the criteria; disagreements were settled through discussion. Since none of the studies were designed with our criteria in mind, however, it is certainly possible, in spite of our efforts, that we have misinterpreted some elements of them. Given the complex variety of the studies, our overall aim is heuristic, not definitive: we are searching for signs of new directions, not attempting to describe all that went before. The results of our search are presented in Table 1.1 (pages 36–39). We examine the contents of Table 1.1 in two ways: first, to describe our set of 45 studies as depicted by our classification scheme and, second, to describe the relations between trust, risk perception and cooperation. Finally, we interpret these results within the framework of a unifying theory.

Classification criteria

The results for the individual criteria are presented first, followed by a description of the relations among the criteria.

Type of target

A large majority of the studies (32) presented respondents with judgement targets that were agents. An additional seven studies presented targets that included both agent-related and object-related elements (note: in Table 1.1, *combined* means that the elements were combined into a single measure, and *separate* indicates multiple measures). Only six studies presented objects as targets. These results indicate that most researchers associated trust with agency, with the (overt or implied) ascription of beliefs, desires and intentions to the judgement target. The minority of studies that presented objects as targets interpreted 'trust' differently, more in the sense of confidence in abilities or performance.

Tangibility of target
The number of studies with concrete targets (19) was slightly less those with abstract targets (22). Four studies included both concrete and abstract targets. These results indicate that there is strong support for both approaches, with neither laying claim to exclusive association with trust. Whether to study trust in concrete or abstract targets appears to be a matter, first of all, of project goals: of what the researcher wants to learn. But it also seems to be a matter of continuity across studies, with some researchers using only one form of tangibility or the other (Sjöberg, for example, is always abstract). Since respondents have very different sorts of information available to them in studies based on concrete targets, compared with studies based on abstract targets, comparability between the two types of studies may be problematical.

Type of judgement
Attribute judgements were used in 26 studies and ratings were used in 18 studies (one study, based on a wide variety of data, included both attribute judgements and ratings). Of the 18 studies that employed concrete targets, however, most (14) used attribute judgements. The 23 studies with abstract targets, in contrast, were evenly split between attribute judgements and ratings, with one study using both. The greater use of attribute judgements with concrete targets reflects the richer information environments in those studies, compared with studies of abstract targets.

Referent of judgement
This criterion applied only to the 26 studies that employed attribute judgements. Values (morality information) were referred to in 14 studies, performance in 5 studies, and both values and performance in 7 studies. As noted earlier, *referent of judgement* is linked to *target of judgement*. Thus, all of the studies with value referents had agent targets, and all of the studies with performance referents had object targets. All but one of the studies with both value and performance referents had agent and object targets.

Relations among the criteria
Some of the relations among the classification criteria have been noted above: concrete targets were associated with attribute judgements, agent targets with value referents, and object targets with performance referents. When all four criteria are applied to the 38 studies for which we have complete and unambiguous information (that is, no combined measures), three patterns emerge. First, there were 13 studies that asked respondents for attribute judgements based on values. For most of these studies (8), the targets were concrete agents. Second, there were 17 studies that asked

respondents for trust ratings. The targets for these studies were about evenly split between abstract and concrete agents. Third, there were six studies in which the targets were objects. Five of these studies, three with concrete targets and two with abstract targets, asked for attribute judgements based on performance; the sixth study asked for ratings of an abstract target. In summary, our review of a diverse set of studies identified two main styles of trust research in risk management contexts: one that asks for attribute judgements based on the values of (mainly) concrete agents, and one that asks for ratings of trust for either concrete or abstract agents. An additional, less prominent, style focuses on the abilities and performance of objects.

Dependent variables

Our set of studies included a wide variety of dependent variables; but we were able to classify all of them as being measures of either *risk perception* or *cooperation*. Although specific wording varies across studies, many measures of risk perception are straightforward, asking respondents, for example, to judge whether 'I personally am very much at risk from (food additives)' or 'The average person is at risk from (food additives)' (Eiser et al, 2002). Some measures of risk perception do not mention the word 'risk', but instead ask about particular negative effects. For example, Greenberg and Williams (1999) asked respondents to judge whether 'Abandoned houses, factories and businesses are a neighborhood problem.' What all measures of risk perception have in common is that they (implicitly or explicitly) ask respondents to indicate the degree to which they believe that certain negative future events will occur as expected. Thus, we can offer the following general definition: risk perception is the belief, based on experience or evidence, that certain negative future events will occur as expected.

Although not used as a variable name in any of the studies examined, we use the term *cooperation* for our second set of dependent variables because it is a general term that characterizes any form of positive relation between respondents and the projects of the targets judged. For example, Kunreuther and colleagues (1990) asked of their Nevada-dwelling respondents: 'If a vote were held today on building a permanent repository, where would you vote to locate the repository (Yucca Mountain, Nevada; Hanford, Washington; or Deaf Smith, Texas)?' In this case, a vote for Yucca Mountain was an indicator of cooperation. In contrast to the simplicity of Kunreuther et al, some researchers, such as Hoban and colleagues (1992), constructed multi-indicator measures of cooperation. Hoban et al used three items to measure opposition to the genetic engineering of plants (for example, 'Genetic engineering is being used to produce new kinds of crops that are not damaged by herbicides. Do you think this is (1) very desirable, (2) somewhat desirable or (3) undesirable?'), and another three items to measure opposition to genetic engineering of animals (for example, 'Genetic engineering is being used to produce a growth hormone that will increase the amount of milk dairy cows can produce. Do you think this is (1) very desirable, (2) somewhat desirable or (3) undesirable?')

Relations between classification criteria and dependent variables

We examined the studies in Table 1.1 (pages 36–39) to determine whether there were any reliable relations between styles of trust research and quantitative results for our dependent variables (due to space limitations, not all quantitative results for all studies are presented in Table 1.1, although they were included in our analyses). As noted above, our review identified two main styles of trust research in risk management contexts: one which asks for attribute judgements based on the values of (mainly) concrete agents (style A), and one which asks for ratings of trust for either abstract or concrete agents (style B). Quantitative analyses revealed large within-group variation and no reliable differences between the two styles of trust research. For style A, the beta weights for trust and perceived risk, for example, varied from $-.13$ to $-.82$; for style B, the correlations between trust and perceived risk varied from insignificant to $-.64$. For the other styles, the numbers of studies were simply too small to permit meaningful comparisons. These results indicate that, to the extent that we can define methodological agreement among researchers, the way in which the relations between trust, risk perception and cooperation are studied may not have significant effects on quantitative results. What is most remarkable about those results is their variability within methods.

Relations between trust, risk perception and cooperation

When we examined the studies in Table 1.1 without regard to research style, the most significant aspect of the quantitative results remained their variability. With regard to risk perception, some studies found it to be strongly related to trust, others found a moderate relation, and still others only a weak one. A similar pattern held for the relation between trust and cooperation. In brief, our review of empirical studies supports neither the strong relationship position presented in some of Slovic's writings, nor the weak relationship position promoted by Sjöberg. Instead, our results suggest that the relation between trust and risk perception (as well as cooperation) is contingent upon certain contextual factors.

What factors might affect the strength of the relations between trust, risk perception and cooperation? Several studies in our review suggest that knowledge of the hazard may be an important factor. Katsuya (2002), for example, found that persons knowledgeable about nuclear power based their judgements of cooperation on trust ($\beta = .27$), perceived risk ($\beta =$

−.38) and perceived benefit (β = .28), while less knowledgeable individuals based their judgements only on trust (β = .54) Thus, the relation between trust and cooperation was stronger in the absence of knowledge, and the relation between perceived risk and cooperation was stronger in the presence of knowledge. Similarly, Siegrist and Cvetkovich (2000) found that the relation between trust and perceived risk was stronger when knowledge was weaker. A number of studies also suggest that the degree of agreement/disagreement on hazard-related values between respondents and targets may also be an important factor in the relation between trust and risk perception. Both Groothuis and Miller (1997) and Pijawka and Mushkatel (1991/1992), for example, found that, for mainly anti-development respondents, the relation between trust and perceived risk was negative for pro-development targets and positive for anti-development targets, with the negative relations generally stronger than the positive.

Conclusion

Although a good deal of rhetoric in the debate over the relation between trust and risk perception has focused on the simple issue of strength (is trust a strong determinant of risk perception or not?), our review of empirical studies has revealed that the relation between trust and risk perception is complex and contextual. Whether the relation between trust and risk perception is strong or weak, positive or negative, depends upon contextual factors, two of which appear to be hazard knowledge and agreement on hazard-related values. In the following section, we describe a general model that integrates these contextual factors with judgements of trust, risk perception and cooperation.

Understanding the relation between risk perception and trust

Drawing on previous work on the relation between trust and confidence (described earlier in this chapter in the section on 'The trust, confidence and cooperation (TCC) model') and on the associated distinction between morality information and performance information (Earle and Siegrist, 2006), we can, in Figure 1.2, present a model of the core relation between trust and risk perception (for simplicity and clarity, we do not include in this figure elements in the full model that are not central to our current argument, such as general trust and general confidence). The model shows two pathways to cooperation: one through trust and one through risk perception. Trust is shown to depend upon the judged agreement between a respondent's hazard-related values and the values attributed by the respondent to the target being judged. Risk perception is shown to depend not only upon the respondent's knowledge of the hazard, but also, in two ways, upon the respondent's trust in the target. First, trust affects risk perception directly; second, trust affects risk perception indirectly

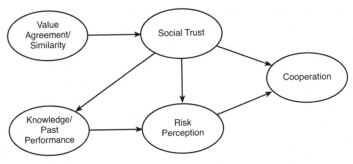

Source: Chapter authors

Figure 1.2 *Model of the relation between risk perception and trust*

through its effects on knowledge. In short, our model depicts risk perception as being thoroughly embedded in social values.

Whereas some models of risk perception (for example Sjöberg, 2000b) portray abstracted individuals encountering abstracted hazards, our model presents persons as members of groups (individuals who share important values with others); the hazards they encounter are interpreted and given meaning within the context of their group's values. Our approach is in accord with Slovic's recent contextualist analysis (Slovic, 2001), in which he compares risk perception to game playing: just as games vary according to specific rules, risk perception varies in accordance with social contexts. In our model, social contexts are represented by two distinct sorts of information, that which has morality or value implications and that which has performance implications. Contexts will vary in the degree to which information of either type is available. Given morality information, an individual can judge whether the values of the target (person or institution) agree with his/her own, thus providing a basis for trust. Given performance information: an individual can judge whether the behaviour of the target (hazard) has been safe, thus providing a basis for risk perception – a judgement of future behaviour. But, as our model shows, the meaning of performance information is not universally self-evident; it must be interpreted within a context of social relations, according to the rules of the game being played. Those rules are inherent in relations of trust. For example, if an individual trusts the manager of a hazard, then performance information about that hazard will be interpreted more positively than if he or she distrusted the manager. In this way, morality information is said to dominate performance information (De Bruin and Van Lange, 2000; Earle and Siegrist, 2006). But the domination of knowledge by trust is not limited to biased interpretation of past behaviour; trust relations also define the legitimacy or quality of performance information (Koehler, 1993). Thus, within a particular social context, certain information about past behaviour may be considered legitimate, while other information may be rejected as bogus.

Since humans appear to be programmed to seek and detect signs of similarity (Harris, 2002), concrete contexts can only rarely be said to be devoid of morality information. In a pinch, individuals can base their judgements on such signals as facial resemblance (DeBruine, 2002) or a smile (Scharlemann et al, 2001). Thus, people are almost always ready to interpret performance information, if it is available. In this meta-sense, risk perception is almost always (especially if we include self-trust) strongly related to trust. But most studies of the relation between trust and risk perception are conducted at less lofty levels, where knowledge-based risk perception provides an alternative route to cooperation. This route, however, is open only when performance information is available and can be transformed into knowledge. To become knowledge, performance information must be seen to be legitimate and must prove to be usable. These criteria are most strongly met by one's personal experience. A great deal of performance information, however, such as technical risk management data, may be granted legitimacy, but turn out to be unusable. When legitimate, usable performance information is unavailable, risk perception must be based on trust. When knowledge is available, it is interpreted within the social relations of trust, and risk perception becomes a joint product of both knowledge and trust. In relation to risk perception, then, trust is always in the background; in the absence of knowledge, it takes centre stage.

Summary

In this section, we used a review of 45 empirical studies to explore the relation between trust and risk perception. We approached each of the studies in our review from the point of view of the respondents: what information were they given? What were they asked to do? A preliminary examination of the studies suggested four criteria by which they could be classified:

1 *target of judgement* (agent or object);
2 *tangibility of target* (concrete or abstract);
3 *type of judgement* (attribute or rating); and
4 *referent of judgement* (morality or performance).

When we applied these criteria to our diverse set of studies, two main styles of trust and risk perception research emerged: one in which respondents are asked to make attribute judgements based on the values of (mainly) concrete agents, and one in which respondents are asked to rate their levels of trust for either abstract or concrete agents. In a third, less prominent, style, respondents are asked to judge the abilities and performance of objects. Although a wide variety of terms were used, all of the dependent variables in the set of studies were judged to be measures of either risk perception or cooperation. Quantitative analyses indicated that there were no reliable differences among styles of research (or any of the criteria singly applied) in the magnitude of the relations between trust and risk perception or trust and cooperation. The outstanding characteristic of our results was their variability overall and within methods of study.

Although the studies in our review were small in number and diverse, our results suggest that the relation between trust and risk perception is affected by factors unattended to in most studies to date. We presented a model that identifies two contextual factors – knowledge and value agreement – that may account for the relation between trust and risk perception, and its variability across contexts. According to our model, the impact of trust on risk perception is greatest when knowledge is lacking. But our model also indicates that knowledge itself is affected by trust, so that risk perception can never escape its web of social relations. The key contribution of our model is that it shows that the ultimate basis for both trust and risk perception is shared values.

Implications for trust in risk management

Our review and analysis does not point to any truly new directions in trust and risk perception research. Instead, we have described a pathway that has been under construction for a decade or two, and that has made considerable progress towards the goal of useful understanding. The direction of the pathway, as we see it, has been away from abstract universalist models and towards concrete contextualist models. Significantly, this has been the direction taken by Slovic (2000, 2001), the most prominent researcher in the field. Much work remains to be done, however, to further explicate the contextual factors that affect trust and risk perception, and to use that knowledge to develop effective risk management tools.

SUMMARY OF LITERATURE REVIEW

We have surveyed a wide variety of both conceptual and empirical studies of trust, drawn from fields of study ranging from personality and social psychology, to political science, to economics. These studies were selected to represent a broad spectrum of trust research, spanning diverse approaches to measurement and methodology and a wide range of application contexts. Our review was motivated by the belief that, varied as trust research certainly is, there is a common core to it. If our belief is well founded, researchers could learn a good deal from each other by crossing disciplinary lines; in particular, risk management researchers could learn from trust researchers in other related domains.

Our aim was to identify commonalities, and we can summarize the results of our review in eight points:

1 There are two separate, but related, pathways to cooperation: trust and confidence.

2   Trust is based on information with value or morality implications, leading to judgements of shared values or common group membership.
3   Confidence is based on information with performance (but not morality) implications, leading to judgements of past performance and/or expectations for future performance.
4   Interpersonal and social (that is, within-group) trust serves to preserve the status quo by conditioning the interpretation of new information. This is the process by which trust builds confidence, which, generally, is the preferred normal state of being.
5   General trust serves to modify the status quo by altering group boundaries – that is, by redefining the meaning of 'us' and 'them'.
6   Situations characterized by uncertainty and vulnerability signal trust. Persons evoking positive affect signal trustworthiness.
7   Confidence is signalled by sanctions, contracts and other attempts at control. Confidence is also associated with feelings of familiarity.
8   The relative importance of trust and confidence varies across contexts. Although confidence is generally preferred, it is sometimes, as when the quality of past performance is difficult to assess, not easy to achieve.

A final, general point should be added to these eight. Contrary to the simple abstractions suggested in many conceptual discussions of trust, the empirical studies that we reviewed demonstrate that trust, in practice, is highly context specific and is manifested in a multitude of ways. Consequently, glib advice on how to build or train trust – advice based on some simple set of principles – is unlikely to produce the desired results. Instead, detailed local knowledge – in concert with a set of empirical generalizations such as those that we have outlined here – is essential to understanding trust in any given context.

GENERAL DISCUSSION

We have presented a new dual-mode model of cooperation based on trust and confidence, the TCC model, and we have used that model in an attempt to improve our understanding of the relation between trust and risk perception. The TCC model addresses the question of how individuals predict cooperation in others. There are two general ways of doing this. One, *trust*, is defined as the willingness, in expectation of beneficial outcomes, to make oneself vulnerable to another based on a judgement of similarity of intentions or values; the other, *confidence*, is defined as the belief, based on experience or evidence, that certain future events will occur as expected. We used brief but representative reviews of the trust and confidence literature in several research domains to evaluate the hypothesis that the existing confusing array of concepts and measures could be accommodated within our simple scheme. The reviews supported our model. But our model, and the conceptual framework on which it is based, does more than simply distinguish trust from confidence. It clarifies and resolves a number of long-standing sources of confusion associated with these concepts. It places trust and confidence within larger contexts of enquiry, relating our model to models in other domains of research. And it provides a basis for a more integrated, productive and useful approach to trust and confidence research in the future.

Distinctive contributions of the TCC model

Our TCC model of cooperation based on trust and confidence makes three distinctive contributions. First, it provides a simple way of integrating a large body of empirical research drawn from a number of disciplines. Our conceptual framework is based on two distinctions. The categorical distinction between agent and object provides the basis for the distinction between trust (the target of which is an agent) and confidence (in which the target is an object). The distinction in degrees between freedom and restriction provides the basis for distinguishing among trust relations that vary in degrees of inclusivity, broadness of scope and porosity of boundaries. General trust is the freest form of trust; social trust within an exclusive, narrow, closed group is the most restricted. Confidence also varies in degrees of freedom, as expressed in degrees of objectivity. All forms of trust and confidence that have been studied in the empirical literature can be located within this scheme and related to one another. By providing a simple, summary scheme, our model makes available a basis for cross-disciplinary integration and exchange of knowledge.

Second, our conceptual framework and model of cooperation identify links between trust and confidence research and other domains of research within social psychology and related disciplines. These connections, a number of which are outlined below, could lead to mutually enriching exchanges across domains and disciplines.

The third and most distinctive contribution of our model is that it provides a general account of the relation between trust and confidence. The relation between trust and confidence is the relation between a group or community and its means of predicting and controlling behaviour. Predictions and controls are accepted and justified only within communities. In general, members of a community endorse predictions that seem to support their favoured way of life; they discount and oppose predictions that seem to undermine that way of life. According to this model, confidence, and a stable, happy life, can only be achieved through trust. This point is broadly important both theoretically and practically. On the practical side, consider the following representative situation. A person claims to be an environmentalist and to be opposed to plans to drill for oil in the Arctic National Wildlife Refuge. In discussions with proponents of drilling, the environmentalist is presented with extensive scientific studies, conducted by well-qualified experts, which claim to show that the wildlife in the refuge will not be harmed by drilling; in fact, the wildlife will benefit. In the face of all this science and logic, the environmentalist, not surprisingly, is unmoved. The scientist's arguments failed to increase the environmentalist's confidence. Why? Opposition to drilling, in this case, defines the environmentalist's community. Thus, the association between a pro-drilling conclusion and scientific arguments (which, all else equal, should lead to increased confidence) negates the logic of the latter. In this case, the scientific arguments would be confidence enhancing only within groups that are either pro-drilling or undefined by attitudes towards drilling. As our model shows, confidence presupposes trust, a community based on shared values. We will try to demonstrate the theoretical importance of this point by examining the relations between our model and other dual-mode models.

Relations to other dual-mode models

Our TCC model is a dual-mode model of cooperation based on trust and confidence. Unlike most other dual-mode models, it is not a model of two systems of reasoning (Sloman, 1996) or two modes of information processing (Smith and DeCoster, 2000; see also Chaiken and Trope, 1999) In their review and integration of a number of dual-mode models, Smith and DeCoster (2000) use the labels 'associative' and 'rule based' to distinguish two models of processing, which, in turn, are linked to different memory systems. Associative processing draws on the slow-learning memory system and is characterized by similarity- and contiguity-based structure, automaticity and lack of awareness. Rule-based processing (which also makes use of the slow-learning system) uses the fast-binding memory system and is distinguished by language- and logic-based structure, optional employment given need and capacity, and the presence of awareness. Smith and DeCoster's associative/rule-based distinction corresponds to our freedom/restriction distinction – that is, the associative/rule-based distinction is perpendicular to the trust/confidence distinction.

The implications of dual-mode models of information processing for our understanding of trust and confidence have not been fully exploited, and this should be a goal of future research. For example, Chaiken and colleagues have conducted a number of studies (Chen and Chaiken, 1999) that have demonstrated how the use of a heuristic (associative) cue can bias systematic (rule-based) processing. In our TCC model (see Figure 1.1), this biasing effect is indicated by the path from social trust to past performance. Also relevant are studies that have demonstrated a 'disconfirmation bias', according to which information that is contrary to one's existing attitude is actively resisted, while preference-consistent information is quickly accepted (see, for example, Ditto et al, 1998). These studies, carried out in a wide variety of contexts, all point to the key role of prior affect in biasing subsequent information processing.

Due to their narrow focus on information processing, most dual-mode models do not account for the general relation between trust and confidence that we have referred to as 'confidence presupposing trust', and which is depicted by the path from social trust to confidence in Figure 1.1. For example, Smith and DeCoster (2000) claim, as in our example of scientific arguments above, that logical reasoning is generally judged to be more valid than reasoning based on intuition. Although we agree with the basis for this claim that judged validity increases with degree of objectivity or exposure to public critique, we disagree on the scope of its application (and Smith and DeCoster also seem to qualify their general claim in their discussion of how rules become associations). According to our model, validity is the product of critique within communities – communities formed on the basis of trust relations. No trust relations, no validity: confidence is produced within communities, not generally across communities or generally across an undifferentiated society. This view is congruent with the large and varied body of theory and research on the social influence of groups (for example, Turner, 1991; Wood, 1999, 2000; Iyengar and Simon, 2000; Tetlock, et al, 2000).

Trust and confidence research could also benefit from Smith and DeCoster's (2000) discussion, mentioned above, of the process whereby rules become associations. Repeated use of a rule creates the conditions for associative learning; rules can thereby become embodied in associations and 'acquire the phenomenologically given quality that the preconscious associative (or experiential) system affords' (Smith and DeCoster, 2000, p123). Within our model, this is the process whereby techniques for achieving confidence within a community (such as science-based arguments) become invisible to community members, accepted as given, unquestioned ways of describing the world. This, the blindness to technique and the universalization of the local is the central and all but unexamined core of trust and confidence research.

Is the distinction merely semantic?

A fundamental objection to our TCC model of cooperation is that trust and confidence are simply minor variations on a single theme – the subjective belief that the target will act according to expectations. But our use of the agent/object distinction makes the difference between trust and confidence both clear and significant. Although some theorists (for example, Hardin, 2001b) have long argued to the contrary, within our framework, one does not trust the other to perform some specific behaviour; one's trust is placed in the freedom of the other, that the other will act (whatever that turns out to be) as the trusting person would in that situation. Confidence, in contrast, has a specific performance criterion that the object will behave in a certain predicted way (based on a thorough review of studies of trust and confidence in organizational settings, Dirks and Ferrin (2001, 2002) make a similar argument). This distinction explains why poor performance can destroy confidence but not trust; trust is not based on expected performance.

The dependence of confidence on a performance criterion, and trust's freedom from such a criterion, also shows why trust is associated with risk and vulnerability, and confidence is not. It is true that both trust and confidence are mechanisms for reducing uncertainty; the difference between them lies in how this is done. Trust reduces uncertainty in one domain (a problem requiring cooperation) by shifting it to another, more restricted domain – the freedom of the other. Thus, a risk is taken, and one becomes vulnerable to the other. Confidence, in contrast, reduces uncertainty by attempting to predict and control the behaviour of the other. The aim of confidence is to eliminate risk and vulnerability, and to replace them with controlled exposure to danger.

In our model, trust is based on 'shared values'. We claim that this label subsumes such terms as morality, benevolence, integrity, fairness and caring. It could be objected that these are qualities of the other – qualities that are not necessarily shared by the trusting person. This can be a concern only if one grants existence to those qualities that is judgement independent. In our model, however, these qualities are inherent in the relation between judge and the other. The morality of the other is at the mercy of the judge. And our model explicitly claims that morality is a product, an epiphenomenon, of judgements of agreement with the other on specific important matters.

Conclusions: Implications for trust in risk management

The TCC model, with its central distinction between trust based on morality information and confidence based on performance information, allows us to see the error of the trust asymmetry hypothesis, which is based on the less useful distinction between positive and negative information. Since morality information and trust dominate performance information and confidence, trust tends to be resilient and confidence tends to be fragile. The TCC model allows us to understand why trust is resilient in the face of seemingly damning negative performance information and why, given certain morality-relevant signals, trust can be instantly established. With regard to risk perception, the TCC model shows how trust dominates both knowledge and risk perception, with clear consequences for risk communication efforts: the establishment of trust is primary and essential.

The TCC model also shows the strengths and weaknesses of general trust. Since general trust is trust in strangers, its contributions to cooperative risk management should be greatest when little is known about the issue in question and arguments in support or opposition to it carry little moral weight. In such cases, the relative presence or absence of general trust may be crucial to accepting a course of action. With regard to issues that generate moral implications – that signal a certain sets of values or ways of life – general trust should be overpowered by trust in a specific entity espousing a particular policy. When an issue is morally important, trust shows its resilience – its ability to absorb and survive negative performance information that does not carry moral significance.

Trust is most resilient in matters of greatest moral importance. Depletion of natural resources, for example, is highly morally important for many people. In such a case, the issue-related intentions or values of the entity responsible for managing the issue strongly dominate the entity's specific behaviour in non-issue-related matters. This can be illustrated by examining the effects of so-called fair procedures or fair treatment. The standard model of fairness, which has been very influential in risk management practice, holds that being treated fairly by an entity leads to trust in that entity, regardless of the outcomes of the entity's actions (see, for example, Tyler, 2001). Fairness, in the standard model, is objective, following certain universal norms such as the encouragement of full public participation in the process. According to the TCC model, however, fair treatment leads to trust (and, subsequently, to cooperation) only when the issue in question is of low personal (moral) importance to those involved. When the issue is morally important, then the value implications of outcomes (for example, the policy advocated by the trusted entity) should dominate process.

For example, persons who believe an issue to be morally important should judge the fairness of the procedures used by decision-makers – and their trust in the decision-makers – based on what the decision-makers do regarding the issue and on the moral implications of the policy adopted. In contrast, individuals who do not believe that the issue is morally important should judge the fairness of procedures on the basis of objective criteria (norms), and that judgement should affect their trust in the decision-makers. In the TCC model, the moral significance of outcomes, ignored in the standard model of fairness, is of central importance.

In addition to substantive implications for understanding trust in risk management, our review of trust studies in several domains and our discussion of the TCC model suggest ways in which studies of trust in risk management could be made more useful. Most such studies either use simple ratings as measures or employ *ad hoc* measures that vary across studies and research programmes. Although one might suppose that simple, consistent measures would lead to more useful results over time, this is not necessarily the case. Consistent measures are, at best, uninformative – or, at worst, as confusing as *ad hoc* measures – when they cannot be interpreted within a coherent theoretical framework.

Whereas the number of empirical studies of trust in risk management has increased in recent years, the development of conceptual frameworks, models and theories that would enable us to make some coherent sense of those studies has not kept pace. In this chapter we have attempted to describe an initial effort towards a general framework that could accommodate a variety of approaches to the study of trust in risk management and integrate the results of those studies into useful knowledge for managing environmental and technological risks.

Table 1.1 Empirical studies of the relation between risk perception and trust

Study	Type of target	Tangibility of target	Type of judgement	Referent of judgement	Dependent variable/ results
Baird (1986)	Agent	Abstract	Attribute	Values	Perceived risk (r = –.50)
Bassett et al (1996)	Agent	Concrete and abstract (separate)	Rating	N/A	Cooperation (r = .16)
Biel and Dahlstrand (1995)	Object	Abstract	Attribute	Performance	Perceived risk (r = –.63)
Bord and O'Connor (1990)	Agent	Abstract	Rating	N/A	Cooperation (r = .60)
Bord and O'Connor (1992)	Agent	Abstract	Attribute	Values	Perceived risk (r = –.22)
Cvetkovich (1999)	Agent	Abstract	Rating	N/A	Perceived risk (r = –.40)
Drottz-Sjöberg and Sjöberg (1991)	Agent and object (combined)	Abstract	Attribute	Values and performance (combined)	Perceived risk (β = –.14)
Eiser et al (2002)	Object	Abstract	Rating	N/A	Perceived risk (Study 1: personal risk, r = –.43; population risk, r = –.44. Study 2: r = –.44. Study 3: r = –.64)
Ferguson et al (2004)	Agent	Abstract	Rating	N/A	Perceived risk (weak, non-significant correlations)
Figueiredo and Drottz-Sjöberg (2000)	Agent	Concrete	Rating	N/A	Perceived risk (r = –.13)
Flynn et al (1992)	Agent	Concrete	Attribute	Values	Perceived risk (β = –.82) and cooperation (no direct effect, but perceived risk did affect cooperation; β = –.68)
Freudenburg (1993)	Object	Abstract	Attribute	Performance	Perceived risk (negative effects shown by c2 analyses)
Frewer et al (2003)	Agent and object (combined)	Abstract	Attribute	Values and performance	Perceived risk (weak, non-significant effects)
Greenberg and Williams (1999)	Agent and object (separate)	Concrete and abstract (separate)	Attribute	Values and performance (separate)	Perceived risk (negative effects shown by discriminant function analysis)

Study	Type	Concreteness	Measure	Outcome	Findings
Grobe et al (1999)	Agent	Concrete	Rating	N/A	Perceived risk (negative effects shown by probit analyses)
Groothuis and Miller (1997)	Agent	Concrete and abstract	Rating	N/A	Perceived risk (regression analyses show negative effects for pro-development targets and positive effects for anti-development targets) and cooperation (positive effects for pro-development targets; negative effects for anti-development targets)
Hine et al (1997)	Agent and object (combined)	Concrete	Attribute	Values and performance (combined)	Perceived risk ($\beta = -.50$) and cooperation ($\beta = .36$; perceived risk also had an effect, $\beta = -.37$)
Hoban et al (1992)	Agent	Abstract	Rating	N/A	Cooperation ($\beta = 27$)
Ibitayo and Pijawka (1999)	Agent	Concrete	Rating	N/A	Cooperation (t-test shows positive relation)
Jungermann et al (1996)	Agent	Abstract	Attribute	Values and performance (separate and combined)	Perceived risk ($\beta = -.22$ for combined measure of values and performance)
Katsuya (2002)	Agent	Concrete	Attribute	Values	Cooperation (high knowledge: affected by trust, $\beta = .27$, perceived risk, $\beta = -.38$ and perceived benefit, $\beta = .28$; low knowledge: affected only by trust, $\beta = .54$)
Kunreuther et al (1990)	Agent	Concrete	Attribute	Values	Cooperation (logistic regression shows positive effect)
McComas (2003)	Agent	Concrete	Attribute	Values	Perceived risk ($r = -.37$) and cooperation ($r = .41$, $\beta = .19$)
Moon and Balasubramanian (2004)	Object	Concrete	Attribute	Performance	Perceived risk ($\beta = -.63$)
Pijawka and Mushkatel (1991/1992)	Agent	Concrete	Attribute	Values	Perceived risk (negative correlations for pro-development targets; positive correlations for anti-development targets)

Table 1.1 continued

Study	Type of target	Tangibility of target	Type of judgement	Referent of judgement	Dependent variable/results
Priest (2001)	Agent	Abstract	Attribute	Values	Cooperation (partial r = .28)
Priest et al (2003)	Agent	Abstract	Attribute	Values	Cooperation (variety of strong, positive effects)
Sapp and Bird (2003)	Agent	Concrete	Rating	N/A	Perceived risk (LISREL analysis: for agencies, negative effects on worry; for consumer advocates, positive effects on both worry and concern)
Siegrist (1999)	Agent and object (combined)	Concrete	Attribute	Values and performance (combined)	Perceived risk ($\beta = -.69$), perceived benefit ($\beta = .54$) and cooperation (indirectly, through perceived risk and benefit)
Siegrist (2000)	Agent	Abstract	Attribute	Values	Perceived risk ($\beta = -.46$), perceived benefit ($\beta = .44$) and cooperation (indirectly, through perceived risk and benefit)
Siegrist and Cvetkovich (2000)	Agent	Abstract	Rating	N/A	Perceived risk (negative correlations for a variety of targets – for example, $r = -.64$) and perceived benefit (positive correlations for a variety of targets), particularly in the absence of knowledge
Siegrist et al (2000)	Agent and object (combined)	Concrete	Attribute	Values	Perceived risk ($\beta = -.77, -.84$ and $-.81$ for three hazards) and perceived benefits ($\beta = .78, .72$ and $.54$ for three hazards)
Sjöberg (1999)	Agent	Abstract	Rating	N/A	Perceived risk ($\beta = -.14$)
Sjöberg (2000a)	Agent	Abstract	Rating	N/A	Perceived risk ($\beta = -.10$)
Sjöberg (2000b)	Agent	Abstract	Rating	N/A	Perceived risk (no significant effect)

Study	Target	Abstract/Concrete	Type	Basis	Results
Sjöberg (2001)	Agent	Abstract	Rating	N/A	Perceived risk (Study 1: r ≈ −.30; Study 2: β = −.22 and β = −.20)
Sjöberg (2002)	Agent	Abstract	Rating	N/A	Perceived risk (accounts for an average of 10% of variance)
Sparks et al (1994)	Agent	Concrete and abstract	Rating	N/A	Cooperation (negative correlations for pro-technology targets; positive correlation for anti-technology target)
Spies et al (1998)	Object	Concrete	Attribute	Performance	Cooperation (logistic regression shows positive effect for community residents, but no effect for community leaders)
Tanaka (2004)	Agent	Concrete	Attribute	Values	Cooperation (β = .26 for siting of nuclear facilities)
Titchener and Sapp (2002)	Agent	Concrete	Attribute	Values	Cooperation (positive effect for pro-technology targets on intentions, β = .12, and willingness, β = .07)
Trumbo and McComas (2003)	Agent	Concrete	Attribute	Values	Risk perception (negative relation for risk management targets, β = −.18 and β = −.13; positive relation for risk critic target, β = .23)
Viklund (2003)	Agent and object (separate and combined)	Abstract	Attribute and rating	Value and performance	Risk perception (varying, mostly significant negative effects)
Williams et al (1999)	Agent	Concrete	Attribute	Not given	Risk perception (begative relation: r = .49, with a function that discriminates between low and high risk perception)
Williams and Hammitt (2001)	Object	Concrete	Attribute	Performance	Risk perception (regression analyses show negative relation to several forms of risk)

Note

1 This chapter is based upon work supported by the National Science Foundation under Grants No 0099278 and 0323451. Concise reviews of empirical studies of interpersonal trust, trust within organizations and trust in government, evaluated from the point of view of the TCC model, are available from the authors.

References

Adler, P. S. and Kwon, S-W. (2002) 'Social capital: Prospects for a new concept', *Academy of Management Review*, vol 27, pp17–40

Bacharach, M. and Gambetta, D. (2001) 'Trust in signs', in K. S. Cook (ed) *Trust in Society*, Russell Sage Foundation, New York, pp148–184

Baird, B. N. R. (1986) 'Tolerance for environmental health risks: The influence of knowledge, benefits, voluntariness, and environmental attitudes', *Risk Analysis*, vol 6, pp425–435

Bassett, G. W., Jenkins-Smith, H. C. and Silva, C. (1996) 'On-site storage of high level nuclear waste: Attitudes and perceptions of local residents', *Risk Analysis*, vol 16, pp309–319

Becker, L. C. (1996) 'Trust as noncognitive security about motives', *Ethics*, vol 107, pp43–61

Bella, D. A., Mosher, C. D. and Calvo, S. N. (1988) 'Establishing trust: Nuclear waste disposal', *Journal of Professional Issues in Engineering*, vol 114, pp40–50

Biel, A. and Dahlstrand, U. (1995) 'Risk perception and the location of a repository for spent nuclear fuel', *Scandinavian Journal of Psychology*, vol 36, pp25–36

Bohnet, I., Frey, B. S. and Huck, S. (2001) 'More order with less law: On contract enforcement, trust, and crowding', *American Political Science Review*, vol 95, pp131–144

Bord, R. J. and O'Connor, R. E. (1990) 'Risk communication, knowledge, and attitudes: Explaining reactions to a technology perceived as risky', *Risk Analysis*, vol 10, pp499–506

Bord, R. J. and O'Connor, R. E. (1992) 'Determinants of risk perceptions of a hazardous waste site', *Risk Analysis*, vol 12, pp411–416

Brehm, J. and Rahn, W. (1997) 'Individual-level evidence for the causes and consequences of social capital', *American Journal of Political Science*, vol 41, pp999–1023

Chaiken, S. and Trope, Y. (1999) *Dual-process Theories in Social Psychology*, The Guilford Press, New York

Chen, S. and Chaiken, S. (1999) 'The heuristic-systematic model in its broader context', in S. Chaiken and Y. Trope (eds) *Dual-Process Theories in Social Psychology* Guilford Press, New York, pp73–96

Chiles, T. H. and McMackin, J. F. (1996) 'Integrating variable risk preferences, trust, and transaction cost economics', *Academy of Management Review*, vol 21, pp73–99

Cook, K. S. and Cooper, R. M. (2003) 'Experimental studies of cooperation, trust, and social exchange', in E. Ostrom and J. Walker (eds) *Trust and Reciprocity: Interdisciplinary Lessons from Experimental Research*, Russell Sage Foundation, New York, pp209–244

Cvetkovich, G. (1999) 'The attribution of social trust', in G. Cvetkovich and R. Löfstedt (eds) *Social Trust and the Management of Risk*, Earthscan, London

Cvetkovich, G. and Löfstedt, R. E. (eds) (1999) *Social Trust and the Management of Risk*, Earthscan, London

Das, T. K. (2001) 'Relational risk and its personal correlates in strategic alliances', *Journal of Business and Psychology*, vol 15, pp449–465

Das, T. K. and Teng, B.-S. (1998) 'Between trust and control', *Academy of Management Review*, vol 23, pp491–512

Das, T. K. and Teng, B.-S. (2004) 'The risk-based view of trust: A conceptual framework', *Journal of Business and Psychology*, vol 19, pp85–116

Dawes, R. M. and Messick, D. M. (2000) 'Social dilemmas', *International Journal of Psychology*, vol 35, pp111–116

De Bruin, E. N. M. and Van Lange, P. A. M. (1999a) 'The double meaning of a single act: Influences of the perceiver and the perceived on cooperative behaviour', *European Journal of Personality*, vol 13, pp165–182

De Bruin, E. N. M. and Van Lange, P. A. M. (1999b) 'Impression formation and cooperative behaviour', *European Journal of Social Psychology*, vol 29, pp305–328

De Bruin, E. N. M. and Van Lange, P. A. M. (2000) 'What people look for in others: Influences of the perceiver and the perceived on information selection', *Personality and Social Psychology Bulletin*, vol 26, pp206–219

DeBruine, L. M. (2002) 'Facial resemblance enhances trust', *Proceedings of the Royal Society of London. Series B*, vol 269, pp1307–1312

Deutsch, M. (1958) 'Trust and suspicion', *Journal of Conflict Resolution*, vol 2, pp265–279

Deutsch, M. (1973) *The Resolution of Conflict*, Yale University Press, New Haven, CT

Dirks, K. T. and Ferrin, D. L. (2001) 'The role of trust in organizational settings', *Organization Science*, vol 12, pp450–467

Dirks, K. T. and Ferrin, D. L. (2002) 'Trust in leadership: Meta-analytic findings and implications for research and practice', *Journal of Applied Psychology*, vol 87, pp611–628

Ditto, P. H., Scepansky, J. A., Munro, G. D. and Apanovitch, A. M. (1998) 'Motivated sensitivity to preference-inconsistent information', *Journal of Personality and Social Psychology*, vol 75, pp53–69

Drottz-Sjöberg, B.-M. and Sjöberg, L. (1991) 'Adolescents' attitudes to nuclear power and radioactive wastes', *Journal of Applied Social Psychology*, vol 21, pp2007–2036

Earle, T. C. and Cvetkovich, G. (1995) *Social Trust: Toward a Cosmopolitan Society*, Praeger, Westport, CT

Earle, T. C. and Cvetkovich, G. (1997) 'Culture, cosmopolitanism, and risk management', *Risk Analysis*, vol 17, pp55–65

Earle, T. C. and Cvetkovich, G. (1999) 'Social trust and culture in risk management', in G. Cvetkovich and R. Löfstedt (eds) *Social Trust and the Management of Risk*, Earthscan, London

Earle, T. E. and Siegrist, M. (2006) 'Morality information, performance information, and the distinction between trust and confidence', *Journal of Applied Social Psychology*, vol 36, no 2, pp383–416

Eckel, C. C. and Wilson, R. K. (2003) 'The human face of game theory: Trust and reciprocity in sequential games', in E. Ostrom and J. Walker (eds) *Trust and*

Reciprocity: Interdisciplinary Lessons from Experimental Research, Russell Sage Foundation, New York

Eiser, J. R., Miles, S. and Frewer, L. J. (2002) 'Trust, perceived risk, and attitudes toward food technologies', *Journal of Applied Social Psychology*, vol 32, pp2423–2433

Ferguson, E., Farrell, K., James, V. and Lowe, K. C. (2004) 'Trustworthiness of information about blood donation and transfusion in relation to knowledge and perceptions of risk: An analysis of UK stakeholder groups', *Transfusion Medicine*, vol 14, pp205–216

Figueiredo, C. J. and Drottz-Sjöberg, B.-M. (2000) 'Perceived control, voluntariness and emotional reactions in relocated areas of Russia, Ukraine and Belarus', *Risk: Health, Safety and Environment*, vol 11, pp233–242

Flynn, J., Burns, W., Mertz, C. K. and Slovic, P. (1992) 'Trust as a determinant of opposition to a high-level radioactive waste repository: Analysis of a structural model', *Risk Analysis*, vol 12, pp417–429

Frank, R. H. (1988) *Passions within Reason*, W. W. Norton, New York

Frank, R. H., Gilovich, T. and Regan, D. T. (1993a) 'The evolution of one-shot cooperation: An experiment', *Ethology and Sociobiology*, vol 14, pp247–256

Frank, R. H., Gilovich, T. and Regan, D. T. (1993b) 'Does studying economics inhibit cooperation?', *Journal of Economic Perspectives*, vol 7, pp159–171

Freudenberg, W. R. (1993) 'Risk and recreancy: Weber, the division of labor, and the rationality of risk perceptions', *Social Forces*, vol 71, pp909–932

Frewer, L. J., Scholderer, J. and Bredahl, L. (2003) 'Communicating about the risks and benefits of genetically modified foods: The mediating role of trust', *Risk Analysis*, vol 23, pp1117–1133

Geyskens, I., Steenkamp, J.-B. E. M. and Kumar, N. (1998) 'Generalizations about trust in marketing channel relationships using meta-analysis', *International Journal of Research in Marketing*, vol 15, pp223–248

Geyskens, I., Steenkamp, J.-B. E. M. and Kumar, N. (1999) 'A meta-analysis of satisfaction in marketing channel relationships', *Journal of Marketing Research*, vol 36, pp223–238

Giddens, A. (1990) *The Consequences of Modernity*, Stanford University Press, Stanford, CA

Granovetter, M. (1985) 'Economic action and social structure: The problem of embeddedness', *American Journal of Sociology*, vol 91, pp481–510

Greenberg, M. R. and Williams, B. (1999) 'Geographical dimensions and correlates of trust', *Risk Analysis*, vol 19, pp159–169

Grobe, D., Douthitt, R. and Zepeda, L. (1999) 'A model of consumers' risk perceptions toward recombinant bovine growth hormone (rbGH): The impact of risk characteristics', *Risk Analysis*, vol 19, pp661–673

Groothuis, P. A. and Miller, G. (1997) 'The role of social distrust in risk-benefit analysis: A study of the siting of a hazardous waste disposal facility', *Journal of Risk and Uncertainty*, vol 15, pp241–257

Gurabardhi, Z., Gutteling, J. M. and Kuttschreuter, M. (2004) 'The development of risk communication', *Science Communication*, vol 25, pp323–349

Hall, M. A. (2002) 'Law, medicine, and trust', *Stanford Law Review*, vol 55, pp463–527

Hardin, R. (1991) 'Trusting persons, trusting institutions', in R. Zeckhauser (ed) *Strategy and Choice*, MIT Press, Cambridge, MA, pp185–209

Hardin, R. (1993) 'The street-level epistemology of trust', *Politics and Society*, vol 21, pp505–529

Hardin, R. (1995) *One for All: The Logic of Group Conflict*, Princeton University Press, Princeton

Hardin, R. (1998) 'Trust in government', in V. Braithwaite and M. Levi (eds) *Trust and Governance*, Russell Sage Foundation, New York, pp9–27

Hardin, R. (1999) 'Do we want trust in government?', in M. E. Warren (ed) *Democracy and Trust*, Cambridge University Press, Cambridge, UK, pp22–41

Hardin, R. (2000) 'The public trust', in S. J. Pharr and R. D. Putnam (eds) *Disaffected Democracies: What's Troubling the Trilateral Countries?*, Princeton University Press, Princeton, NJ, pp31–51

Hardin, R. (2001a) 'Distrust', *Boston University Law Review*, vol 81, pp495–522

Hardin, R. (2001b) 'Conceptions and explanations of trust', in K. S. Cook (ed) *Trust in Society*, Russell Sage Foundation, New York, pp3–39

Harris, P. L. (2002) 'Checking our sources: The origins of trust in testimony', *Studies in History and Philosophy of Science*, vol 33, pp315–333

Hayashi, N. and Yamagishi, T. (1998) 'Selective play: Choosing partners in an uncertain world', *Personality and Social Psychology Review*, vol 2, pp276–289

Hine, D. W., Summers, C., Prystupa, M. and McKenzie-Richer, A. (1997) 'Public opposition to a proposed nuclear waste repository in Canada: An investigation of cultural and economic effects', *Risk Analysis*, vol 17, pp293–302

Hoban, T., Woodrum, E. and Czaja, R. (1992) 'Public opposition to genetic engineering', *Rural Sociology*, vol 57, pp476–493

Hovland, C. I., Janis, I. L. and Kelley, H. H. (1953) *Communication and Persuasion: Psychological Studies of Opinion Change*, Yale University Press, New Haven

Ibitayo, O. O. and Pijawka, K. D. (1999) 'Reversing NIMBY: An assessment of state strategies for siting hazardous-waste facilities', *Environment and Planning C: Government and Policy*, vol 17, pp379–389

Iyengar, S. and Simon, A. F. (2000) 'New perspectives and evidence on political communication and campaign effects', *Annual Review of Psychology*, vol 51, pp149–169

Jones, W. H., Couch, L. and Scott, S. (1997) 'Trust and betrayal', in R. Hogan, J. Johnson and S. Briggs (eds) *Handbook of Personality Psychology*, Academic Press, New York, pp465–482

Jones, G. R. and George, J. M. (1998) 'The experience and evolution of trust: Implications for cooperation and teamwork', *Academy of Management Review*, vol 23, pp531–546

Jungermann, H., Pfister, H.-R. and Fischer, K. (1996) 'Credibility, information preferences, and information interests', *Risk Analysis*, vol 16, pp251–261

Kahan, D. M. (2001) 'Trust, collective action and law', *Boston University Law Review*, vol 81, pp333–347

Kasperson, R. E., Golding, D. and Tuler, S. (1992) 'Siting hazardous facilities and communicating risks under conditions of high social distrust', *Journal of Social Issues*, vol 48, pp161–187

Katsuya, T. (2002) 'Difference in the formation of attitude toward nuclear power', *Political Psychology*, vol 23, pp191–203

Koehler, J. J. (1993) 'The influence of prior beliefs on scientific judgements of evidence quality', *Organizational Behaviour and Human Decision Processes*, vol 56, pp28–55

Kollock, P. (1993) 'An eye for an eye leaves everyone blind: Cooperation and accounting systems', *American Sociological Review*, vol 58, pp768–786

Kollock, P. (1994) 'The emergence of exchange structures: An experimental study of uncertainty, commitment, and trust', *American Journal of Sociology*, vol 100, pp313–345

Kollock, P. (1998a) 'Social dilemmas: The anatomy of cooperation', *Annual Review of Sociology*, vol 24, pp183–214

Kollock, P. (1998b) 'Transforming social dilemmas: Group identity and cooperation', in P. A. Danielson (ed) *Modeling Rationality, Morality and Evolution*, Oxford University Press, New York, pp186–210

Kramer, R. M. (1999) 'Trust and distrust in organizations', *Annual Review of Psychology*, vol 50, pp569–598

Kramer, R. M., Hanna, B. A., Su, S. and Wei, J. (2001) 'Collective identity, collective trust, and social capital: Linking group identification and group cooperation', in M. E. Turner (ed) *Groups at Work: Theory and Research*, Lawrence Erlbaum, Mahwah, NJ, pp173–196

Kunreuther, H., Easterling, D., Desvousges, W. and Slovic, P. (1990) 'Public attitudes toward siting a high-level nuclear waste repository in Nevada', *Risk Analysis*, vol 10, pp469–484

Kydd, A. (2000) 'Trust, reassurance, and cooperation', *International Organization*, vol 54, pp325–357

Larrick, R. P. and Blount, S. (1997) 'The claiming effect: Why players are more generous in social dilemmas than in ultimatum games', *Journal of Personality and Social Psychology*, vol 72, pp810–825

Levi, M. (1998) 'A state of trust', in V. Braithwaite and M. Levi (eds) *Trust and Governance*, Russell Sage Foundation, New York, pp77–101

Levi, M. and Stoker, L. (2000) 'Political trust and trustworthiness', *Annual Review of Political Science*, vol 3, pp475–508

Lewicki, R. J. and Bunker, B. B. (1995) 'Trust in relationships: A model of development and decline', in B. B. Bunker, J. Z. Rubin and Associates (eds) *Conflict, Cooperation, and Justice*, Jossey-Bass, San Francisco, pp133–173

Lipshitz, R., Gilad, Z., and Suleiman, R. (2001) 'The one-of-us effect in decision evaluation', *Acta Psychologica*, vol 108, pp53–71

Luhmann, N. (1979) *Trust and Power*, John Wiley and Sons, Chichester

Luhmann, N. (1988) 'Familiarity, confidence, trust: Problems and alternatives', in D. Gambetta (ed) *Trust: Making and Breaking Cooperative Relations*, Basil Blackwell, Oxford, pp94–107

Luhmann, N. (1993) *Risk: A Sociological Theory*, Aldine de Gruyter, New York

Malle, B. F. (2001) 'Folk explanations of intentional action', in B. F. Malle, L. J. Moses and D. A. Baldwin (eds) *Intentions and Intentionality: Foundations of Social Cognition*, The MIT Press, Cambridge, MA, pp265–286

Malle, B. F. and Knobe, J. (1997a) 'The folk concept of intentionality', *Journal of Experimental Social Psychology*, vol 33, no 2, pp101–121

Malle, B. F. and Knobe, J. (1997b) 'Which behaviors do people explain? A basic actor-observer asymmetry', *Journal of Personality and Social Psychology*, vol 72, no 2, pp288–304

March, J. G. (1994) *A Primer on Decision Making*, Free Press, New York

McComas, K. A. (2003) 'Citizen satisfaction with public meetings used for risk communication', *Journal of Applied Communications Research*, vol 31, pp164–184

Messick, D. M. (1999) 'Alternative logics for decision making in social settings', *Journal of Economic Behaviour and Organization*, vol 39, pp11–28

Messick, D. M. (2000) 'Context, norms, and cooperation in modern society: A postscript', in M. Van Vugt, M. Snyder, T. B. Tyler and A. Biel (eds) *Cooperation in Modern Society: Promoting the Welfare of Communities, States and Organizations*, Routledge, New York, pp231–240

Messick, D. M. and Kramer, R. M. (2001) 'Trust as a form of shallow morality', in K. S. Cook (ed) *Trust in Society*, Russell Sage Foundation, New York, pp89–118

Misztal, B. (1996) *Trust in Modern Societies*, Polity Press, Cambridge, UK

Molm, L. D., Takahashi, N. and Peterson, G. (2000) 'Risk and trust in social exchange: An experimental test of a classical proposition', *American Journal of Sociology*, vol 105, pp1396–1427

Moon, W. and Balasubramanian, S. K. (2004) 'Public attitudes toward agrobiotechnology: The mediating role of risk perceptions on the impact of trust, awareness, and outrage', *Review of Agricultural Economics*, vol 26, pp186–208

Morrison, B. (1999) 'Interdependence, the group, and social cooperation', in M. Foddy, M. Smithson, S. Schneider and M. Hogg (eds) *Resolving Social Dilemmas: Dynamic, Structural, and Intergroup Aspects*, Psychology Press, Philadelphia, pp295–308

O'Neill, O. (2004) 'Accountability, trust, and informed consent in medical practice and research', *Clinical Medicine*, vol 4, pp269–276

Ostrom, E. (1998) 'A behavioural approach to rational choice theory of collective action', *American Political Science Review*, vol 92, pp1–22

Ostrom, E. (2003) 'Toward a behavioural theory linking trust, reciprocity, and reputation', in E. Ostrom and J. Walker (eds) *Trust and Reciprocity: Interdisciplinary Lessons from Experimental Research*, Russell Sage Foundation, New York, pp19–79

Oswald, M. E. and Fuchs, T. (1998) 'Readiness to trust in complex situations', *Swiss Journal of Psychology*, vol 57, pp248–254

Peeters, G. and Czapinski, J. (1990) 'Positive–negative asymmetry in evaluations: The distinction between affective and informational negativity effects', in W. Stroebe and M. Hewestone (eds) *European Review of Social Psychology*, vol 1, Wiley, New York, pp33–60

Pijawka, K. D. and Mushkatel, A. H. (1991/1992) 'Public opposition to the siting of the high-level nuclear waste repository: The importance of trust', *Policy Studies Review*, vol 10, pp180–194

Pillutla, M. M. and Chen, X.-P. (1999) 'Social norms and cooperation in social dilemmas: The effects of context and feedback', *Organizational Behaviour and Human Decision Processes*, vol 78, pp81–103

Plous, S. (1991) 'Biases in the assimilation of technological breakdowns: Do accidents make us safer?', *Journal of Applied Social Psychology*, vol 21, no 13, pp1058–1082

Portes, A. (1998) 'Social capital: Its origins and applications in modern sociology', *Annual Review of Sociology*, vol 24, pp1–24

Priest, S. H. (2001) 'Misplaced faith', *Science Communication*, vol 23, pp97–110

Priest, S. H., Bonfadelli, H. and Rusanen, M. (2003) 'The "trust gap" hypothesis: Predicting support for biotechnology across national cultures as a function of trust in actors', *Risk Analysis*, vol 23, pp751–766

Renn, O. and Levine, D. (1991) 'Credibility and trust in risk communication', in R. E. Kasperson and P. J. M. Stallen (eds) *Communicating Risks to the Public*, Kluwer, Dordrecht, The Netherlands, pp175–218

Ribstein, L. E. (2001) 'Law v. trust', *Boston University Law Review*, vol 81, pp553–590

Ring, P. S. (1996) 'Fragile and resilient trust and their roles in economic exchange', *Business and Society*, vol 35, pp148–175

Rosenberg, S., Nelson, C. and Vivekananthan, P. S. (1968) 'A multidimensional approach to the structure of personality impressions', *Journal of Personality and Social Psychology*, vol 9, pp283–294

Rotter, J. B. (1971) 'Generalized expectancies for interpersonal trust', *American Psychologist*, vol 26, pp443–452

Rotter, J. B. (1980) 'Interpersonal trust, trustworthiness, and gullibility', *American Psychologist*, vol 35, pp1–7

Rousseau, D. M., Sitkin, S. B., Burt, R. S. and Camerer, C. (1998) 'Not so different after all: A cross-discipline view of trust', *Academy of Management Review*, vol 23, pp393–404

Sapp, S. G. and Bird, S. R. (2003) 'The effects of social trust on consumer perceptions of food safety', *Social Behaviour and Personality*, vol 31, pp413–422

Scharlemann, J. P. W., Eckel, C. C., Kacelnik, A. and Wilson, R. K. (2001) 'The value of a smile: Game theory with a human face', *Journal of Economic Psychology*, vol 22, pp617–640

Seligman, A. B. (1997) *The Problem of Trust*, Princeton University Press, Princeton

Seligman, A. B. (1998) 'Trust and sociability: On the limits of confidence and role expectations', *American Journal of Economics and Sociology*, vol 57, pp391–404

Seligman, A. B. (2001) 'Role complexity, risk, and the emergence of trust', *Boston University Law Review*, vol 81, pp619–634

Siegrist, M. (1999) 'A causal model explaining the perception and acceptance of gene technology', *Journal of Applied Social Psychology*, vol 29, pp2093–2106

Siegrist, M. (2000) 'The influence of trust and perceptions of risks and benefits on the acceptance of gene technology', *Risk Analysis*, vol 20, pp195–203

Siegrist, M. and Cvetkovich, G. (2000) 'Perception of hazards: The role of social trust and knowledge', *Risk Analysis*, vol 20, pp713–719

Siegrist, M., Cvetkovich, G. and Roth, C. (2000) 'Salient value similarity, social trust, and risk/benefit perception', *Risk Analysis*, vol 20, pp353–362

Siegrist, M., Earle, T. C. and Gutscher, H. (2003) 'Test of a trust and confidence model in the applied context of electromagnetic field (EMF) risks', *Risk Analysis*, vol 23, pp705–716

Siegrist, M., Earle, T. C. and Gutscher, H. (2005) 'Perception of risk: The influence of general trust and general confidence', *Journal of Risk Research*, vol 8, pp145–156

Simon, B., Loewy, M., Stümer, S., Weber, U., Freytag, P., Habig, C., Kampmeier, C. and Spahlinger, P. (1998) 'Collective identification and social movement participation', *Journal of Personality and Social Psychology*, vol 74, pp646–658

Simon, B., Stümer, S. and Steffens, K. (2000) 'Helping individuals or group members? The role of individual and collective identification in AIDS volunteerism', *Personality and Social Psychology Bulletin*, vol 26, pp497–506

Sitkin, S. B. and Roth, N. L. (1993) 'Explaining the limited effectiveness of legalistic "remedies" for trust/distrust', *Organization Science*, vol 4, pp367–392

Sjöberg, L. (1999) 'Perceived competence and motivation in industry and government as factors in risk perception', in G. Cvetkovich and R. Löfstedt (eds) *Social Trust and the Management of Risk*, Earthscan, London, pp89–99

Sjöberg, L. (2000a) 'Perceived risk and tampering with nature', *Journal of Risk Research*, vol 3, pp353–367

Sjöberg, L. (2000b) 'Factors in risk perception', *Risk Analysis*, vol 20, pp1–11

Sjöberg, L. (2001) 'Limits of knowledge and the limited importance of trust', *Risk Analysis*, vol 21, pp189–198

Sjöberg, L. (2002) 'Attitudes toward technology and risk: Going beyond what is immediately given', *Policy Sciences*, vol 35, pp379–400

Skowronski, J. J. and Carlston, D. E. (1989) 'Negativity and extremity biases in impression formation: A review of explanations', *Psychological Bulletin*, vol 105, pp131–142

Sloman, S. (1996) 'The empirical case for two systems of reasoning', *Psychological Bulletin*, vol 119, pp3–22

Slovic, P. (1993) 'Perceived risk, trust and democracy', *Risk Analysis*, vol 13, pp675–682

Slovic, P. (2000) *The Perception of Risk*, Earthscan, London

Slovic, P. (2001) 'The risk game', *Journal of Hazardous Materials*, vol 86, pp17–24

Slovic, P., Flynn, J. and Layman, M. (1991) 'Perceived risk, trust, and the politics of nuclear waste', *Science*, vol 254, pp1603–1607

Smith, C. (2001) 'Trust and confidence: Possibilities for social work in "high modernity"', *British Journal of Social Work*, vol 31, pp287–305

Smith, E. R. and DeCoster, J. (2000) 'Dual-process models in social and cognitive psychology: Conceptual integration and links to underlying memory systems', *Personality and Social Psychology Review*, vol 4, pp108–131

Sniezek, J. A. and Van Swol, L. M. (2001) 'Trust, confidence, and expertise in a judge-advisor system', *Organizational Behaviour and Human Decision Processes*, vol 84, pp288–307

Sparks, P., Shepherd, R. and Frewer, L. J. (1994) 'Gene technology, food production, and public opinion: A UK study', *Agriculture and Human Values*, vol 11, pp19–28

Spies, S., Murdock, S. H., White, S., Krannich, R., Wulfhorst, J. D., Wrigley, K., Leistritz, F. L., Sell, R. and Thompson, J. (1998) 'Support for waste facility siting: Differences between community leaders and residents', *Rural Sociology*, vol 63, pp65–93

Stolle, D. (1998) 'Bowling together, bowling alone: The development of generalized trust in voluntary associations', *Political Psychology*, vol 19, pp497–525

Stolle, D. (2001a) 'Clubs and congregations: The benefits of joining an association', in K. S. Cook (ed) *Trust in Society*, Russell Sage Foundation, New York, pp202–244

Stolle, D. (2001b) '"Getting to trust": An analysis of the importance of institutions, families, personal experiences and group membership', in P. Dekker and E. M. Uslaner (eds) *Social Capital and Participation in Everyday Life*, Routledge, London, pp118–133

Sullivan, J. L. and Transue, J. E. (1999) 'The psychological underpinnings of democracy: A selective review of research on political tolerance, interpersonal trust, and social capital', *Annual Review of Psychology*, vol 50, pp625–650

Tanaka, Y. (2004) 'Major psychological factors determining public acceptance of the siting of nuclear facilities', *Journal of Applied Social Psychology*, vol 34, pp1147–1165

Tenbrunsel, A. E. and Messick, D. M. (1999) 'Sanctioning systems, decision frames, and cooperation', *Administrative Science Quarterly*, vol 44, pp684–707

Tetlock, P. E., Kristel, O. V., Elson, S. B., Green, M. C. and Lerner, J. S. (2000) 'The psychology of the unthinkable: Taboo trade-offs, forbidden base rates, and heretical counterfactuals', *Journal of Personality and Social Psychology*, vol 78, pp853–870

Titchener, G. D. and Sapp, S. G. (2002) 'A comparison of two approaches to understanding consumer opinions of biotechnology', *Social Behaviour and Personality*, vol 30, pp373–382

Tonkiss, F. and Passey, A. (1999) 'Trust, confidence and voluntary organisations: Between values and institutions', *Sociology*, vol 33, pp257–274

Trumbo, C. W. and McComas, K. A. (2003) 'The function of credibility in information processing for risk perception', *Risk Analysis*, vol 23, pp343–353

Turner, J. C. (1991) *Social Influence*, Brooks/Cole, Pacific Grove, CA

Tyler, T. R. (2001) 'Trust and law abidingness: A proactive model of social regulation', *Boston University Law Review*, vol 81, pp361–406

Uslaner, E. M. (2001) *Trust as a Moral Value*, Paper prepared for the conference Social Capital: Interdisciplinary Perspectives, University of Exeter, UK, January 2001

Van Lange, P. A. M. and Sedikides, C. (1998) 'Being honest but not necessarily more intelligent than other: Generality and explanations for the Muhammad Ali effect', *European Journal of Social Psychology*, vol 28, pp675–680

Viklund, M. J. (2003) 'Trust and risk perception in western Europe: A cross-national study', *Risk Analysis*, vol 23, pp727–738

Weber, J. M., Kepelman, S. and Messick, D. M. (2004) 'A conceptual review of decision making in social dilemmas: Applying a logic of appropriateness', *Personality and Social Psychology Review*, vol 8, no 3, pp281–307

Williams, B. L., Brown, S., Greenberg, M. and Kahn, M. A. (1999) 'Risk perception in context: The Savanna River Site stakeholder study', *Risk Analysis*, vol 19, pp1019–1035

Williams, P. R. D. and Hammitt, J. K. (2001) 'Perceived risks of conventional and organic produce: Pesticides, pathogens, and natural toxins', *Risk Analysis*, vol 21, pp319–330

Wojciszke, G. (1994) 'Multiple meanings of behaviour: Construing actions in terms of competence or morality', *Journal of Personality and Social Psychology*, vol 67, pp222–232

Wojciszke, B., Bazinska, R. and Jaworski, M. (1998) 'On the dominance of moral categories in impression formation', *Personality and Social Psychology Bulletin*, vol 24, pp1251–1263

Wood, W. (1999) 'Motives and modes of processing in the social influence of groups', in S. Chaiken and Y. Trope (eds) *Dual-Process Theories in Social Psychology*, Guilford Press, New York, pp547–570

Wood, W. (2000) 'Attitude change: Persuasion and social influence', *Annual Review of Psychology*, vol 51, pp539–570

Yamagishi, T. (1986) 'The provision of a sanctioning system as a public good', *Journal of Personality and Social Psychology*, vol 51, pp110–116

Yamagishi, T. (1988) 'The provision of a sanctioning system in the United States and Japan', *Social Psychology Quarterly*, vol 51, pp265–271

Yamagishi, T. (2001) 'Trust as a form of social intelligence', in K. S. Cook (ed) *Trust in Society*, Russell Sage Foundation, New York, pp121–147

Yamagishi, T. and Sato, K. (1986) 'Motivational bases of the public goods problem', *Journal of Personality and Social Psychology*, vol 50, pp67–73

Yamagishi, T. and Yamagishi, M. (1994) 'Trust and commitment in the United States and Japan', *Motivation and Emotion*, vol 18, pp129–166

Yamagishi, T., Cook, K. S. and Watabe, M. (1998) 'Uncertainty, trust, and commitment formation in the United States and Japan', *American Journal of Sociology*, vol 104, pp165–194

Zucker, L. (1986) 'Production of trust: Institutional sources of economic structure, 1840–1920', in B. M. Staw and L. L. Cummings (eds) *Research in Organizational Behaviour*, vol 8, JAI Press, Greenwich, CT, pp53–111

2 Social Identity and the Group Context of Trust: Managing Risk and Building Trust through Belonging[1]

Michael A. Hogg

As a species, humans are relatively weak and vulnerable, and depend upon others for physical and social survival. We have an overwhelming need to form bonds with and attachments to others (Baumeister and Leary, 1995) – a need that is laid down in very early childhood (Bowlby, 1988). Because this fundamental social dependency, based upon personal vulnerability, puts us at potential risk of physical and psychological exploitation from those we depend upon, we need to assess the degree of risk associated with particular relationships in order to decide whom to trust. Trust, therefore, plays a key role in social dependency and social connections. Trust is the bond that nourishes close personal relationships with partners, friends and family, and a sense of belonging and attachment to social groups and larger social categories. According to Marar (2003), it is personal vulnerability that lies at the heart of the need to trust – if we did not feel vulnerable we would not feel the need to trust people.

At the interpersonal level, trust is usually vested in people we know very well (for example, friends, relatives, parents and partners) and is closely associated with self-disclosure, such that violation of trust, and exploitation of vulnerability and self-disclosure, is usually the death knell for a personal relationship, and often transforms love and loyalty into hatred and vindictiveness (see, for example, Yamagishi and Yamagishi, 1994; Boon and Holmes, 1999; Holmes, 2002). Not surprisingly, in most societies trust violation within the family is considered abhorrent and attracts severe punishments.

At the group level, trust is often associated with people whom we do not know very well, or at all. We trust people simply because they are in the same group as us – they are 'one of us' (Brewer, 1981; Macy and Skvoretz, 1998; Yamagishi and Kiyonari, 2000; Buchan et al, 2002). We expect loyalty from group members, and trust others to work for the good of the group and, thus, of ourselves as group members. Indeed, in most societies betrayal of the trust that a group places in its members can be labelled treason and is often a capital offence, attracting, in medieval times, particularly gruesome punishment.

Group trust, relative to interpersonal trust, has an extra facet – there is an intra-group and an inter-group dimension, and the two dimensions may be dynamically related such that the less one trusts specific outgroups (inter-group distrust), the more one has to, or needs to, trust the in-group (intra-group trust). Indeed, as we will see below, out-group suspicion and distrust may be a strategic device that leaders use to build in-group trust and solidarity. Lack of trust between groups may underpin some of the world's greatest social and environmental catastrophes. A history of inter-group distrust can make it extremely difficult for conflict to be resolved or avoided – there is, between groups, what Prentice and Miller (1999) call a cultural divide based on past atrocities and betrayals. In the environmental context, if two groups who share a replenishable resource, such as a fishery, do not trust one another to conserve the resource, then both groups vigorously pursue their own short-term gain. This inevitably depletes the resource and often destroys it.

This chapter focuses on the group dimension of trust. Specifically, it outlines a social identity analysis of the social psychology of group and inter-group trust and risk management. The social identity perspective has developed over the past 30 years to become an influential mid-range social psychological theory of the relationship between collective self-conception and group and inter-group behaviour, and yet it does not explicitly and directly discuss the role of trust. This chapter says a little about trust research in social psychology, describes the social identity perspective, and then draws out social identity implications for trust and the role of trust in group life.

The simple take-home message is that when a specific group membership is an important part of who one is – one's collective self-concept – one treats fellow in-group members as 'extensions' of one's self. Their feelings, attitudes, intentions and behaviours are largely one's own, and so they can, of course, be trusted not to do harm to the group or oneself. Identification with a group generates trust among group members. However, where there is an in-group there are out-groups, and in-group trust is often associated with deep distrust between groups.

SOCIAL PSYCHOLOGY OF TRUST

Social psychologists investigate the causes, course and consequences of human interaction. In doing this, the discipline has shown a strong preference for theories that elaborate upon the role of social cognitive processes in the mind of the individual – how people perceive and think about other people (Devine et al, 1994; Taylor, 1998). A social psychology that is so cognitively oriented has also found it rather challenging to theorize about more affective, 'hot', aspects of social life, such as emotion, mood, love and so forth. What tends to happen is that theories of these aspects adopt a rational cognitive approach – for example, Forgas's (1995) affect-infusion model of the role of mood in social thought, and Kelley and Thibaut's (1978) social exchange/cost-benefit theory of interpersonal relations. Even the social identity perspective, which is elaborated upon below, has a strong cognitive dimension (Turner et al, 1987).

The study of trust is likewise problematic for social psychologists. Perhaps this is one reason why there is rather less explicit research in social psychology on trust than one would expect – significant recent exceptions include research on social trust by Cvetkovich and his colleagues (see, for example, Earle and Cvetkovich, 1995; Cvetkovich and Löfstedt, 1999) and research on trust in organizational settings by Kramer and Tyler (for example, Kramer and Tyler, 1996; Kramer, 1999; Tyler and Huo, 2002). Although trust is largely a feeling, social psychological analyses of trust, or rather of phenomena that involve trust, almost always view people as rational beings who make careful cost-benefit analyses before deciding whether to risk engaging in particular behaviours. *Homo oeconomicus* reigns, and affect is error variance.

Game theory and the study of trust

The principle domain for trust research in social psychology is the study of cooperation and competition among individuals whose goals are in conflict. The classic scenario is Luce and Raiffa's (1957) anecdote about the prisoner's dilemma. Two obviously guilty suspects are questioned separately by detectives who have only enough evidence to convict them of a lesser offence. The suspects are separately offered a chance to confess, knowing that if one confesses but the other does not, the confessor will be granted immunity and the confession will be used to convict the other of the more serious offence. If both confess, each will receive a moderate sentence. If neither confesses, each will receive a very light sentence. If the prisoners trust one another, then neither will confess; but if, as is overwhelmingly the case, there is no trust, they almost always both confess (see, for example, Dawes, 1991).

The model, introduced by Von Neumann and Morgenstern (1944), used mainly during the 1960s and 1970s to study dilemmas of trust, is called *decision theory*, *game theory* or *utility theory*. It is highly abstract and rational, and its relevance to the real world of conflict has been questioned (see, for example, Nemeth, 1970) so that since the 1980s its popularity in social psychology has declined dramatically. Its treatment of trust can also be criticized on the grounds that, in fact, it ultimately leaves trust out of the equation and instead talks about cost-benefit analyses and risk-management (Rothstein, 2000). As Marar puts it: 'Game theorists and managers look for the best move, while in reality we choose to trust or distrust precisely when we cannot know the best move' (Marar, 2003, p157).

Social dilemmas

Although game theory research has focused on interpersonal trust rather than what may happen within and between groups, other research does look at trust/distrust-related behaviour at the group level – in particular, in the context of social dilemmas such as the commons dilemma (see, for example, Liebrand et al, 1992; Dawes and Messick, 2000; Kerr and Park, 2001). In the commons dilemma, a shared pasture for grazing is destroyed by overgrazing because people who have access to the common pasture do not trust each other to show restraint, and so each pursues his or her own personal short-term interest, rather than the long-term shared interest of the group. The commons dilemma is a replenishable resource dilemma.

Another social dilemma is a public goods dilemma, where people let others contribute to the good that benefits all, but they themselves do not contribute – this is the free-rider problem that describes tax avoidance, fare evasion and so forth. The trust aspect is less obvious here. However, it is more obvious in another manifestation of free-riding – participation in social movements (Klandermans, 1997; Stürmer and Simon, 2004). Social movements, if successful, benefit large numbers of people. However, engagement in protest and struggle can be costly for the individual in terms of time and money, and can sometimes carry physical risk. People are unlikely to take part unless their cost-benefit analysis is favourable, and they can trust others to also take part.

The key question, then, is how can one build trust and resolve social dilemmas? Research suggests that when self-interest is pitted against the collective good, the usual outcome is competition and resource destruction (see, for example, Sato, 1987). Often the most effective way of resolving a dilemma is to by-pass trust altogether and, instead, impose a structural solution (see Foddy et al, 1999) that regulates utilization of, or contribution to, the resource. This usually involves having a leader manage the resource (see, for example, van Vugt and de Cremer, 1999). For this to work, the leader needs to be an enlightened and powerful authority that can implement measures, manage the bureaucracy and police violations. This can be difficult to bring about.

Another line of research finds that people who identify strongly with a group that accesses a resource develop strong bonds of mutual trust and obligation, and they utilize or contribute to the resource in ways that benefit the entire group (see Brewer and Schneider, 1990; de Cremer and van Vugt, 1999, 2002). Self-interest is transformed into group interest. It is this idea, of group belonging transforming relationships among people, which is the core of the social identity perspective that is the focus of the rest of this chapter.

Social identity perspective

The social identity perspective has its roots in research by Tajfel during the late 1960s and the early 1970s on social categorization, inter-group relations, discrimination and cognitive aspects of prejudice (see Tajfel, 1969; Tajfel and Turner, 1979). Since then, it has developed into an integrative social psychological analysis of the relationship between collective self-construal (social identity) and group processes and inter-group relations (for recent conceptual and empirical overviews, see Turner, 1999; Abrams and Hogg, 2001; Hogg, 2001a, 2003, 2005a; see also Hogg and Abrams, 1988). The social identity perspective has a number of compatible and integrated conceptual components. These deal, for example, with the structure of self (for example, Turner, 1982); processes of social comparison (see Hogg, 2000a); self-enhancement and uncertainty reduction motives (Abrams and Hogg, 1988; Hogg, 2000b); social influence and conformity (see Turner, 1991); inter-group relations (Tajfel and Turner, 1979); and the generative role of social categorization (Turner et al, 1987).

At the heart of the social identity perspective is the view that group phenomena are associated with people identifying and evaluating themselves in terms of the defining attributes of a group that are shared by fellow members. Group phenomena are associated with collective self-conception in terms of a specific group membership – social identity (we/us versus them). Social identity is distinguished from personal identity – self-definition in terms of close personal relationships and/or attributes that are thought to be uniquely or idiosyncratically self-defining (I/me versus you). The social identity perspective does not elaborate upon personal identity or personal identity-related phenomena.

Categories, prototypes and salience

People represent social categories (groups) as prototypes. Prototypes are fuzzy sets of attributes (beliefs, attitudes, feelings, perceptions and behaviours) that capture similarities within a group at the same time as they also capture differences between groups. Prototypes are governed by the principle of meta-contrast (they are composed in such a way as to maximize the ratio of inter-group differences to intra-group differences), and so accentuate category distinctiveness and entitativity (the extent to which a category has the properties of a distinct and bounded entity). As such, prototypes have two other important features. The prototype usually describes the ideal, not the typical or average group member (the prototype is often polarized away from a specific out-group and is therefore more extreme than the average in-group member), and prototypes can vary from context to context as a function of the out-group that the in-group is compared against.

In a given social context, people recruit social categorizations to render the context and their place within it meaningful. To do this people cognitively 'try out' categorizations that are 'accessible' in their mind and in the context itself. If the categorization does not account for similarities and differences among people, and does not explain why people behave as they do, then the category has poor 'comparative fit' and poor 'normative fit' to the social context. Other available categorizations will be tried out until one has adequate fit.

Depersonalization, self-categorization and self-extension

The specific social categorization that best fits becomes the psychologically salient basis of social perception and self-construal in that particular context. The categorization becomes the basis of social perception of other people – in-group members are viewed in terms of the in-group prototype and out-group members in terms of the out-group prototype. Idiosyncratic differences among people are perceptually submerged so that people are viewed through the lens of the relevant prototype. People are depersonalized in terms of the relevant prototype so that they are seen only as embodiments of the prototype and differ only in how well they appear to fit the prototype. Another way to put this is that social categorization leads to stereotypic perceptual homogenization within groups and differentiation between groups.

The pivotal point for the social identity perspective is that people also categorize themselves. What this produces is depersonalization of self in terms of the in-group prototype. Self-categorization transforms self-conception so that one views oneself in terms of the social category – one's social identity as a member of that category is the psychologically salient basis of self-construal. A specific social identity governs one's perceptions, beliefs, attitudes, feelings and behaviours. One feels prototypically similar to fellow in-group members and prototypically different from out-group members, and one's perceptions, thoughts, feelings and behaviours conform to those prescribed by the in-group prototype.

Another way of thinking about the consequences of self-categorization for self-construal is in terms of the notion of self-extension. Research on interpersonal relationships by Aron and his colleagues (see, for example, Aron et al, 1991) has shown that people are able to extend their self-conception to incorporate features of another person; indeed, in close relationships people often describe themselves in terms of attributes that are actually possessed by their partner. This idea has been extended to show that in group contexts, people can include the in-group, or in-group attributes, as part of the self (for example, Smith and Henry, 1996; Smith et al, 1999). Recently, self-extension has been offered as an alternative way of characterizing the process and strength of group identification (for example, Tropp and Wright, 2001; Wright et al, 2002); this characterization is not inconsistent with the self-categorization process described above.

Self-enhancement and uncertainty reduction

Social identity processes are underpinned by two major motivations. Perhaps the most fundamental is uncertainty reduction (Hogg, 2000b, 2001b, 2004). Identifying with a group provides you with a self-definition that prescribes how you should feel, perceive, think and behave. It tells you what sort of person you are and how you should act. It also tells you about other people, out-group members and fellow in-groupers, and how they will interact with you. Identification reduces social and self-uncertainty and allows you to predict and, thus, act effectively in a complex social world. Identification reduces risk and produces in-group trust.

The other motivation is self-enhancement (for example, Abrams and Hogg, 1988; Rubin and Hewstone, 1998). People generally like to feel good about themselves and for others to view them positively. Identifying with a group provides, as self-evaluation, all the evaluative properties of the group: its social valence, its social standing and its status. Thus, people strive to be members of groups that furnish an evaluatively positive social identity. If the group's social evaluation is favourable, they will protect it. If the group's social evaluation is unfavourable, they may try to leave the group and join better groups, or they may work in a variety of ways to improve the group's social evaluation (see Tajfel and Turner, 1979; Hogg and Abrams, 1988).

Cohesion, social attraction and inter-subjectivity

Because people like to view their group favourably, the in-group prototype is generally evaluatively positive. This, working in conjunction with depersonalization, produces favourable inter-individual evaluations and feelings within the group. Within the group, people tend to like one another 'as group members', and this 'liking' is especially governed by how prototypical others are perceived to be. There is prototype-based depersonalized social attraction among members that can be experienced as a sense of group solidarity and cohesion (Hogg, 1993).

Self-categorization or self-extension enhances, through depersonalization, perceived prototypical similarity among in-group members and therefore produces conditions that favour inter-subjectivity. This means that within a group, people are likely to be able to see the world through fellow group members' eyes, and effectively take their perspective (see, for example, Batson et al, 1997; Maner et al, 2002). In addition, they may feel their feelings, experience their internal states and processes, empathize with them, and so forth – there is a degree of inter-subjectivity (see Mackie et al, 2000; Norton et al, 2003).

SOCIAL IDENTITY, TRUST AND RISK MANAGEMENT

Having briefly overviewed some key features of the social identity perspective, two things emerge – little or nothing is explicitly said about trust and risk management, and a lot could be said. This section explicitly draws out the implications of the social identity perspective for an analysis of trust and risk management.

Social identity reduces risk and builds in-group trust. Although interpersonal relations also do this among people who know one another, social identity provides a basis for trust among strangers simply because they are categorized as fellow in-group members. Group membership-based trust, therefore, has a much wider and more immediate benefit for society: categorization of a complete stranger as a fellow in-group member immediately and automatically reduces risk and increases trust.

The reduction in risk comes about through self-categorization-based uncertainty reduction. As we saw above, self-categorization defines self, prescribes one's conduct and renders relations among people predictable – you know who you are, who others are, and how people will relate to, and interact with, you. The risk that something unpredictable may happen, or that fellow in-groupers will do you harm, is minimized. This alone lays the groundwork for you to trust fellow in-group members. Indeed, going back to Hirschmann (1970), we see that in-group loyalty is one powerful response to the perception of uncertainty and risk – the other responses are to exit the group or to voice concerns.

Identification quite possibly functions as an effective heuristic for the management of risk under uncertainty. This idea is consistent with Tversky and Kahneman's (1974) argument that when we make judgements under uncertainty, we tend to take cognitive short cuts (heuristics) that deliver the best solution in the shortest time and with minimum cognitive effort (see also Kahneman et al, 1982). Perhaps it is when we have the luxury of more deliberation time and are less self-conceptually uncertain that we make more careful and deliberate risk assessments as the basis of whom we trust and how much we trust them (compare Chapter 4 in this volume).

More generally, social identification may be the process that mediates the relationship between value similarity and trust. Research has shown that in the absence of reliable diagnostic information about a specific person, people base trust and subsequent cooperation on the perception of value similarity (Earle and Cvetkovich, 1995; Siegrist et al, 2000). Shared values, attitudes and practices map out the contours of social identities; thus, the perception of value or attitude similarity between self and other is a potent prime for social identification, and the process of social identification produces trust. Indeed, Bianco (1998) has stated that identity cues are the most diagnostic cues for trust.

Trust is also strengthened by social attraction, social identity itself and inter-subjectivity. Self-categorization is associated with social attraction – you like fellow group members, as group members (Hogg, 1993). It is cognitively inconsistent to like someone and to harm them or for you to expect them to harm you (see, for example, Abelson et al, 1968). Thus, in salient groups, social attraction (cohesion and solidarity) is associated with a web of inter-member trust.

Social identity is evaluative self-definition in terms of the group prototype – thus, group members are unlikely to behave towards fellow members in ways that will damage the integrity and valence of the prototype. In groups, you can trust fellow members (particularly those who value membership highly) not to do you harm if you, too, are 'one of us' and behave in ways that protect and promote the group.

Finally, the notion of inter-subjectivity increases the probability that you will not harm fellow members because you would 'feel their pain' – therefore, for the same reason, you can trust fellow members not to do you harm. Inter-subjectivity also reduces risk because it allows you to take the perspective of the other and to know, therefore, what that person is thinking, feeling and planning to do.

PATTERNS OF TRUST WITHIN GROUPS

Social identification reduces risk and generates trust among people who belong to the same group. However, things may not be quite so straightforward. Within groups, members are highly attuned to prototypicality – they pay close attention to information about the context-specific prototype and tend to differentiate between members who are highly prototypical and those who are not (Hogg, 2005b). Groups are not homogeneous – they are differentiated in many ways, among which, perhaps, the most critical is prototypicality.

Highly prototypical members are more likely to be trusted than are marginal members. This is because the former are more likely to identify strongly with the group and to have a greater investment in the group. Their membership credentials are unquestioned, and they are expected and trusted to treat the group and fellow members fairly, and in ways that protect and promote the group and its members. Marginal members are not taken on trust as they still need to validate their membership credentials. There is greater risk of harm from a marginal than a core member; thus, correspondingly less trust is invested in marginal members.

There are two relatively recent social identity literatures that connect with this analysis: leadership and deviance.

Leadership

The social identity theory of leadership (Hogg, 2001c; Hogg and van Knippenberg, 2003; van Knippenberg and Hogg, 2003) argues, and provides evidence, that as people identify more strongly with a group, effective leaders are more likely to be high than low prototypical members. This is because members pay more attention to prototypical leaders (they are informative about the prototype), are more likely to be persuaded by the position they occupy (conformity to the prototype), consensually like them more (social attraction) and assign them high status within the group, and tend to view them as more charismatic.

In addition, prototypical leaders are trusted more precisely because they are prototypical and, therefore, trustworthy and likely to be doing the best for the group (Tyler, 1997). This trust in the leader generates a paradox. Being prototypical actually allows the leader to act non-prototypically and to be innovative in remodelling the group's identity, norms and prototype, and in steering the group in new directions – an idea that is consistent with Hollander's (1958) earlier notion that leaders who conform to group norms on the way up earn idiosyncrasy credits that can be spent when they reach the top. To a great extent, it is trust that allows leaders of high salience groups to actually lead effectively rather than merely to manage.

Leaders of high salience groups who are not highly prototypical have their work cut out for them. They find it difficult to be innovative because they need to behave prototypically in order to validate their membership credentials and to be trusted to be 'doing it for the group' (for example, Platow and van Knippenberg, 2001). Once again it is trust that underpins effective leadership. As an aside – it should be noted that the other, non-prototype-based, way of steering a salient group in new directions is by exercising coercive power over the group. However, by most definitions of leadership (for example, Chemers, 2001), coercive power is not leadership.

This analysis of the role of trust in effective leadership is consistent with the wider literature on leadership. For example, many management and industrial/organizational psychology theories of leadership maintain that leaders need to be innovative. In order to be innovative they need to transform subordinates so that they think of themselves as 'we' not 'I' – for this to happen, subordinates need to trust the leader (see Yukl, 1998). Leadership is largely an exercise in building trust so that you can be innovative. As Marar (2003, p152) puts it: 'If you want to lead ... then you had better be someone people trust.' In pursuit of trust, prototypical leaders are also well placed to strategically build in-group trust by focusing the group's attention on threats posed by an enemy out-group that cannot be trusted. The social identity perspective provides a social cognitive account of the generative role of group membership in trust-based leadership of salient groups.

Deviance

In many respects, in salient groups, the mirror image of leadership is deviance – leaders are highly prototypical members and deviants are highly non-prototypical members (Hogg, 2005b; Hogg et al, 2005). One line of social identity research on deviance focuses on 'black sheep' – in-group members who are so marginally prototypical that they pretty much sit right on the inter-group boundary (see, for example, Marques and Páez, 1994; Marques et al, 2001a). These members are strongly disliked and often rejected entirely as members – indeed, they are often disliked more than out-group members who occupy the same position.

Although a trust analysis is not offered, one readily flows from our preceding discussion. 'Black sheep' muddy inter-group boundaries and thus threaten the clarity and focus of the in-group prototype and raise uncertainty. They put the group and its members at self-definitional risk. 'Black sheep' are also dangerously close to being out-group members and they are disliked in social attraction terms. Together, these perceptions make 'black sheep' appear untrustworthy and a threat to the group. Reactions to 'black sheep' are often very extreme, which suggests that there is more going on than simply low social attraction and the perception of group entitativity and valence threat. 'Black sheep' violate in-group trust, pose the possibility of betrayal, and therefore deeply threaten the affective and self-definitional base of group solidarity.

More broadly, in groups there is often a dynamic of deviance that rests heavily on a discourse of trust. In particular, leaders often identify members or sub-groups as marginal in order to bolster their own leadership position (for example, Reicher and Hopkins, 2003). Such prototypically marginal members are vilified and characterized as distrustful, disloyal and actively working to betray the group and all it stands for. Stalin's purges of 'dissidents' are a good example of this.

What happens when people are non-prototypical, but in a positive way – in other words, they have attributes that make them even more different from the out-group than is the in-group prototype itself? Social identity research on such 'positive deviants' suggests that they can invite the same approbation as 'black sheep' because, after all, they violate group norms, threaten group integrity and solidarity, and are low in prototypicality (see, for example, Marques et al, 2001b). Other research qualifies this – where group valence or status (rather than entitativity and distinctiveness) is under threat and the positive deviants publicly give credit for their behaviour to the group, and they are embraced rather than rejected by the group (Fielding et al, 2006; Hogg et al, 2005). Presumably such deviants are viewed as acting for the group by helping the group redefine itself in more positive terms – as such, they are trusted in much the same way as are highly prototypical leaders.

Self-construal

There is a further qualification to the social identity analysis of intra-group trust. Research has shown that people can differ in their preferred mode of self-construal, some preferring the independent self, autonomous and separate from others, and others the relational self, defined by relationships with specific other people (Markus and Kityama, 1991) – a difference that reflects cultural differences between the individualist West and collectivist East (Oyserman et al, 2002), and between men and women (Cross and Madson, 1997).

Brewer and Gardner (1996) have argued that this difference can be reflected in different bases for, and forms of, social identification; for some people (such as Westerners and men), group membership and social identity rest more on the sense of collective and depersonalized self that we have discussed above, whereas for others (Easterners and women), it may rest more on networks of relationships within the group (see, for example, Seeley et al, 2003). The implications for group membership-based trust have not been fully explored; but, in principle, they are clear. When group membership is salient, Westerners and men will behave much as described above, whereas Easterners and women will show patterns of intra-group and group membership-based trust that are more personalized (Maddux and Brewer, in press).

TRUST BETWEEN GROUPS

Thus far we have focused on trust and risk management among members of the same group. Group identification reduces risk and builds trust within groups, but this is patterned by how prototypical members are. Nevertheless, identification seems to be a good way of building trust in society. There is, however, a catch – a rather serious catch. As we have seen above, in-groups, by definition, imply out-groups, and the mechanisms of cohesion, solidarity, social attraction, inter-subjectivity, common bonds and so forth that prevail within groups certainly do not hold between groups. There is an oft-quoted passage on ethnocentrism from Sumner (1906, p13) that describes inter-group relations rather accurately:

> ... one's own group is the centre of everything, and all others are scaled and rated with reference to it... Each group nourishes its own pride and vanity, boasts itself superior, exalts its own divinities and looks with contempt on outsiders. Each group thinks its own folkways the only right one... Ethnocentrism leads a people to exaggerate and intensify everything in their own folkways which is peculiar and which differentiates them from others.

There is, therefore, a very real risk that building in-group trust through identification simply recreates (extreme) distrust at a superordinate intergroup level. Indeed, social identity processes actually generate inter-group distrust – out-groups are disliked, there is no inter-subjectivity to promote trust, and out-groups are largely out to protect or raise their own status at the expense of the in-group. Out-groups certainly cannot be trusted not to do you harm.

Inter-group distrust, particularly at the global level, is a significant obstacle to social harmony (see, for example, Kramer and Messick, 1998; Kramer and Jost, 2002) – there are countless examples, which include Northern Ireland (Murphy, 1978; Hewstone et al, 2005) and Israel and Palestine (Kelman, 1990). In addition, many minorities who are in a subordinate power relationship to the social majority not only distrust the majority, but also the entire social system that the majority has put in place (see Steele et al, 2002) – a distrust that can sometimes be buffered by elaborate cognitive acrobatics to try to justify the system (Jost and Hunyadi, 2002; see also Jost et al, 2003).

Since so much social harm lies at the feet of destructive and distrustful inter-group relations, what can be done to improve inter-group relations? Specifically, how can one build trust between different groups? This is one of social psychology's most important research questions (Pettigrew, 1998). From a social identity point of view, the immediate answer is that one should create conditions in which the two groups categorize themselves as one superordinate group – inter-group relations are transformed into intra-group relations and the trust problem is solved. Sadly, although this can be done in controlled laboratory experiments, it is very difficult to do in the real world.

Relations between real groups are often characterized by a long and entrenched history of distrust, hatred and violence – there is a yawning cultural divide between groups (Prentice and Miller, 1999), often with large power and status differentials. Even in the somewhat less charged atmosphere of organizational mergers, successful recategorization can be very difficult to achieve (see, for example, Terry et al, 2001).

One important reason why recategorization does not easily work is that people have a self-definitional vested interest in retaining the distinctiveness of their existing groups. Recategorization threatens distinctiveness and, ultimately, self-definition, and it also raises self-definitional uncertainty. Recategorization implies that one should cut one's bonds of solidarity and trust with one's former group and its members, and, perhaps, develop bonds of trust with one's erstwhile foe – a very risky business. There are, therefore, strong forces at play that work against successful recategorization.

The art, then, of increasing trust between groups is to assiduously avoid methods that, along the way, threaten the distinctiveness of the groups. One possibility that has been proposed in different guises by a number of social identity researchers is to create what are, effectively, positive role relations between groups such that each group fundamentally depends upon the distinctive existence of the other group (see Hewstone and Brown, 1986; Hornsey and Hogg, 2000; Hogg and Hornsey, 2006). Under these circumstances, there is a superordinate identity that does not require superordinate recategorization and sub-group decategorization, but rather specifies mutually enhancing and additive inter-group role relations. Role relations such as these require inter-group trust. Indeed, this arrangement may cause people to define themselves at the subordinate level in terms of these positive inter-group relations.

The argument is that this creates a self-concept that celebrates and defines self in terms of diversity. The self-concept extends to include positive relations with specific out-groups as part of self-definition (Roccas and Brewer, 2002; Wright et al, 2002; Hogg and Hornsey, 2006), and this may encourage one to take the perspective of specific out-groups and see the world the way they do (for example, Galinsky and Moskowitz, 2000; Galinsky, 2002). In this way, inter-group trust is established without uncertainty and without any threat to sub-group social identity. The best real world examples of an arrangement like this probably come from societies that endorse multiculturalism or cultural pluralism – ethnic sub-groups retain their distinctiveness and integrity within a superordinate nation that explicitly celebrates ethnic diversity as a component of self-definition.

CONCLUSIONS

The aim of this chapter has been to explore ways in which the social identity perspective may be able to comment on risk management and trust. Although, with some exceptions, the social identity perspective does not explicitly or systematically build trust and risk management into its analysis of collective self-construal and group and inter-group behaviour, it can readily do so.

Group identification reduces subjective uncertainty and builds depersonalized social attraction and inter-subjectivity among in-group members. These processes reduce the perception of being at risk of harm from fellow in-group members. However, all in-group members are not trusted equally. All things being equal, there is greater trust for highly prototypical in-group members than marginally prototypical members. In addition, in-group trust is balanced by inter-group distrust. People are quite certainly at risk of harm from out-groups that are in competition with their group over status and prestige. Hence, there is often deep and implacable distrust between groups. One of the major conceptual challenges for social psychology, and practical challenges for society, is to understand the conditions under which inter-group distrust can be ameliorated.

Hopefully this chapter shows there is a clear place for an analysis of trust and risk management in the social identity perspective. Indeed, the social identity perspective probably needs to systematically draw out this potential. This chapter gives some initial ideas of where we might start.

NOTE

1 This chapter was written while spending a sabbatical year at the University of California, Santa Barbara, US.

REFERENCES

Abelson, R. P., Aronson, E., McGuire, W. J., Newcomb, T. M., Rosenberg, M. J. and Tannenbaum, P. H. (eds) (1968) *Theories of Cognitive Consistency: A Sourcebook*, Rand McNally, Chicago

Abrams, D. and Hogg, M. A. (1988) 'Comments on the motivational status of self-esteem in social identity and intergroup discrimination', *European Journal of Social Psychology*, vol 18, pp317–334

Abrams, D. and Hogg, M. A. (2001) 'Collective identity: Group membership and self-conception', in M. A. Hogg and R. S. Tindale (eds) *Blackwell Handbook of Social Psychology: Group Processes*, Blackwell, Oxford, UK, pp425–460

Aron, A., Aron, E. N., Tudor, M. and Nelson, G. (1991) 'Close relationships as including other in the self', *Journal of Personality and Social Psychology*, vol 60, pp241–253

Batson, C. D., Early, S. and Salvarani, G. (1997) 'Perspective taking: Imagining how another feels versus imagining how you would feel', *Personality and Social Psychology Bulletin*, vol 23, pp751–758

Baumeister, R. F. and Leary, M. R. (1995) 'The need to belong: Desire for interpersonal attachments as a fundamental human motivation', *Psychological Bulletin*, vol 117, pp497–529

Bianco, W. T. (1998) 'Uncertainty, appraisal, and common interest: The roots of constituent trust', in V. Braithwaite and M. Levi (eds) *Trust and Governance*, Russell Sage Foundation, New York, pp246–266

Boon, S. D. and Holmes, J. G. (1999) 'Interpersonal risk and the evaluation of transgressions in close relationships', Personal Relationships, vol 6, pp151–168

Bowlby, J. (1988) *A Secure Base: Parent–Child Attachment and Healthy Human Development*, Basic Books, New York

Brewer, M. B. (1981) 'Ethnocentrism and its role in interpersonal trust', in M. B. Brewer and B. Collins (eds) *Scientific Inquiry and the Social Sciences*, Jossey-Bass, San Francisco, CA, pp345–360

Brewer, M. B. and Gardner, W. (1996) 'Who is this "We"? Levels of collective identity and self representation', *Journal of Personality and Social Psychology*, vol 71, pp83–93

Brewer, M. B. and Schneider, S. (1990) 'Social identity and social dilemmas: A double-edged sword', in D. Abrams and M. A. Hogg (eds) *Social Identity Theory: Constructive and Critical Advances*, Harvester Wheatsheaf, London, pp169–184

Buchan, N. R., Croson, R. T. A. and Dawes, R. M. (2002) 'Swift neighbors and persistent strangers: A cross-cultural investigation of trust and reciprocity in social exchange', *American Journal of Sociology*, vol 108, pp168–206

Chemers, M. M. (2001) 'Leadership effectiveness: An integrative review', in M. A. Hogg and R. S. Tindale (eds) *Blackwell Handbook of Social Psychology: Group Processes*, Blackwell, Oxford, UK, pp376–399

Cross, S. E. and Madson, L. (1997) 'Models of the self: Self-construals and gender', *Psychological Bulletin*, vol 122, pp5–37

Cvetkovich, G. and Löfstedt, R. E. (eds) (1999) *Social Trust and the Management of Risk*, Earthscan, London

Dawes, R. M. (1991) 'Social dilemmas, economic self-interest, and evolutionary self-interest', in D. R. Brown, and J. E. Keith-Smith (eds) *Frontiers of Mathematical Psychology: Essays in Honour of Clyde Coombs*, Springer-Verlag, New York, pp53–79

Dawes, R. M. and Messick, D. M. (2000) 'Social dilemmas', *International Journal of Psychology*, vol 35, pp111–116

de Cremer, D. and van Vugt, M. (1999) 'Social identification effects in social dilemmas: A transformation of motives', *European Journal of Social Psychology*, vol 29, pp871–893

de Cremer, D. and van Vugt, M. (2002) 'Intergroup and intragroup aspects of leadership in social dilemmas: A relational model of cooperation', *Journal of Experimental Social Psychology*, vol 38, pp126–136

Devine, P. G., Hamilton, D. L. and Ostrom, T. M. (eds) (1994) *Social Cognition: Impact on Social Psychology*, Academic Press, San Diego, CA

Earle, T. C. and Cvetkovich, G. T. (1995) *Social Trust: Toward a Cosmopolitan Society*, Praeger, Westport, CT

Fielding, K. S., Hogg, M. A. and Annandale, N. (2006) 'Reactions to positive deviance: Social identity and attribution dimensions', *Group Processes and Intergroup Relations*, vol 9, pp199–218

Foddy, M., Smithson, M., Schneider, S. and Hogg, M. A. (eds) (1999) *Resolving Social Dilemmas: Dynamic, Structural and Intergroup Aspects*, Psychology Press, Philadelphia, PA

Forgas, J. P. (1995) 'Mood and judgment: The affect infusion model', *Psychological Bulletin*, vol 117, pp39–66

Galinsky, A. D. (2002) 'Creating and reducing intergroup conflict: The role of perspective-taking in affecting out-group evaluations', in H. Sondak (ed) *Toward Phenomenology of Groups and Group Membership: Research on Managing Groups and Teams*, vol 4, Elsevier, New York, pp85–113

Galinsky, A. D. and Moskowitz, G. B. (2000) 'Perspective-taking: Decreasing stereotype expression, stereotype accessibility, and in-group favoritism', *Journal of Personality and Social Psychology*, vol 78, pp708–724

Hewstone, M. and Brown, R. (1986) 'Contact is not enough: An intergroup perspective', in M. Hewstone and R. Brown (eds) *Contact and Conflict in Intergroup Encounters*, Blackwell, Oxford, UK, pp1–44

Hewstone, M., Cairns, E., Voci, A., Paolini, S., McLernon, F., Crisp, R., Niens, U. and Craig, J. (2005) 'Intergroup contact in a divided society: Challenging segregation in Northern Ireland', in D. Abrams, M. A. Hogg and J. M. Marques (eds) *The Social Psychology of Inclusion and Exclusion*, Psychology Press, New York, pp265–292

Hirschmann, A. (1970) *Exit, Voice, and Loyalty: Responses to Decline in Firms, Organizations, and States*, Harvard University Press, Cambridge, MA

Hogg, M. A. (1993) 'Group cohesiveness: A critical review and some new directions', *European Review of Social Psychology*, vol 4, pp85–111

Hogg, M. A. (2000a) 'Social identity and social comparison', in J. Suls and L.

Wheeler (eds) *Handbook of Social Comparison: Theory and Research*, Kluwer/Plenum, New York, pp401–421

Hogg, M. A. (2000b) 'Subjective uncertainty reduction through self-categorization: A motivational theory of social identity processes', *European Review of Social Psychology*, vol 11, pp223–255

Hogg, M. A. (2001a) 'Social categorization, depersonalization, and group behaviour', in M. A. Hogg and R. S. Tindale (eds) *Blackwell Handbook of Social Psychology: Group Processes*, Blackwell, Oxford, UK, pp56–85

Hogg, M. A. (2001b) 'Self-categorization and subjective uncertainty resolution: Cognitive and motivational facets of social identity and group membership', in J. P. Forgas, K. D. Williams and L. Wheeler (eds) *The Social Mind: Cognitive and Motivational Aspects of Interpersonal Behaviour*, Cambridge University Press, New York, pp323–349

Hogg, M. A. (2001c) 'A social identity theory of leadership', *Personality and Social Psychology Review*, vol 5, pp184–200

Hogg, M. A. (2003) 'Social identity', in M. R. Leary and J. P. Tangney (eds) *Handbook of Self and Identity*, Guilford Press, New York, pp462–479

Hogg, M. A. (2004) 'Uncertainty and extremism: Identification with high entitativity groups under conditions of uncertainty', in V. Yzerbyt, C. M. Judd and O. Corneille (eds) *The Psychology of Group Perception: Perceived Variability, Entitativity, and Essentialism*, Psychology Press, New York, pp401–418

Hogg, M. A. (2005a) 'The social identity approach', in S. A. Wheelan (ed) *The Handbook of Group Research and Practice*, Sage, Thousand Oaks, CA

Hogg, M. A. (2005b) 'All animals are equal but some animals are more equal than others: Social identity and marginal membership', in K. D. Williams, J. P. Forgas and W. von Hippel (eds) *The Social Outcast: Ostracism, Social Exclusion, Rejection, and Bullying*, Psychology Press, New York

Hogg, M. A. and Abrams, D. (1988) *Social Identifications: A Social Psychology of Intergroup Relations and Group Processes*, Routledge, London

Hogg, M. A., Fielding, K. S. and Darley, J. (2005) 'Fringe dwellers: Processes of deviance and marginalization in groups', in D. Abrams, M. A. Hogg and J. M. Marques (eds) *The Social Psychology of Inclusion and Exclusion*, Psychology Press, New York, pp191–210

Hogg, M. A. and Hornsey, M. J. (2006) 'Self-concept threat and differentiation within groups', in R. J. Crisp and M. Hewstone (eds) *Multiple Social Categorization: Processes, Models, and Applications*, Psychology Press, New York

Hogg, M. A. and van Knippenberg, D. (2003) 'Social identity and leadership processes in groups', in M. P. Zanna (ed) *Advances in Experimental Social Psychology*, vol 35, Academic Press, , San Diego, CA, pp1–52

Hollander, E. P. (1958) 'Conformity, status, and idiosyncracy credit', *Psychological Review*, vol 65, pp117–127

Holmes, J. G. (2002) 'Interpersonal expectations as the building blocks of social cognition: An interdependence theory perspective', *Personal Relationships*, vol 9, pp1–26

Hornsey, M. J. and Hogg, M. A. (2000) 'Assimilation and diversity: An integrative model of subgroup relations', *Personality and Social Psychology Review*, vol 4, pp143–156

Jost, J. T., Glaser, J., Kruglanski, A. W. and Sulloway, F. J. (2003) 'Political conservatism as motivated social cognition', *Psychological Bulletin*, vol 129, pp339–375

Jost, J. T. and Hunyadi, O. (2002) 'The psychology of system justification and the palliative function of ideology', *European Review of Social Psychology*, vol 13, pp111–153

Kahneman, D., Slovic, P. and Tversky, A. (1982) *Judgment under Uncertainty: Heuristics and Biases*, Cambridge University Press, Cambridge, UK

Kelley, H. H. and Thibaut, J. (1978) *Interpersonal Relations: A Theory of Interdependence*, Wiley, New York

Kelman, H. C. (1990) 'Applying a human needs perspective to the practice of conflict resolution: The Israeli–Palestinian case', in J. Burton (ed) *Conflict: Human Needs Theory*, St Martin's Press, New York, pp283–297

Kerr, N. L. and Park, E. S. (2001) 'Group performance in collaborative and social dilemma tasks: Progress and prospects', in M. A. Hogg and R. S. Tindale (eds) *Blackwell Handbook of Social Psychology: Group Processes*, Blackwell, Oxford, UK, pp107–138

Klandermans, B. (1997) *The Social Psychology of Protest*, Blackwell, Oxford, UK

Kramer, R. M. (1999) 'Trust and distrust in organizations', *Annual Review of Psychology*, vol 50, pp569–598

Kramer, R. M. and Jost, J. T. (2002) 'Close encounters of the suspicious kind: Outgroup paranoia in hierarchical trust dilemmas', in D. M. Mackie and E. R. Smith (eds) *From Prejudice to Intergroup Emotions: Differentiated Reactions to Social Groups*, Psychology Press, New York, pp173–189

Kramer, R. M. and Messick, D. M. (1998) 'Getting by with a little help from our enemies: Collective paranoia and its role in intergroup relations', in C. Sedikides and J. Scholpler (eds) *Intergroup Cognition and Intergroup Behaviour*, Erlbaum, Mahwah, NJ, pp233–255

Kramer, R. M. and Tyler, T. R. (eds) (1996) *Trust in Organizations: Frontiers of Theory and Research*, Sage, Thousand Oaks, CA

Liebrand, W., Messick, D. and Wilke, H. (eds) (1992) *Social Dilemmas: Theoretical Issues and Research Findings*, Pergamon, Oxford, UK

Luce, R. D. and Raiffa, H. (1957) *Games and Decisions*, Wiley, New York

Mackie, D. M., Devos, T. and Smith, E. R. (2000) 'Intergroup emotions: Explaining offensive action tendencies in an intergroup context', *Journal of Personality and Social Psychology*, vol 79, pp602–616

Macy, M. W. and Skvoretz, J. (1998) 'The evolution of trust and cooperation between strangers: A computational model', *American Sociological Review*, vol 63, pp638–660

Maddux, W., and Brewer, M. B. (in press) 'Gender differences in the relational and collective bases for trust', to be published in *Group Processes and Intergroup Relations* journal

Maner, J. K., Luce, C. L., Neuberg, S. L., Cialdini, R. B., Brown, S. and Sagarin, B. J. (2002) 'The effects of perspective taking on motivations for helping: Still no evidence for altruism', *Personality and Social Psychology Bulletin*, vol 28, pp1601–1610

Marar, Z. (2003) *The Happiness Paradox*, Reaktion, London

Markus, H. R. and Kitayama, S. (1991) 'Culture and the self: Implications for cognition, emotion, and motivation', *Psychological Review*, vol 98, pp224–253

Marques, J. M., Abrams, D., Páez, D. and Hogg, M. A. (2001a) 'Social categorization, social identification and rejection of deviant group members', in M. A.

Hogg and R. S. Tindale (eds) *Blackwell Handbook of Social Psychology: Group Processes*, Blackwell, Oxford, UK, pp400–424

Marques, J. M., Abrams, D. and Serôdio, R. (2001) 'Being better by being right: Subjective group dynamics and derogation of in-group deviants when generic norms are undermined', *Journal of Personality and Social Psychology*, vol 81, pp436–447

Marques, J. M. and Páez, D. (1994) 'The black sheep effect: Social categorization, rejection of ingroup deviates, and perception of group variability', *European Review of Social Psychology*, vol 5, pp37–68

Murphy, D. (1978) *A Place Apart*, Penguin, Harmondsworth, UK

Nemeth, C. (1970) 'Bargaining and reciprocity', *Psychological Bulletin*, vol 74, pp297–308

Norton, M. I., Monin, B., Cooper, J. and Hogg, M. A. (2003) 'Vicarious dissonance: Attitude change from the inconsistency of others', *Journal of Personality and Social Psychology*, vol 85, pp47–62

Oyserman, D., Coon, H. M. and Kemmelmeier, M. (2002) 'Rethinking individualism and collectivism: Evaluation of theoretical assumptions and meta-analyses', *Psychological Bulletin*, vol 128, pp3–72

Pettigrew, T. F. (1998) 'Intergroup contact theory', *Annual Review of Psychology*, vol 49, pp65–85

Platow, M. J. and van Knippenberg, D. (2001) 'A social identity analysis of leadership endorsement: The effects of leader ingroup prototypicality and distributive intergroup fairness', *Personality and Social Psychology Bulletin*, vol 27, pp1508–1519

Prentice, D. A. and Miller, D. T. (eds) (1999) *Cultural Divides: Understanding and Overcoming Group Conflict*, Russell Sage Foundation, New York

Reicher, S. and Hopkins, N. (2003) 'On the science and art of leadership', in D. van Knippenberg and M. A. Hogg (eds) *Leadership and Power: Identity Processes in Groups and Organizations*, Sage, London, pp197–209

Roccas, S. and Brewer, M. B. (2002) 'Social identity complexity', *Personality and Social Psychology Review*, vol 6, pp88–109

Rothstein, B. (2000) 'Trust, social dilemmas and collective memories', *Journal of Theoretical Politics*, vol 12, pp477–501

Rubin, M. and Hewstone, M. (1998) 'Social identity theory's self-esteem hypothesis: A review and some suggestions for clarification', *Personality and Social Psychology Review*, vol 2, pp40–62

Sato, K. (1987) 'Distribution of the cost of maintaining common resources', *Journal of Experimental Social Psychology*, vol 23, pp19–31

Seeley, E., Gardner, W., Pennington, G. and Gabriel, S. (2003) 'Circle of friends or members of a group? Sex differences in relational and collective attachment to groups', *Group Processes and Intergroup Relations*, vol 6, pp251–263

Siegrist, M., Cvetkovich, G. T and Roth, C. (2000) 'Salient value similarity, social trust, and risk/benefit perception', *Risk Analysis*, vol 20, pp353–362

Smith, E. R., Coats, S. and Walling, D. (1999) 'Overlapping mental representations of self, in-group, and partner: Further response time evidence and a connectionist model', *Personality and Social Psychology Bulletin*, vol 25, pp873–882

Smith, E. and Henry, S. (1996) 'An in-group becomes part of the self: Response time evaluation', *Personality and Social Psychology Bulletin*, vol 22, pp635–642

Steele, C. M., Spencer, S. J. and Aronson, J. (2002) 'Contending with group image: The psychology of stereotype and social identity threat', in M. P. Zanna (ed) *Advances in Experimental Social Psychology*, vol 34, Academic Press, San Diego, CA, pp379–440

Stürmer, S. and Simon, B. (2004) 'Collective action: Towards a dual-pathway model', *European Review of Social Psychology*, vol 15, pp59–99

Sumner, W. G. (1906) *Folkways*, Ginn, Boston, MA

Tajfel, H. (1969) 'Cognitive aspects of prejudice', *Journal of Social Issues*, vol 25, pp79–97

Tajfel, H. and Turner, J. C. (1979) 'An integrative theory of intergroup conflict', in W. G. Austin and S. Worchel (eds) *The Social Psychology of Intergroup Relations*, Brooks/Cole, Monterey, CA, pp33–47

Taylor, S. E. (1998) 'The social being in social psychology', in D. T. Gilbert, S. T. Fiske and G. Lindzey (eds) *The Handbook of Social Psychology*, fourth edition, McGraw-Hill, New York, pp58–95

Terry, D. J., Carey, C. J. and Callan, V. J. (2001) 'Employee adjustment to an organizational merger: An intergroup perspective', *Personality and Social Psychology Bulletin*, vol 27, pp267–280

Tropp, L. R. and Wright, S. C. (2001) 'Ingroup identification as inclusion of ingroup in the self', *Personality and Social Psychology Bulletin*, vol 27, pp585–600

Turner, J. C. (1982) 'Towards a cognitive redefinition of the social group', in H. Tajfel (ed) *Social Identity and Intergroup Relations*, Cambridge University Press, Cambridge, UK, pp15–40

Turner, J. C. (1991) *Social Influence*, Open University Press, Milton Keynes, UK

Turner, J. C. (1999) 'Some current issues in research on social identity and self-categorization theories', in N. Ellemers, R. Spears and B. Doosje (eds) *Social Identity*, Blackwell, Oxford, UK, pp6–34

Turner, J. C., Hogg, M. A., Oakes, P. J., Reicher, S. D. and Wetherell, M. S. (1987) *Rediscovering the Social Group: A Self Categorization Theory*, Blackwell, Oxford, UK

Tversky, A. and Kahneman, D. (1974) 'Judgement under uncertainty: Heuristics and biases', *Science*, vol 185, pp1124–1131

Tyler, T. R. (1997) 'The psychology of legitimacy: A relational perspective on voluntary deference to authorities', *Personality and Social Psychology Review*, vol 1, pp323–345

Tyler, T. R. and Huo, Y. J. (2002) *Trust in the Law: Encouraging Public Cooperation with the Police and Courts*, Russell Sage Foundation, New York

van Knippenberg, D. and Hogg, M. A. (2003) 'A social identity model of leadership in organizations', in R. M. Kramer and B. M. Staw (eds) *Research in Organizational Behaviour*, vol 25, JAI Press, Greenwich, CT, pp243–295

van Vugt, M. and de Cremer, D. (1999) 'Leadership in social dilemmas: The effects of group identification on collective actions to provide public goods', *Journal of Personality and Social Psychology*, vol 76, pp587–599

Von Neumann, J. and Morgenstern, O. (1944) *Theory of Games and Economic Behaviour*, Princeton University Press, Princeton, NJ

Wright, S. C., Aron, A. and Tropp, L. R. (2002) 'Including others (and groups) in the self: Self-expansion and intergroup relations', in J. P. Forgas and K. D. Williams (eds) *The Social Self: Cognitive, Interpersonal, and Intergroup Perspectives*, Psychology Press, New York, pp343–363

Yamagishi, T. and Kiyonari, T. (2000) 'The group as the container of generalized reciprocity', *Social Psychology Quarterly*, vol 63, pp116–132

Yamagishi, T. and Yamagishi, M. (1994) 'Trust and commitment in the United States and Japan', *Motivation and Emotion*, vol 18, pp129–166

Yukl, G. (1998) *Leadership in Organizations*, Prentice Hall, New York

3 Trust and Risk: Implications for Management[1]

Eric M. Uslaner

Trust presupposes risk. Yet, people who trust others minimize risk. More precisely, one form of trust, which I shall call 'strategic trust', manages risk. Another form of trust, which I call 'moralistic' or 'generalized' trust, discounts risk. It is this second form of trust that opens up the promise of trust to make our social and political life more cooperative and less confrontational. Moralistic trust waves away risk by doing something most economists and business people might find odd: downplaying evidence.

Why should we discount risk? Generalized trusters don't dismiss risk. Rather, they interpret evidence in a more positive light and are less prone to see high levels of risk than mistrusters. They look at interactions with strangers as opportunities for mutual advantage, rather than as tripwires. They look at people who are different from themselves as presenting opportunities for forming new bridging relationships. So, they see immigrants and open markets as positive forces in promoting growth, rather than as threats to cultural and economic hegemony. They see new technologies as ways of making life easier, rather than as perils to privacy.

Trusters are more tolerant of minorities; they are more likely to donate to charity and to give of their time in volunteering. People with faith in others are more likely to agree on a strong system of legal norms. When trust is high, it is easier to reach agreement across ideological divides. High-trusting countries are less corrupt and have better functioning governments. They spend more on programmes to help those less well off. High-trusting countries have more open markets and greater economic growth. And they have lower crime rates and more effective judiciaries (Uslaner, 2002, Chapters 7 and 8).

Trust becomes an alternative to risk when the world seems less risky. Perhaps the most important way in which we protect ourselves from risk is through the legal system. Yet, we have less need for the strong arm of the law when fewer people have malevolent attitudes and intentions. James Madison, one of the Founding Fathers of the American Republic, remarked: 'If men were angels, there would be no need for government.'

Dasgupta (1988, p53) argued more than two centuries later: 'The problem of trust would ... not arise if we were all hopelessly moral, always doing what we said we would do in the circumstances in which we said we would do it.' If everyone were a truster, we would have less need for a strong legal system to protect us from scofflaws. Yet, if only a small share of people trust their fellow citizens, the foundation of a legal system will be too weak to dole out justice to miscreants.

Trust is important for management because a more trusting environment makes for less conflict in the firm and between firms. Trust also promotes diversity and better relations among different groups. Trust makes it easier to work in a globalized economy. And, perhaps most critically, we know that countries where more people trust each other have higher rates of economic growth (Uslaner, 2002, Chapter 8). So, it also seems likely that companies with more trusters would have higher growth rates. Most critically, trust is important for management because a generalized faith in others makes us less likely to worry about risks – and opens up new opportunities for innovation.

The general argument of this chapter is laid out below; evidence from surveys that supports it is then presented.

VARIETIES OF TRUST

A bond of trust lets us put greater confidence in other people's promises that they mean what they say when they promise to cooperate (compare Elster, 1989, pp274–275; Putnam, 1993, p170).

Yamigishi and Yamigishi (1994) call it 'knowledge-based trust'. Offe (1999) states: 'Trust in persons results from past experience with concrete persons.' If Jane trusts Bill to keep his word, and if Bill trusts Jane to keep her word, they can reach an agreement to cooperate and thus make both of them better off. Even without some external enforcement mechanism (such as an arbitrator, the police or the courts), they will keep to their agreements.

If Jane and Bill did not know each other, they would have no basis for trusting each other. Moreover, a single encounter will not suffice to develop trust. Jane and Bill have to interact over time to develop reputations for keeping their word. And even when they get to know each other better, their mutual trust will be limited to what they know about each other (Hardin, 1992, p154; Misztal, 1996, p121).

The decision to trust another person is essentially *strategic*. Strategic (or knowledge-based) trust presupposes risk (Misztal, 1996, p18; Seligman, 1997, p63). Jane is at risk if she does not know whether Bill will pay her back. And she is at risk if she knows that Bill intends to default on the loan. Trust helps us to solve collective action problems by reducing transaction costs – the price of gaining the requisite information that Bill and Jane need to place confidence in each other (Putnam, 1993, p172;

Offe, 1996, p27). It is a recipe for telling us *when* we can tell whether other people are trustworthy (Luhmann, 1979, p43).[2]

Beyond the strategic view of trust is another perspective. I call it moralistic trust (Mansbridge, 1999, favours 'altruistic trust'). It is based upon the idea that trust has a moral dimension. Moralistic trust is a moral commandment to treat people *as if* they were trustworthy. The central idea behind moralistic trust is the belief that most people share your fundamental moral values. To put it another way, a wide range of people belong to your moral community. They need not share your views on policy issues or even your ideology. They may have different religious beliefs. Yet, despite these differences, we see deeper similarities. Fukayama (1995, p153) states the central idea behind moralistic trust: 'trust arises when a community shares a set of moral values in such a way as to create regular expectations of regular and honest behaviour'. When others share our basic premises, we face fewer risks when we seek agreement on collective action problems. Moralistic trust is based upon 'some sort of belief in the goodwill of the other' (Seligman, 1997, p43; compare Yamigishi and Yamigishi, 1994, p131). We believe that others will not try to take advantage of us (Silver, 1989, p276).

The moral dimension of trust is important because it answers questions that the strategic view cannot. Bill and Jane may develop confidence in each other as they learn more about each other. Each successive cooperative decision that Bill makes increases Jane's faith in him – and vice versa. But why would Bill or Jane decide to cooperate with each other in the first place? If Bill were a Scrooge and Jane were a Bob Cratchit, Jane's confidence in Bill would be misplaced. And this sour experience might lead Jane not to trust other people in the future. The strategic view of trust would lead us to expect that both Bill and Jane would be far more likely to be Scrooges than Cratchits. As Dasgupta (1988) argues, in a world of Cratchits, you wouldn't need strategic trust.

Beyond the distinction between moralistic and strategic trust is a continuum from particularized to generalized trust. Generalized trust is the perception that *most* people are part of your moral community. The difference between generalized and particularized trust is similar to the distinction that Putnam (2000, p22) draws between 'bonding' and 'bridging' social capital. We bond with our friends and people like ourselves. We form bridges with people who are different from ourselves. *The central idea distinguishing generalized from particularized trust is how inclusive your moral community is.* When you only trust your own kind, your moral community is rather restricted. And you are likely to extend trust only to people you think you know. So particularized trusters rely heavily upon their experiences (strategic trust) – or stereotypes that they believe to be founded in knowledge in deciding whom to trust. Particularized trusters fear the unknown – and people who are different from themselves seem risky at best. So beyond people we know from our places of work and worship, we are most likely to trust people from our race, our ethnic group or our religious denomination – or any other group with which we strongly identify.

A summary of these arguments is presented in Table 3.1. The key distinction between generalized and particularized trust is how wide the scope of your moral community is. The belief that 'most people can be trusted' reflects generalized trust. When you only have faith in people like yourself, you are a particularized truster. Generalized and particularized trust form a continuum, not a dichotomy. At one end are people who trust *most others* – it makes no sense to place your faith in *all people*. At the other extreme, some people have little faith in anyone (their own families are generally exceptions; see Banfield, 1958, p110). Particularized trusters lie at this end – although the scope of their moral community may range from small (one's own family) to large (an ethnic, religious or regional community).

Table 3.1 *Dimensions of trust*

Type of trust	Basis of distinction
Moralistic versus strategic trust	Trust based upon values versus trust based upon experience
Generalized versus particularized trust	Scope of trust: belief that 'most people can be trusted' versus only trusting people like yourself

Note: Generalized trust is similar to moralistic trust, but there are some determinants of generalized trust (education, age, race, economic inequality) that do reflect experience rather than values (see Uslaner, 2002, Chapter 2).

Moralistic and strategic trust differ in their foundations: the latter is based upon experience, the former largely not. They are not mutually exclusive, but rather largely independent of each other. Moralistic trusters do not abjure strategic trust. Indeed, moralistic trusters are not so naive that they would hire just *any* contractor or loan a perfect stranger money. Moralistic trust mostly does not reflect how we interact with any *particular person* (which is the domain of strategic trust), but rather how we approach strangers, especially people who may be different from ourselves, more generally. Am I open to risks or am I risk averse? Strategic trust comes into play in deciding *which risks to take*. Moralistic and strategic trust are largely independent of each other. Almost everyone has faith in their own family and close associates, their friends, co-workers, members of their houses of worship and fellow club members (Uslaner, 2002, pp29–30). With such little variation, it is hardly surprising that placing faith in people like oneself has little effect on generalized trust in people who may be different (Uslaner, 2002, pp145–148).

Generalized trust and moralistic trust are largely, though not completely, overlapping. The roots of generalized trust are largely moral since neither form of trust depends heavily upon experience. However, there are some foundations of generalized trust that *do* reflect life experiences – age, education, economic inequality and race – so generalized trust is more than *simply* moralistic trust (Uslaner, 2002, Chapters 4, 6 and 8). Particularized trust is based largely upon experience – with close associates or with people very much like oneself – so it has similar foundations to strategic trust. Yet, particularized trust may also be based upon stereotypes with little underpinning in actual experiences. As a result, the dimensions of generalized versus particularized trust, on the one hand, and moralistic versus strategic trust, on the other, largely overlap, but do not completely do so.

STRATEGIC AND MORALISTIC TRUST

Moralistic trust is a value that rests on an optimistic view of the world and one's ability to control it. It differs from strategic trust in several crucial respects. Moralistic trust is not a relationship between specific persons for a particular context. Jane doesn't trust Bill to repay a $20 loan. Jane just 'trusts' (other people in general, most of the time, for no specific purpose). If the grammar of strategic trust is 'A trusts B to do X' (Hardin, 1992, p154), the etymology of moralistic trust is simply 'A trusts.'[3]

Strategic trust reflects our expectations about how people *will* behave. For strategic trust, Bill must trust Jane *and* Jane must trust Bill. Otherwise there is no deal. Moralistic trust is a statement about how people *should* behave. *People ought to trust each other*. The Golden Rule (which is the foundation of moralistic trust) does *not* demand that you do unto others as they do unto you. Instead, you do unto others *as you would have them* do unto you. The Eighth Commandment is *not* 'Thou shalt not steal unless somebody takes something from you.' Nor does it state: 'Thou shalt not steal from Bill.' Moral dictates are absolutes. Adam Seligman (1997) makes a telling distinction: 'Were the trusting act to be dependent (i.e. conditional) upon the play of reciprocity (or rational expectation of such), it would not be an act of trust at all but an act predicated on *one's expectations of how others will behave*' (compare Mansbridge, 1999).

Strategic trust is not predicated upon a negative view of the world, but rather upon uncertainty. Levi (1997, p3) argues: 'The opposite of trust is not distrust; it is the lack of trust' (compare Hardin, 1992, p154; Offe, 1999). Strategic trust is all about reducing transaction costs by gaining additional information – whether it is positive or negative. But moralistic trust must have positive feelings at one pole and negative ones at the other. It would be strange to have a moral code with good juxtaposed against undecided. So, we either trust most people or we distrust them.

Moralistic trust is predicated upon a view that the world is a benevolent place with good people (compare Seligman, 1997, p47), that things are going to get better, and that you are the master of your own fate. The earliest treatments of interpersonal trust put it at the centre of an upbeat world view (Rosenberg, 1956). People who believe that others can be trusted have an optimistic view of the world. They believe that things will get better *and that they can make the world better by their own actions* (Rosenberg, 1956; Lane, 1959, pp163–166).

All but the most devoted altruists will recall – and employ – the Russian maxim (adopted by President Ronald Reagan in dealing with the Soviets): trust but verify. When dealing with specific people, we use strategic trust. It is hardly contradictory for someone who places great faith in *people* to check out the qualifications and honesty of *specific persons*, such as contractors, mechanics and doctors. *Strategic trust is all about finding ways of minimizing the risk inherent in social life*. Moralistic trust is *not* faith in specific people; rather, it is faith in the 'generalized other'. On the other hand, people who are *not* generalized trusters can only rely on strategic trust. For them, 'trust' means experiences with specific persons.

Strategic trust is fragile since new experiences can change one's view of another's trustworthiness (Bok, 1978, p26; Hardin, 1998, p21). Trust, Levi (1998, p81) argues, may be 'hard to construct and easy to destroy' (compare Dasgupta, 1988, p50). Values are not divorced from experience, but they are largely resistant to the ups and downs of daily life. Moralistic trust is thus *not fragile at all, but quite stable over time*. It is more difficult to build than to destroy because trust is not so easily transferable from one person to another.

Optimists are prone to discount bad news and to give too much credence to good tidings. Pessimists overemphasize adversity and dismiss upbeat messages. Both groups look at evidence selectively. Their reasoning is a 'cognitive "leap" beyond the expectations that reason and experience alone would warrant' (Lewis and Weigart, 1985, p970; compare Baron, 1998, p409; Mansbridge, 1999). It may be a good thing that moralistic trusters are not concerned with reciprocity, for they might well make erroneous decisions on who is trustworthy and who is not. Orbell and Dawes (1991, pp521, 526) report results from an experimental game showing that trusters are overly optimistic about the motivations of others. They use their own good intentions (rather than life experiences) to extrapolate about whether strangers would cooperate in experimental games.

Moralistic trusters are also significantly more likely than mistrusters to say that other people trust them.[4] People who feel good about themselves interpret ambiguous events in a positive light, while people who have a poor self-image and who look at life pessimistically interpret the same experiences negatively (Diener et al, 1997). Since moralistic trusters look at the world with (at least partial) blinkers on, it should not be surprising that this type of trust is not at all fragile – and that *moralistic trust is largely dismissive of risk*.

It would be easier to monitor peoples' trustworthiness if we could simply look at people and determine whether we should trust them. Their appearance would send us a signal that they would not betray us. In a world where knowledge is costly and sometimes scarce, we often find this tactic a useful device to reduce uncertainty.

One fail-safe solution to the problem would be for all trusters to wear outfits with a 'T' and mistrusters to wear clothes marked with an 'M' (compare Frank, 1988). Clearly, this is infeasible. So, for good or ill, we are likely to trust people who look and think most like ourselves. People who look like us are most likely to share our values.

Particularized trust offers a solution to the problem of signalling. The Maghribi of Northern Africa relied on their extended Jewish clan – and other Jews in the Mediterranean area – to establish a profitable trading network in the 12th century. Maghribis and other Jews did not wear clothing with a 'J' (for Jew) or 'T' (for trustworthy). But, as a small enough minority group, Jews could identify each other. They believed that others in their in-group were more likely to deal honestly with them, so they could minimize being exploited when trading with people whom they did not know (Greif, 1993). As long as members of an in-group can identify each other, they can limit their interactions to people whom they expect to be trustworthy.

THE WORLD VIEWS OF GENERALIZED AND PARTICULARIZED TRUST

When you feel good about yourself and others, it is easy to have an expansive moral community. Generalized trusters have positive views towards both their own in-group and out-groups. But they rank their own groups less highly than do particularized trusters (Uslaner, 2002, Chapter 4). If you believe that things are going to get better – and that you have the capacity to control your life – trusting others isn't so risky. Generalized trusters are happier in their personal lives and believe that they are the masters of their own fate (Rosenberg, 1956, pp694–695; Lane, 1959, pp165–166; Brehm and Rahn, 1997, p1015). They are tolerant of people who are different from themselves and believe that dealing with strangers opens up opportunities more than it entails risks (Rotter, 1980, p6; Sullivan et al, 1981, p155).

When you are optimistic about the future, you can look at encounters with strangers as opportunities to be exploited. Optimists believe that they control their own destinies. Perhaps you can learn something new from the outsider or exchange goods so that you both become better off. Even if the encounter turns out to be unprofitable, you can minimize any damage by your own actions. For pessimists, a stranger is a competitor for what little you have. She may also represent the sinister forces that control your life (as pessimists believe). They suspect that outsiders are trying to exploit them. And, given their long-term history, they might be right. But pessimists might also overestimate the likelihood of a bad experience with a stranger, depriving themselves of the opportunities of mutual exchange. Just as some bad experiences are not going to turn optimists into misanthropes, a few happy encounters with strangers will not change long-term pessimists into trusters. Change is possible, but it is likely to occur slowly.

TRUST, RISK AND THE LAW

All of our interactions have some element of risk. Social ties with close friends are hardly a gamble. Hiring a new contractor is more hazardous, while interacting with strangers (including businesses without a national reputation) may seem fraught with uncertainty and possible danger. Where information is plentiful, we can rely upon strategic trust. We can check with consumer bureaus or neighbours about contractors. But when information is scarce, we must fall back on moralistic or generalized trust.

Yet, there *is* an instrument of strategic trust that can reduce, if not minimize, risk: the legal system. If someone takes advantage of us, we can take them to court. A strong legal system will reduce transaction costs, making trust less risky. On one view, a strong system of laws can generate more trust: the more experience people have with compliance, the more likely they are to have confidence in others' goodwill (Brehm and Rahn, 1997, p1008; Levi, 1998; Offe, 1999).

So, Bill knows that if he hires Jane to paint his house and she accepts his payment and does a poor job, he can take her to court for redress. Thus, he won't worry so much if he has to look for a new painter. My own family benefited from this very type of protection: we hired a contractor to repave our driveway and he used an inferior grade of concrete. After a year or more, the Maryland Home Improvement Commission ruled in our favour and we recovered our initial investment. Cohen (1997, p19) argues that 'legal norms of procedural fairness, impartiality and justice that give structure to state and some civil institutions, limit favouritism and arbitrariness, and protect merit are the *sine qua non* for society-wide "general trust", at least in a modern social structure'.

There is plenty of evidence that people are more likely to obey laws and pay taxes if they believe that laws are enforced fairly and if people trust government (Tyler, 1990; Scholz and Pinney, 1995). Rothstein (2001a) argues (compare Misztal, 1996, p251; Offe, 1996, p27; Seligman, 1997, p37; Levi, 1998):

> Political and legal institutions that are perceived as fair, just and (reasonably) efficient, increase the likelihood that citizens will overcome social dilemmas... In a civilized society, institutions of law and order have one particularly important task: to detect and punish people who are 'traitors' – that is, those who break contracts, steal, murder and do other such non-cooperative things and therefore should not be trusted. Thus, if you think that particular institutions do what they are supposed to do in a fair and efficient

manner, then you also have reason to believe ... that people will refrain from acting in a treacherous manner and you will therefore believe that 'most people can be trusted'.

Rothstein (2001b, p21) argues in favour of the linkage between trust in the legal system and faith in people by citing correlations between the trust in different governmental institutions and generalized trust, in Swedish surveys conducted from 1996 to 2000. Of 13 governmental institutions, the correlations with trust in people are highest (though barely) for the police and the courts.

There is little reason to presume that government enforcement of laws will build trust. Yes, coercion can increase *compliance* with the law. Obeying the law because you fear the wrath of government will not make you more trusting – no matter how equally the heavy hand of the state is applied. Generalized trusters are, in fact, less likely than mistrusters to endorse unconditional compliance. In the US General Social Survey, just 35 per cent of trusters say that you should *always* obey the law, even if it is unjust, compared to 48 per cent of mistrusters.[5] Simply getting people to obey laws will not produce trust. Perhaps this is a caricature of the argument on building trust; but it is easy to confuse compliance with voluntary acceptance, to confuse the law-abiding people of Singapore with those of Sweden (compare Rothstein, 2001a). Even in high-trusting countries such as Sweden, the linkage between confidence in the legal system and the police and trust in people is not very strong (Rothstein, 2001a).[6]

Courts can save us from rascals only if there are few rascals (compare Sitkin and Roth, 1993). Law-abiding citizens, not rogue outlaws, create constitutions that work. You may write any type of constitution that you wish, but statutes alone won't create generalized trust. Macaulay (1963, pp58, 61–63) argues that business executives and lawyers prefer transactions based upon trust – and handshake seals the deal – to those based upon contracts and threats of legal sanctions. Most executives and even lawyers have faith that other people will keep their end of a bargain. Resorting to formal documents might undo the goodwill that undergirds business relationships (Macauley, 1963, p63; Mueller, 1999, p96). Coercion, Gambetta (1988, p220) argues, 'falls short of being an adequate alternative to trust... It introduces an asymmetry which disposes of *mutual* trust and promotes, instead, power and resentment' (compare Baier, 1986, p234; Knight, 2000, p365). Generalized trust does *not* depend upon contracts. Indeed, trusting others is sometimes said to be a happy substitute for monitoring their standing (Putnam, 2000, p135).[7]

There is a linkage between confidence in the legal system and trust in people; the direction of causality goes from trust to confidence in the legal system. Trusting societies have strong legal systems, able to punish the small number of scofflaws. Rothstein (2001a) argues that Russians have low levels of trust in each other because they don't have faith in the law. It seems more likely that this direction of causality runs the other way: Russians have a weak legal system because not enough people have faith in each other.

In a cross-national study of trust and corruption, Gabriel Badescu and I found that the correlation between trust and perceptions of corruption were uniformly *low* in countries with high levels of corruption (Uslaner and Badescu, 2004). People in highly corrupt countries – especially the former Communist nations of Central and Eastern Europe – don't blame fellow citizens for corruption. The elites are to blame, so that any successful attempts to control corruption *will not increase trust*. When corruption is high, people learn how to cope. They know who must be paid off (and how much) and they are reasonably certain that the scofflaws will not face the strong arm of the law (which is corrupt itself). Where corruption is low (as in the Nordic countries), people are more likely to make a link between malfeasance and the trustworthiness of their fellow citizens. These high-trusting societies have well-functioning legal systems (see Uslaner, 2002, Chapter 8) and there is little reason to believe that if the transgressors are caught, they will face severe punishment.

Seeking to instil generalized trust from the top down (by reforming the legal system) misses the mark in most cases. If courts or government, more generally, can build up any type of trust at all, it is strategic trust: I can protect myself from an occasional contractor (or corrupt business) who takes advantage of me by seeking legal redress. When most contractors are dangerous, there is little reason to believe that the courts will be able to protect me from risk.

A strong legal system depends upon trust – in other words, upon people who minimize the risks involved in daily life. People who trust others are the strongest supporters of the fundamental norms that make for a civil and cooperative society, according to data from the World Values Survey. Trusters are more likely to say that it is wrong to purchase stolen goods, to claim government benefits that you are not entitled to, to keep money you have found, and to hit someone else's car without making a report. Trust and one's own moral code lead people to endorse strong standards of moral behaviour – and not expectations of others' morality. Trust matters most on moral questions when the stakes are highest (in terms of real monetary costs) and when there is the least consensus on what is moral. When everyone agrees that something is wrong – say, on joyriding – or when violating a norm has small consequences – say, on avoiding a fare on public transportation – trust doesn't matter so much. Trust also matters most when a specific person bears the brunt of breaching a norm. Trust is not quite so important for actions affecting the government – say, cheating on taxes or avoiding fares – as it is when we can point to a specific, though unknown, victim, such as keeping money you have found or hitting someone's car without making a report.[8]

Trusting people are more supportive of *the legal order* as well as of legal norms. They are substantially more willing to serve on a jury – where they not only help to run the system of laws but also are likely to interact with people unlike themselves. Generalized trusters are more likely to say that they are willing to serve on a jury. And particularized trust matters even more: people who rank their own in-groups highly are much less likely to say that they would serve, while those who give more favourable ratings to out-groups are much *more* willing to do their jury duty.[9] The foundation of the legal system that helps to reduce risk is a populace who does not see the world as full of risk.

RISK AND NEW OPPORTUNITIES

Mistrusters look at people who are different from themselves (out-groups) with suspicion. A deep-seated pessimism makes people view outsiders as threats to what little they have.[10] Minorities and immigrants are seeking to take jobs away from the majority population; open markets take jobs away from native citizens. Protecting yourself and others like you against these risks becomes paramount (Uslaner, 2002, pp193–199). When people see little hope for the future and believe that others control their fate, they naturally develop a sense of fatalism and mistrust. Perhaps one of the best descriptions came from Edward Banfield's (1958, p110) description of the social distrust in the Italian village of Montegrano in the 1950s, where poverty was rife and people had little hope for the future: 'any advantage that may be given to another is necessarily at the expense of one's own family. Therefore, one cannot afford the luxury of charity, which is giving others more than their due, or even justice, which is giving them their due.' Banfield's discussion is controversial – not everyone agrees that Montegrano was marked by such mistrust. However, the picture that Banfield drew is a dramatic portrayal of the misanthrope, who sees risk and danger at every corner. Racists and authoritarians fit the portrait just as well.

In contrast, generalized trusters look at people who are different from themselves as members of their moral community. Interacting with them broadens your vistas. So, it is hardly surprising that moralistic trusters have warm feelings towards people who are different from themselves: in the US, white trusters admire African-Americans and favour programmes such as affirmative action that might tear down racial barriers. As with African-Americans, trusters don't see illegal immigrants taking jobs from natives. And they have far more favourable views of *legal* immigrants than mistrusters: immigrants don't increase crime rates, generally help the economy, don't take jobs away from people who were born in America, and make the country more open to new ideas. And trusters don't believe that immigrants can readily work their way up the economic ladder, any more than African-Americans can, without government assistance (Uslaner, 2002, pp193–197).

Trusters also have more positive evaluations of other groups in the society that have faced discrimination. They rate gays and lesbians more highly than mistrusters. Generalized trusters are much more supportive of gays and lesbians serving in the military and adopting children. In each case – general affect, military service and adopting children – particularized trusters (as measured by the difference in feeling thermometers of out- and in-groups in the 1992 American National Election Studies (ANES)) are far less supportive of homosexuals. Particularized trust is by far the strongest determinant of overall affect and it is also more powerful for military service. Trusters are far more supportive of gays' and lesbians' right to teach and speak in schools, and for the right of libraries to have books by gay and lesbian authors. Since trusters don't fear strangers – or even people they don't like or agree with – they are willing to extend the same rights to atheists and racists.

Trusters want to let more immigrants come to America since they are more likely to believe that newcomers share the basic values of people already here. And trusters also favour free trade as a means of boosting economic growth. People with faith in others are less afraid that trading with other countries will permit other countries to take unfair advantage of the US.

People who trust others do not believe that people who are different from themselves constitute a risk. We are all in this together, trusters believe, because Americans share a common set of values.

Similarly, trusters and mistrusters *both* use the new technology of the internet, with little difference in how often they go online. However, there are differences, according to surveys conducted by the Pew Center for the Internet and American Life in 1998 and 2000, in how they go online.[11] First, the new innovation of the internet – chat rooms – offers some hope that people of different backgrounds might get together and learn to trust one another. But here, of all places, we see some evidence of misanthropy. People who visit chat rooms or who make new friends online are *less trusting* than others (compare Shah et al, 2001, p149). Perhaps people who make friends online, often anonymously, feel uncomfortable with meeting 'real' strangers. And many, maybe most, chat rooms are marked by a dominant world view or ideology – and dissidents often find out rather rudely that they are not welcome (Hill and Hughes, 1997). People who frequent chat rooms seem to trust only people like themselves and fear people with different views.

On matters not related to privacy and security, there is little that separates trusters and mistrusters on the internet. Trusting people are no more likely to go online to get information of any sort – or even to buy products. They are no more prone to go to the web for fun – or to spend lots of time on it. Offline, trusting people overall see the web as a place occupied with lots of trustworthy people and companies. They have no desire to hide their identity. Trusting people are more tolerant of people of different races and religions and of minorities who have faced discrimination. They have more favourable attitudes towards immigrants and are more likely to favour open markets.

Where privacy issues are concerned, there is a marked difference in how people approach risk: online, trusters respond to emails from strangers – and receive fewer offensive missives from people they don't know (either because it takes more to offend them or they get on fewer lists with people who write nasty notes). They worry less about what others might learn about them and don't fear that others will invade their personal lives or spread lies. They are more likely to demand that companies ask permission to get personal information; but they will use their credit card numbers for phone orders (though, surprisingly, there is no difference for internet orders).

DOES TRUST SHAPE ATTITUDES TOWARDS RISK?

The argument I have laid out suggests that trusters *underestimate* risk. Even when the world outside seems threatening, people who trust others are less likely to see danger than mistrusters. There are alternative perspectives: Hardin (1992, p165) argues that trust is merely 'encapsulated experience'. His perspective views *all* trust as strategic:

> Suppose ... that I started life with such a charmed existence that I am now too optimistic about trusting others, so that I often overdo it and get burned. Because I am trusting, I enter into many interactions and I collect data for updating my Bayesian estimates very quickly. My experience soon approaches the aggregate average and I reach an optimal level of trust that pays off well in most of my interactions, more than enough to make up for the occasions when I mistakenly overrate the trustworthiness of another.

And Yamagishi (1998) argues that trusters are more sophisticated than mistrusters – and are especially likely to recognize likely defectors in iterative games. Yamagishi believes that trusters are more likely to perceive risk than mistrusters. We might expect that this reflects the greater levels of education among trusters; but this is controlled in Yamagishi's experiments. So the implication seems to be that trusters are *more alert to the dangers of risk than are mistrusters – and even that trusters may overestimate risk*.

Which of these perspectives is correct? This is not an easy set of claims to test. However, there is an ideal set of data that permits an examination of these three alternative arguments. The Pew Center for the People and the Press 1996 Civic Engagement Survey of metropolitan Philadelphia asked the standard survey question about how safe it is to walk in your neighbourhood at night, with responses ranging from very safe, somewhat safe, not too safe, to not at all safe. This question clearly measures risk, though it clearly does not cover all forms of danger. But it is the only survey

I know that includes both perceptions of risk and the generalized trust question.

My argument would expect that *trusters consistently overestimate the level of safety controlling for the actual level of violence in their neighbourhoods*. Hardin's (1992) thesis would expect no difference between trusters and mistrusters in the level of safety, *controlling for the actual level of violence*, since information on the level of violence should 'wash out' any effects of trust (which is endogenous to information). There is no *a priori* reason to presume that trusters and mistrusters would gain different levels of information once other factors (such as where you live, etc.) are controlled. Yamigishi's (1998) argument suggests that trusters would *underestimate the level of safety controlling for the actual level of violence in their neighbourhoods*.

I tested these arguments in a statistical analysis reported in Uslaner (2004). The analysis controls for factors *other than trust* that might lead people to feel safe or unsafe in their neighbourhoods. The 1996 Pew survey is especially important because it includes an objective measure of risk: the actual level of violence in a neighbourhood, as compiled for voting precincts from official police department statistics. If trust is nothing more than encapsulated experience, then we would expect that the level of violence in your neighbourhood would largely determine perceptions of safety. But perceptions of safety may reflect more than just the levels of violence (here from sexual assaults).

Other questions in the survey may also shape attitudes about the safety of your neighbourhood: whether your parent was a victim of a crime, how much you trust your city government, whether you live in the city centre or a suburb, gender, how much you like both your neighbourhood and Philadelphia, and how much local television news you watch. Experiencing crime – even indirectly (through one's parents) – should make you believe that your neighbourhood is unsafe. If you trust your local government and like both your neighbourhood and your city, you are more likely to feel secure (and protected). Suburban residents correctly feel more secure than inner-city residents, while women are more likely to feel that their neighbourhoods are unsafe (since they may be more vulnerable to attacks). People who watch local television news in the US might be very susceptible to the 'mean world' effect. When you see lots of violence on television, you are more likely to believe that the real world is just as 'mean' as the 'television world' and thus be less likely to trust others (Gerbner et al, 1980, pp17–19). Local television news in the US shows a lot of violent crime. People in the news business say of local television news: 'If it bleeds, it leads' – violent news comes first. These control variables should 'equalize' information levels to the extent possible (education was not significant). Generalized trust is the main variable of interest.

Experiences clearly shape the perceptions of neighbourhood safety. The actual level of violence influences perceptions of neighbourhood safety. People living in the least violent neighbourhoods are almost nine times as likely to say that their neighbourhoods are very safe as are people from the most violent neighbourhoods – and they are half as likely to say that their neighbourhoods are very unsafe. People living in the central city are also far more likely to say that their neighbourhoods are unsafe, as are women and people who watch a lot of local television news. People who trust the city government, who like Philadelphia and especially who like their neighbourhood are far more likely to feel safe. Perhaps suprisingly, if your parents were the victim of a crime – *or even if you were the victim of a crime* – you are no more likely to feel unsafe.

Yet, perceptions of risk are not just about experience. Trust shapes attitudes towards risk. *Trusters are more likely to perceive their neighbourhoods as being safe: 38 per cent of trusters compared to 28 per cent of mistrusters say that their neighbourhood is safe, even controlling statistically for all of the other variables (including the actual level of violent crime). Trusters are also half as likely as mistrusters to say that their neighbourhoods are 'not at all safe', regardless of the actual level of violence.* Even at moderately high levels of rape violence, mistrusters are about twice as likely as trusters to say that their neighbourhoods are 'not at all safe'.

Suburban trusters are generally more likely to perceive their neighbourhoods as being very safe than central city trusters. *Yet, within each area of residence, trusters are substantially more likely to see their neighbourhoods as very safe.* Suburban trusters are about 20 per cent more likely to perceive their neighbourhoods as being very safe compared to suburban mistrusters. There is a smaller advantage for trusters in the central city. At low levels of violence, trusters in the central city comprise about 10 per cent. Only when the level of violence becomes relatively high do we find little difference between central city trusters and mistrusters. We cannot make comparisons for very high levels of violence because we only see such high rates in the central city and not in the suburbs. Central city mistrusters *always see their neighbourhoods as less safe than trusters and their perceptions of insecurity rise exponentially as the crime rate goes up.* In the suburbs, the actual level of violent crime does not seem to make people less secure. Trust in the suburbs shapes perceptions of risk, regardless of the level of crime. Trusters are consistently (across levels of violent crime) 10 per cent more likely to say that their neighbourhoods are very safe in the suburbs.

People's perceptions of the safety of their neighbourhoods do depend upon the actual level of violence (though the simple correlation is just –0.292). But trust matters as well. Trust is not simply a sieve for information and context, as Hardin would have it. Nor are trusters ever on the alert for miscreants, as Yamigishi would lead us to believe. Instead, trusters consistently downplay the level of insecurity. Across every comparison – through actual level of violence and where one lives – trusters believe that their neighbourhoods are safer than mistrusters do, *net of any other experiences such as being a victim of crime, having a parent who was a victim, how often you watch local television news (featuring lots of violence, reflected in the saying: 'if it bleeds, it leads'), and how much you like your neighbourhood or the city.* Trusters may not damn contrary evidence, but they discount it.

The Pew measure of safety is not an abberation. Trusters are far less likely to lock their doors. And they are less less likely to feel that they must protect themselves from criminals with a gun (Uslaner, 2002, pp199, 263).[12] Next to living in an urban area, trust has the biggest effect of any variable on whether people think they should lock their doors. Even being attacked or robbed three times in the last five years doesn't matter as much as being a truster. In a variety of circumstances, trusters feel more secure against threats.

Reprise

Optimists underestimate risk. Are they irrational? Hardly: optimism (and trust) pays. Trusters find it easier to work with others in their community and nation. This cooperative spirit leads to higher levels of economic growth and better-functioning legal systems. Trust ultimately pays better than mistrust. This is hardly a novel idea. We know from game theory that being nice is better than being mean (Axelrod, 1984). We know the same thing from Aesop's fable about the tortoise and the hare. So it should hardly be surprising that depreciating risk is a winning strategy. When you fear encounters with strangers, you lose the opportunity to expand your markets. Trust opens up opportunties: it expands the base of people with whom we interact. Trust makes us more likely to embrace new technologies. Trust makes us more likely to take risks in daily life and in business. The trusting person seeks cooperation rather than confrontation, so closing the deal is easier when trust is widespread. To be sure, trusters might be more likely to be taken in by rogues. Yet, over the long haul, they will do better than mistrusters *because they are less likely to be consumed by the fear of failure and they are more likely to search for common ground*.

Trust is important for management for several reasons. First, by fostering tolerance and a cooperative spirit, trust makes for a more inclusive and consensual decision-making style. It leads to broader perspectives and a greater willingness to engage in a globalized world (Uslaner, 2002, Chapters 7 and 8). Second, by minimizing perceptions of risk, trust emboldens entrepreneurs. It makes them less likely to fear privacy (as my internet studies have shown) and less worried about risk, in general. Innovation depends upon a willingness to take risks and if trusters are more open to taking chances, they will also be the ones who are more likely to arrive at new solutions to management problems – and to prosper. We know that trusting countries have higher growth rates and we might also expect that firms with many trusting employees should be more innovative and more successful. Trust is a key factor enhancing a cooperative team spirit, a willingness to take risks and a willingness to work with people who are different from oneself. No one does business on a handshake in a complex world; but many people are very quick to forego negotiations and mediation and head straight to court. A lack of trust means that each search for a cooperative solution (a compromise, if you will) must start from scratch. Assuming that there *is* common ground is far more productive and efficient. Kenneth Arrow (1974, p23) argued:

> Now trust has a very important pragmatic value, if nothing else. Trust is an important lubricant of a social system. It is extremely efficient; it saves a lot of trouble to have a fair degree of reliance on other people's word. Unfortunately, this is not a commodity which can be bought very easily. If you have to buy it, you already have some doubts about what you've bought.

What can business do to build trust? We should not expect a major transformation in society from business since trust is generally learned early in life. However, business can take steps to create trust: some direct and some indirect steps. The direct steps include stressing diversity in the workplace and seeking non-confrontational solutions to negotiations. Indirect strategies may be more likely to have lasting effects. Business can create family-friendly workplaces, giving workers time to spend with their families so that they can transmit these positive values to their children. Business can provide opportunities for furthering education by employees since we know that education broadens people's perspectives on the world and sharply increases trust (Uslaner, 2002, Chapter 4). Businesses can get involved in their communities and set examples.

Sponsoring volunteering programmes and charitable giving programmes can create trust – but only if they are truly voluntary. The *Wall Street Journal* reported (Shellenbarger, 2003) that many American firms are 'forcing' their workers to volunteer (as schools require volunteering of their students) – and many workers resent this (perhaps even reducing trust!). Businesses can make substantial contributions to their communities themselves, realizing that a bit of altruism can yield a better society and a stronger business climate. Swiss Re, the big reinsurance company, is an excellent example of a large firm that has excelled in community service. So are Ben and Jerry's, the ice cream company, and Birkenstock, which makes sandals and other shoes.

Promoting good deeds such as volunteering and charity produces a 'warm glow' among those who help others: benevolence is rewarded with a good feeling, leading to greater trust (Andreoni, 1989; Uslaner, 2002, Chapter 5). Ben and Jerry's contributed a share of profits to charity (at least when it was privately held). Birkenstock provides free time for its employees to do good works and it contributes to many charities. In particular, it has donated considerable sums of money to causes that have little to do with its business, such as the Elizabeth Glaser Pediatric AIDS Foundation. It has also paid employees to volunteer and to give sacks of money to worthy causes without publicity (Lewis, 2004). And each small step in doing 'good' can help to reduce, even by a small amount, the level of inequality in a society. Simply recognizing the importance of trust and its benefits, is a key first step.

NOTES

1. An earlier version of this chapter was presented as a paper to the Conference on Trust, Department of Philosophy, University of California, Riverside, 27–28 February 2004. Some of the material on the implications of trust for business and management came from a talk, 'Generalized trust and why it matters for business in the age of globalization' that I gave at the Caux Conference for Business and Industry, Caux, Switzerland, 21 July 2004. The research assistance of Mitchell Brown is greatly appreciated. I am also grateful to the Russell Sage Foundation and the Carnegie Foundation for a grant under the Russell Sage programme on the social dimensions of inequality (see www.russellsage.org/programs/proj_reviews/social-inequality.htm) and to the General Research Board of the Graduate School of the University of Maryland, College Park. Some of the data reported here come from the Inter-University Consortium for Political and Social Research (ICPSR), which is not responsible for any interpretations. Other data come from the Pew Center for the People and the Press and I am grateful to Andrew Kohut for making them available to me. Most of the arguments here come from Uslaner (2002). I am grateful for the comments of David Levin.
2. The term 'strategic trust' is mine. Most of the people I cite would probably find the terminology congenial. Hardin (1992, p163) emphatically holds that 'there is little sense in the claim of some that trust is a more or less consciously chosen policy'. Trust based on experience can be strategic even if we do not make a deliberate choice to trust on specific occasions.
3. As I note below, it is foolish to trust everyone all of the time. Moralistic trust doesn't demand that. But it does presume that we trust most people under most circumstances (where most is widely defined).
4. This finding comes from the Pew Research Center for the People and the Press's 1996 Trust and Citizen Engagement survey in metropolitan Philadelphia. Ninety-seven per cent of moralistic trusters said that other people trust them, compared to a still very high 86 per cent of mistrusters (tau-b = 0.174; gamma = 0.627). This result may reflect either reality – perhaps we are more likely to trust people who trust us – or it may also be part of the general syndrome of overinterpretation.
5. Phi = –0.128; Yule's Q = –0.269. The question was asked in 1985, 1990 and 1996.
6. The correlation between trust in people and confidence in the legal system in the World Value Survey is modest (tau-c = 0.069; gamma = 0.122). And the country-by-country correlations tend to be higher where trust in people is higher.
7. Others who see trust as knowledge based – notably, Dasgupta (1988, p53), Hardin (1995, pp8–9) and Misztal (1996, pp121–123) – argue that it is *based upon* reputation.
8. See Uslaner (1999) for a more detailed examination of this evidence.

9   People with faith in others are between 7 and 16 per cent more likely to say that they are willing to serve. The effects of in-group and out-group trust are even higher: between 17 and 24 per cent.
10 Most of this section comes from Uslaner (2002, Chapter 7). The databases and the specific statistical analyses (all multivariate) are discussed in that chapter.
11 This section comes from Uslaner (2001).
12 Only three other variables – living in a border state or the south, and whether you or a family member witnessed a crime – has a bigger effect on defending yourself with a gun compared to trust.

REFERENCES

Andreoni, J. (1989) 'Giving with impure altruism: Applications to charity and Ricardian ambivalence', *Journal of Political Economy*, vol 97, pp1447–1458
Arrow, K. J. (1974) *The Limits of Organization*, W.W. Norton, New York
Axelrod, R. (1984) *The Evolution of Cooperation*, Basic Books, New York
Baier, A. (1986) 'Trust and antitrust', *Ethics*, vol 96, pp231–260
Banfield, E. (1958) *The Moral Basis of a Backward Society*, Free Press, New York
Baron, J. (1998) 'Trust: beliefs and morality', in A. Ben-Ner and L. Putterman (eds) *Economics, Values and Organization*, Cambridge University Press, New York, pp408–418
Bok, S. (1978) *Lying*, Pantheon, New York
Brehm, J. and Rahn, W. (1997) 'Individual level evidence for the causes and consequences of social capital', *American Journal of Political Science*, vol 41, pp888–1023
Cohen, J. L. (1997) *American Civil Society Talk*, National Commission on Civic Renewal, Working Paper no 6, College Park, MD
Dasgupta, P. (1988) 'Trust as a commodity', in D. Gambetta (ed) *Trust*, Basil Blackwell, Oxford, UK, pp49–72
Diener, E., Suh, E. and Oishi, S. (1997) 'Recent findings on subjective well-being: University of Illinois manuscript', *Indian Journal of Clinical Psychology*, vol 24, no 1, pp25–41
Elster, J. (1989) *The Cement of Society*, Cambridge University Press, New York
Frank, R. H. (1988) *Passions within Reason*, W. W. Norton, New York
Fukayama, F. (1995) *Trust: The Social Virtues and the Creation of Prosperity*, Free Press, New York
Gambetta, D. (1988) 'Can we trust trust?', in D. Gambetta (ed) *Trust*, Basil Blackwell, Oxford, UK, pp213–237
Gerbner, G., Gross, L., Morgan, M. and Signorielli, N. (1980) 'The "mainstreaming" of America: Violence profile no 11', *Journal of Communication*, vol 30, pp10–29
Greif, A. (1993) 'Contract enforceability and economic institutions in early trade: The Maghribi Traders Coalition', *American Economic Review*, vol 83, pp525–548
Hardin, R. (1992) 'The street-level epistemology of trust', *Analyse and Kritik*, vol 14, pp152–176
Hardin, R. (1995) 'Trust in government', text prepared for presentation at the Pacific Division Meeting of the American Philosophical Association, San Francisco, CA, April

Hardin, R. (1998) *Conceptions and Explanations of Trust*, Russell Sage Foundation Working Paper no 129, New York, April

Hill, K. A. and Hughes, J. E. (1997) 'Computer-mediated political communication: The USENET and political communities', *Political Communication*, vol 14, pp3–27

Knight, J. (2000) 'Social norms and the rule of law: Fostering trust in a socially diverse society', in K. S. Cook (ed) *Trust in Society*, Russell Sage Foundation, New York

Lane, R. E. (1959) *Political Life*, Free Press, New York

Levi, M. (1997) *A State of Trust*, Unpublished manuscript, University of Washington, Washington

Levi, M. (1998) 'A state of trust', in M. Levi and V. Braithwaite (eds) *Trust and Governance*, Russell Sage Foundation, New York, pp77–101

Lewis, J. D. and Weigert, A. (1985) 'Trust as social reality', *Social Forces*, vol 63, pp967–985

Lewis, M. (2004) 'The irresponsible investor', *New York Times Magazine*, 6 June, pp68–71

Luhmann, N. (1979) 'Trust', in N. Luhmann (ed) *Trust and Power*, John Wiley and Sons, New York, pp1–103

Macaulay, S. (1963) 'Non-contractual relations in business: A preliminary study', *American Sociological Review*, vol 28, pp55–67

Mansbridge, J. (1999) 'Altruistic trust', in M. Warren (ed) *Democracy and Trust*, Cambridge University Press, New York, pp290–310

Misztal, B. A. (1996) *Trust in Modern Societies*, Polity Press, Cambridge, UK

Mueller, J. (1999) *Democracy, Capitalism, and Ralph's Pretty Good Grocery*, Princeton University Press, Princeton

Offe, C. (1996) *Social Capital: Concepts and Hypotheses*, Unpublished manuscript, Humboldt University, Germany

Offe, C. (1999) 'Trust and knowledge, rules and decisions: Exploring a difficult conceptual terrain', in M. Warren (ed) *Democracy and Trust*, Cambridge University Press, Cambridge

Orbell, J. and Dawes, R. M. (1991) 'A "cognitive miser" theory of cooperators' advantage', *American Political Science Review*, vol 85, pp513–528

Putnam, R. D. (1993) *Making Democracy Work: Civic Traditions in Modern Italy*, Princeton University Press, Princeton

Putnam, R. D. (2000) *Bowling Alone*, Simon and Schuster, New York

Rosenberg, M. (1956) 'Misanthropy and political ideology', *American Sociological Review*, vol 21, pp690–695

Rothstein, B. (2001a) 'Trust, social dilemmas and collective memories: On the rise and decline of the Swedish model', *Journal of Theoretical Politics*, vol 12, pp477–501

Rothstein, B. (2001b) 'Social capital in the social democratic welfare state', *Politics and Society*, vol 29, pp207–241

Rotter, J. B. (1980) 'Interpersonal trust, trustworthiness, and gullibility', *American Psychologist*, vol 35, pp1–7

Scholz, J. T. and Pinney, N. (1995) 'Duty, fear, and tax compliance: The heuristic basis of citizenship behavior', *American Journal of Political Science*, vol 39, pp490–512

Seligman, A. B. (1997) *The Problem of Trust*, Princeton University Press, Princeton

Shah, D. V., Kwak, N. and Holbert, R. L. (2001) '"Connecting" and "disconnecting" with civic life: Patterns of internet use and the production of social capital', *Political Communication*, vol 18, pp141–162

Shellenbarger, S. (2003) 'Drafted volunteers: Employees face pressure to work on company charities', *Wall Street Journal*, 20 November, pD1

Silver, A. (1989) 'Friendship and trust as moral ideals: An historical approach', *European Journal of Sociology*, vol 30, no 2, pp274–297

Sitkin, S. B. and Roth, N. L. (1993) 'Explaining the limited effectiveness of legalistic remedies for trust/distrust', *Organization Science*, vol 4, pp367–392

Sullivan, J. H., Piereson, J. and Marcus, G. E. (1981) *Political Tolerance and American Democracy*, University of Chicago Press, Chicago

Tyler, T. R. (1990) *Why People Obey the Law*, Yale University Press, New Haven

Uslaner, E. M. (1999) 'Trust but verify: Social capital and moral behavior', *Social Science Information*, vol 38, pp29–56

Uslaner, E. M. (2001) *Trust, Civic Engagement and the Internet*, Unpublished manuscript, University of Maryland, College Park, Maryland

Uslaner, E. M. (2002) *The Moral Foundations of Trust*, Cambridge University Press, New York

Uslaner, E. M. (2004) *Trust as an Alternative to Risk*, Presented at the Conference on Trust, Department of Philosophy, University of California, Riverside, 27–28 February

Uslaner, E. M. and Badescu, G. (2004) 'Honesty, trust, and legal norms in the transition to democracy: Why Bo Rothstein is better able to explain Sweden than Romania', in J. Kornai, S. Rose-Ackerman and B. Rothstein (eds) *Creating Social Trust: Problems of Post-Socialist Transition*, Palgrave, New York

Yamagishi, T. (1998) *The Structure of Trust: The Evolutionary Games of Mind and Society*, Tokyo University Press, Tokyo

Yamigishi, T. and Yamigishi, M. (1994) 'Trust and commitment in the United States and Japan', *Motivation and Emotion*, vol 18, pp129–166

4 A Social Judgement Analysis of Trust: People as Intuitive Detection Theorists[1]

Mathew P. White and J. Richard Eiser

INTRODUCTION

The social amplification and attenuation of technological risks have concerned risk regulators and policy-makers for many years (Kasperson et al, 1988). Moreover it has long been recognized that due to the complexity of many modern technologies, members of the public may have difficulty in making personal assessments of their risks and benefits. A number of commentators have therefore argued that public risk perceptions and acceptance have become largely associated with issues of trust and confidence in risk assessors, managers and communicators (Wynne, 1980; Starr, 1985; Eiser, 1990; Freudenburg and Rursch, 1994). Indeed, the authors of the Social Amplification of Risk Framework (SARF) (Kasperson et al, 2003, p33) have recently suggested that 'understanding how trust is shaped, altered, lost or rebuilt in the processing of risk by social and individual stations of risk is a priority need in social amplification research'.

The aim of this chapter is to introduce a new perspective on understanding public trust in the field of risk management, one we believe can improve our understanding of trust processes. The approach proposes that public decisions regarding whether or not to trust any given risk assessment can be likened to the process whereby signal detection theorists (SDTs) – for example, Green and Swets (1974/1988) – evaluate the decision-making performance of experts under conditions of uncertainty. Specifically, we suggest that the public tends to evaluate the performance of risk managers and communicators in terms of two parameters or dimensions – namely, *competence* and *decision bias*. In the context of risk, competence refers to the decision-maker's ability to discriminate between instances of safety and danger, and decision bias refers to their general predisposition to make more or less risky or cautious decisions. However, unlike formal detection theorists, the public is generally less able and willing to engage in detailed performance analysis, and will tend to use a number of heuristic assessments, such as the degree to which those under scrutiny are open and transparent with respect to their decision-making processes. In line with conceptions of people as naive or *intuitive* psychologists (Heider, 1958), statisticians (Kelley, 1967), politicians (Tetlock, 2002) and toxicologists (Kraus et al, 1992), we refer to members of the lay public as *intuitive* signal detection theorists – or *intuitive detection theorists* (IDTs) for short.

Although there appear to be a number of similarities between the IDT and other approaches at explaining trust dimensionality, one advantage of our approach is that it is based on a clear theoretical background. Frewer (2003), for example, has recently argued that the understanding of the relationships between trust and risk perception has been hampered by the 'atheoretical' nature of much of the work to date; thus, a more theoretical approach appears timely. Conceptually, our approach clarifies a certain amount of ambiguity that exists in the current literature regarding the number of dimensions that exist and the interrelations between them. In terms of policy, it suggests that current conceptions of how trust is gained, maintained and lost may be overly simplistic, and when viewed as intuitive detection theorists, public reactions become clearer.

The chapter continues with a brief overview of the issue of trust dimensionality in the risk perception literature and highlights some of the conceptual confusion that currently exists. We then introduce the intuitive detection theorist account of trust, and argue that this more theoretical approach can resolve much of the ambiguity in the current literature on dimensionality. We then present empirical findings from a public survey carried out to investigate the IDT perspective with regard to trust and mobile phone technology. Finally, we consider some of the overlaps with other perspectives and provide some suggestions with regard to further research.

TRUST DIMENSIONALITY AND RISK PERCEPTION

Numerous dimensions have been proposed to underlie trust in the context of risk management, including care, competence, concern, consensual (or shared) values, consistency, expertise, fairness, faith, honesty, knowledge, objectivity, openness, past performance, predictability, reliability and sympathy (see, for example, Renn and Levine, 1991; Kasperson et al, 1992; Mishra, 1996; Peters et al, 1997; Maeda and Miyahara, 2003; Siegrist et al, 2003). Although the list of possible dimensions seems daunting, a number of factor analytic studies (Frewer et al, 1996; Jungermann et al, 1996; Metlay, 1999; Poortinga and Pidgeon, 2003) and a conceptual review (Johnson, 1999) suggest that this list can be reduced to two (or three) key dimensions.

Jungermann et al (1996), for example, referred to the two dimensions that emerged from their factor analysis as *competence* (including: competence) and *honesty* (including: honesty, credibility, comprehensiveness and one-sidedness). While Frewer et al (1996) did not name their factors, it seems from their account that the two dimensions that emerged could be referred to as *knowledge* (including: distorted, biased, truthful, good track record, accurate, responsible and knowledgeable) and *accountability* (including: accountable, self-protection, vested interest and sensationalization). This latter dimension appears to suggest that a source would attempt to conceal (rather than release) risk-related information. Metlay (1999) talks of two 'components' (rather than dimensions) of trust that emerge from his analysis: a *competence* component (including: competence) and an *affective* component (including: openness, reliability, integrity, credibility, fairness and caring). Poortinga and Pidgeon (2003) also reported two dimensions from their factor analysis, which they labelled *general trust* (including: competence, care, fairness and openness) and *scepticism* (including: credibility, reliability and integrity). Clearly, by factoring together competence with care and fairness, the results of this analysis are somewhat different from the others. Johnson's (1999) more conceptual approach also concluded that there were two key dimensions – *competence* and *care* – although he also recognized the role of *consensual values*, as highlighted by Earle and Cvetkovich (1995), as a possible third dimension with slightly different properties.

In sum, the overall picture to emerge from this literature appears to be both illuminating and yet slightly confusing at the same time. Certainly, at least four of the five attempts to reduce the number of dimensions discussed above see competence as a dimension in its own right, in keeping, it seems, with the entire trust and risk management literature (although see Earle and Cvetkovich, 1995).[2] However, there seems to be less agreement and conceptual clarity as to the nature of the second (or third) dimension. Thus, while Frewer et al (1996) and Johnson (1999) stress the importance of whether or not the other is seen to be acting in the public rather than just personal interest, Jungermann et al (1996) focus on communication issues, such as honesty and openness. Metlay (1999) effectively combines these two aspects into a single dimension and Poortinga and Pidgeon (2003) combine both with competence to form a single trust dimension, with scepticism forming a second dimension quite different from those proposed by other commentators.

Moreover, the aspects of trust that load together in the factor analytic studies do not always appear to agree with each other and thus lack a certain amount of face validity. For example, why should *biased* be on the *knowledgeable* rather than *accountable* dimension (Frewer et al, 1996), and why should *comprehensiveness* load with *honesty* and not *competence* (Jungermann et al, 1996)? Furthermore, if we treat Poortinga and Pidgeon's (2003) analysis at face value, trust dimensionality seems simply a reflection of informational valence. We believe that this ambiguity in the current literature is hampering further understanding of trust-related processes.

In part, this ambiguity seems to derive from the largely descriptive nature of earlier approaches to examining trust dimensionality. Certainly, the use of factor analytic techniques suggests a 'bottom-up' data-driven approach, which, although helpful in describing a phenomenon in the first instance, is limited by the specific nature of the risk context examined. Thus, the differences in factor structures that emerge in the literature could be due to differences in the risk context, rather than a reflection of any underlying differences in the nature of trust. As a result, greater clarity with respect to trust dimensionality requires the development of a theoretical perspective, which can span specific risk contexts.

THE INTUITIVE DETECTION THEORIST APPROACH

Signal detection theory

The intuitive detection theorist (IDT) account of trust is derived from an approach to decision-making under uncertainty called signal detection theory (SDT). Not only does SDT have a long and well-established history in applied diagnostic settings (see, for example, Einhorn, 1974; Swets, 1996; Swets et al, 2000), many of its principles are pervasive in common language, such as the term 'false alarm', suggesting that the ideas are more than just theoretical abstractions.

SDT was originally developed to understand the discriminations made by risk assessors, such as radar operators interpreting blips on radar screens (Swets, 1996). Such blips might represent, for instance, an enemy bomber or some random interference due to faulty transmission – for example, weather conditions. In other words, a blip might be a real 'signal' or it could be meaningless 'noise'. How operators would react would be a function both of the physical properties of the signal, and of their expectation of the ratio of signals to noise and the overlap in their distributions.

SDT proposes a procedure for analysing the performance of any discriminator or discrimination system. This involves first classifying the person (or system's) responses in terms of two dichotomies: whether the response is correct (true) or incorrect (false); and whether it is positive ('signal present') or negative ('signal absent'). What counts as positive or negative obviously varies with the context; but for most purposes considered here it will be helpful to think of 'positive' as referring to the identification of the presence of danger, and 'negative' as referring to a judgement that no danger is present. Within this scheme, then, there are four classes of response (see Table 4.1): true positives (that is, 'hits', or correct identifications of danger); true negatives (that is, correct rejections, 'all clears' or reassurance of absence of danger); false positives (that is, 'false alarms') and false negatives (that is, 'misses').

Table 4.1 *The four possible outcomes of a simple binary detection task*

		Perceiver says signal is	
		Present	Absent
Signal really is	Present	TRUE POSITIVE (hit)	FALSE NEGATIVE (miss)
	Absent	FALSE POSITIVE (false alarm)	TRUE NEGATIVE (all clear)

Naturally, this all requires some objective criterion for establishing whether any response is correct or incorrect. This criterion, however, is something that is applied after the event. At the time of the discrimination, all that is available are signals and noise of different strength. In other words, decisions are being made under conditions of uncertainty.

SDT distinguishes between two independent parameters in terms of which discrimination performance can be described. The first parameter is termed sensitivity or discrimination ability. This refers to the likelihood of responding positively to a signal compared with responding positively to noise. The second parameter is termed criterion or response bias. This refers to the strength of input required before something is categorized as signal rather than noise, and is determined by an evaluation of the perceived benefits associated with correct decisions ('hits' and 'all clears'), the perceived costs associated with incorrect decisions ('misses' and 'false alarms'), and perceptions about the probabilities of the possible outcomes. Critically, what counts as 'good performance' depends both upon a person's ability to discriminate between noise and noise plus signal, and where they set their criterion for saying that a signal is present. In other words, 'good performance' isn't just about reducing uncertainty, it's also about how one reacts when uncertainty remains.

This point can be illuminated by referring to the example of medical staff trying to decide whether or not a tumour is present on an X-ray (see, for example, Swets et al, 2000). Performance can range from very poor to very good and will be determined by how good the member of staff is at distinguishing tumours from benign matter on X-rays (that is, discrimination ability) and their response bias when they are uncertain. Let us assume, for the sake of argument, that the costs of a miss (saying a tumour is absent when it is present) are greater than the costs of a false alarm (saying a tumour is present when it is not), and that the benefits of a hit (which enables effective treatment) are greater than the benefits of an all clear (no unnecessary stress). Although clearly an oversimplification, such a situation would suggest that the response bias should be set so as to increase the chance of hits and decrease the chance of misses. Thus, when uncertain, the X-ray examiner should, on the basis of an understanding of the different costs and benefits of correct and incorrect decisions, assume that a tumour is present rather than absent.

Very good performance would thus be characterized by high discrimination ability and an appropriate response bias. In other words, this person would be able to correctly identify most instances of tumour present or not present; when uncertain, they will adopt a cautious criterion and assume that a tumour is nevertheless present. At the other end of the spectrum, very poor performance would be characterized by someone who was not very good at identifying tumours (low discrimination ability) and yet maintained that most X-rays probably did not show tumours (inappropriate response bias). Moderate performance could be achieved in two ways. First, one might show high discrimination ability but an inappropriate response bias or low discrimination ability but an appropriate response bias.

In sum, SDT argues that performance under conditions of uncertainty can be technically separated into two parameters that are conceptually independent: discrimination ability and response bias. In the context of risk management, a decision-maker who makes a relatively large number of hits and correct rejections compared to misses and false alarms has high discrimination ability. One who makes many hits but, at the same time, many false alarms has a response bias towards caution, and one who makes many correct rejections but also many misses has a relatively risk-tolerant response bias. While high discrimination ability is always likely to be preferred compared to low ability, preferences for a more or less risk-tolerant response bias will depend upon a consideration of the relative costs and benefits of the various outcomes, and these will depend upon context.

Extending the signal detection theory argument to social trust

Although the principles of SDT have been widely applied in assessing the performance of actual risk managers in many different settings, those carrying out the analysis have tended to be experts themselves in the principles of SDT (see, for example, Swets et al, 2000). Indeed, formal use of the theory requires relatively complex statistical procedures for separating performance in terms of the two dimensions, and the complexity of these procedures naturally increases when we move beyond a simple binary state of the world such as we have discussed so far (for greater discussion of the complexities of the SDT, see, for example, MacMillan and Creelman, 1991). Nevertheless, there is now plenty of evidence to suggest that lay decision-makers can make a range of highly functional decisions with the use of relatively simple heuristic judgement processes that still tend to mirror more formal procedures in certain key respects (see Gigerenzer et al, 1999; Gilovich et al, 2002).

Thus, what we are suggesting here, following Eiser's (1990) original proposal, is that members of the public also evaluate the performance of risk managers and communicators using the same two parameters as formal detection theorists, albeit at a much more heuristic level. That is to say, in attempting to assess a decision-maker's performance in risk-related contexts, we argue that people effectively act as *intuitive detection theorists* and make assessments about the decision-maker's abilities to discriminate safe from dangerous situations and their tendencies to be more or less cautious when uncertain.

Moreover, since in many risk contexts members of the public cannot directly assess the risks for themselves, these judgements about the other's performance will also tend to underpin *trust* in them – that is, the conclusions of those who are perceived to show high levels of performance will be trusted more than those who show low levels of performance. In essence, then, we are arguing that trust in the messages about a given risk context (that is, are the risks acceptable or not?) will be higher when discrimination ability is perceived to be high and when perceived response bias appears to be 'appropriate' in any given situation. Of course, what is deemed 'appropriate' will depend upon the perspective of the observer, although there is evidence that for many new technologies, at least, the public may well deem a more cautious response bias to be more appropriate (see Tetlock and Boettger, 1994; White et al, 2003).

Importantly, in terms of our earlier discussion of trust dimensionality in the risk management literature, the IDT perspective highlights two dimensions that appear to have important similarities to those that emerged from more descriptive approaches. Specifically (as stated above), three dimensions, in particular, emerged from earlier analyses – namely, competence, acting in the public versus private interest, and openness and transparency in communication. Although there was not always consistent agreement about these three dimensions in the earlier literature, the competence dimension appears closely related to the IDT aspect of discrimination ability and a perception of public versus private interest appears related to perceptions of response bias. That is to say, when perceived as adopting an appropriate response bias by members of the public, a decision-maker is more likely to also be perceived as acting in the public interest than in personal interest, which, in turn, is more likely to be inferred when an 'inappropriate' response bias is believed to have been adopted. Importantly, therefore, the greater clarity of the IDT approach enables us to recognize that dimensions such as care, concern, shared values and vested interest all appear related to the same fundamental aspect – namely, perceived response bias.

However, the third dimension that is sometimes proposed – openness and transparency – raises an important issue about the IDT perspective. Specifically, while formal detection theorists have open access to performance data and can therefore assess discrimination ability and response bias directly, members of the public rarely have such privileged access. Rather, they often have to infer these aspects of performance from what the decision-maker is prepared to say about their decision-making processes.

When the decision-maker is open and transparent, then it is relatively easy for the lay observer (should they be motivated to do so) to assess their performance as an intuitive detection theorist. Under these conditions, performance evaluation and, thus, trust in future decisions will depend upon perceptions of discrimination ability and response bias as outlined above.

However, when the decision-maker is reticent to reveal the decision processes and is less than fully transparent, the lay observer is hindered in attempts to act as an IDT and is, instead, forced to draw conclusions and attributions about performance from this reluctance to be open. An obvious conclusion to draw from a lack of transparency is that if someone withholds something from you, it's because they think you would not like to see what is revealed. In other words, a lack of transparency may be used to conceal either low discrimination ability, a response bias that it is thought would be inappropriate in the eyes of the public, or, worst of all, both. In other words, according to the IDT approach, a lack of transparency will undermine trust because it implies poor performance on one or both of the other indicators. However, the IDT perspective also highlights why transparency and openness is not a panacea for trust since if transparency reveals poor performance in terms of one or both of the dimensions, then trust may be low anyway. In short, in many risk contexts, transparency is a necessary but not sufficient requirement for trust.

In sum, seeing people as intuitive detection theorists provides a more theoretical account of trust in risk managers and communicators than has been proposed to date. One of the chief advantages of such a perspective is that some of the ambiguity that exists in the current literature on trust dimensionality based on more descriptive approaches can be addressed and explained. In particular, the IDT account is able to highlight the underlying similarity between many of the proposed dimensions relating to perceptions of response bias and to suggest why openness and transparency is important in its own right, yet at the same time is no guarantee of trust.

How is our theoretical perspective supported empirically? Over the last four years we have carried out a number of survey and experimental studies to investigate the IDT approach. In the next section we present summary findings from one of the earliest studies to adopt an IDT perspective: a public survey into attitudes towards mobile phone technologies. While the survey had a number of aims that need not concern us here, one part, in particular, focused on issues of trust in sources of information about any possible health risks from the technology. Importantly, we asked not only about trust in the various sources, but also about perceptions of their knowledge about the risks (that is, discrimination ability) and the amount of evidence of harm they would require before warning the public (response bias).[3] Support for the IDT perspective would be found if trust was positively related to greater perceived discrimination ability and the degree to which the sources exhibited an 'appropriate' response bias. What should be considered as 'appropriate' in the context of mobile phones is dealt with in more detail below.

MOBILE PHONE SURVEY

Introduction

Shortly before the survey was carried out in autumn 2001, the UK's Independent Expert Group on Mobile Phones (IEGMP, 2000) had concluded that although there was little evidence of negative risks to health from the use of mobile phones, there was enough ambiguity to warrant a precautionary approach. Moreover, the report of the group urged more social scientific research into the issue, including the development of greater understanding of public risk perceptions. In part, the survey was a response to this call and, as a consequence, addressed a number of issues not directly relevant here.

In total, the postal survey contained 89 items and an opportunity for additional comments, although of primary concern here were the items relating to trust and some of the risk perception questions (see below). Although stratified sampling techniques were not used, the resulting sample of 1320 residents of the Greater Sheffield region, UK, was broadly representative of the national population in terms of a number of indicators, such as age, gender, employment profiles and mobile phone use (European Commission, 2000; White et al, 2004).

The six potential sources of information about the possible health risks of mobile phones selected for the survey were chosen because prior research suggested that they represented the broad spectrum of trust for a host of other risk-related issues (see Frewer et al, 1996; Jungermann et al, 1996; Langford et al, 2001). Accordingly, we asked about trust in the mobile phone industry, the government, the media, environmental groups, independent scientists and the medical profession. While we recognize that these are rather abstract categories, they were nevertheless sufficient for our purposes, and further work has since been carried out using more specific sources.

Recall that our predictions were for greater trust in sources with greater perceived discrimination ability and more appropriate response biases. While the discrimination ability hypothesis is relatively straightforward, we need to clarify what we mean by appropriate in this context. First, note that for many risks people tend to prefer a more cautious approach. This would suggest that sources who adopted a risk-tolerant response bias and thus increase their chances of making misses would be particularly low in trust. However, does this mean that sources who adopt a particularly risk-averse response bias resulting in an increased chance of false alarms would be greatly trusted? In the case of mobile phones, and unlike many other risk contexts, this is probably not the case because the vast majority of the public, even in the UK in 2001 (IEGMP, 2000), had already purchased a mobile phone and were thus already aware of the benefits. For this reason, it may be that a source who was prepared to offer a large number of false alarms might be trusted rather less than one who adopted a more moderate response bias, and who would warn the public on the basis of significant but not excessive evidence of risk. In short, because most of the public was already happy to use mobile phones, in this particular context it may be that what counts as an 'appropriate' response bias is one which is relatively moderate, not too strict (requiring a large amount of evidence of a problem) and not too lax (requiring the merest hint of a problem).

Moreover, in addition to these general definitions of an 'appropriate' response bias of sources, we also suspected that there would be important individual differences in what members of the public felt to be appropriate. Specifically, those who felt that the probability of negative health effects was relatively high might prefer sources who exhibited a more lax response bias and were prepared to warn the public with relatively little evidence. By contrast, those who thought the possibility of risks was generally low might be reluctant to endorse such a criterion and be more amenable to sources who were not so quick to raise the alarm.

Specific questions

The key perceived discrimination ability, response bias and trust-related questions were, in order, as follows:

1 'How much knowledge do you feel each of the following has about any possible health effects of mobile phone technology?' (0 = 'No knowledge' to 6 = 'Complete knowledge').
2 'How willing do you think each of the following would be to warn the public about any possible health effects of mobile phone technology?' (0 = 'Will warn even if no evidence' to 6 = 'Will warn only if link is proven').
3 'How much do you trust each of the following as sources of information about any possible health effects of mobile phone technology?' (0 = 'No trust' to 6 = 'Complete trust').

In addition, respondents were asked: 'How likely do you think it is that the following would experience any negative health effects due to regular use of a (hand-held) mobile phone:

- you personally;
- the average person of your age and sex; and
- the average child' (0 = 'Not at all' to 6 'Extremely').

For current purposes, these three responses were collapsed (a = .74) to form a perceived probability of harm score that was subsequently split three ways to reflect those with low (0 to < 3; n = 359), moderate (3 to < 4; n = 496) and high (4 to 6; n = 424) perceptions of risk.

Results

Perceived discrimination ability, response bias and trust in the six sources

The means and standard deviations for each of the three questions for each of the six sources can be seen in Table 4.2. Three one-way repeated measures analysis of variance (ANOVA), one for each variable, were carried out to compare the means for the six sources.

Table 4.2 *Perceived discrimination ability, response bias and trust in various sources*

	\multicolumn{12}{c}{Source of information (from most to least trusted)}											
	Independent scientists		Medical profession		Environmental groups		Government		Media		Mobile industry	
	M	SD	M	SD	M	SD	M	SD	M	SD	M	SD
Sensitivity[a]	4.04	1.26	3.74	1.35	3.60	1.46	3.61	1.73	2.64	1.43	4.21	1.47
Response bias[b]	3.48	1.66	3.60	1.48	1.85	1.73	4.60	1.62	1.73	1.85	5.48	1.28
Trust[c]	3.89	1.29	3.79	1.35	3.19	1.69	1.86	1.60	1.53	1.49	0.88	1.26

Notes:
a Seven-point scale from 0 'No knowledge' to 6 'Complete knowledge'.
b Seven-point scale from 0 'Will warn even if no evidence' to 6 'Will warn only if link is proven'.
c Seven-point scale from 0 'No trust' to 6 'Complete trust'.

The first analysis with perceived discrimination ability as the dependent variable found a large main effect of source $F(5, 6280) = 413.14$, $p < 0.001$. *Post-hoc* comparisons found that the means for all sources differed significantly from each other (all $ps < 0.001$) except those for government and environmentalists.[4] That is to say, the mobile phone industry was perceived to have greater discrimination ability with regard to the potential risks than any other source, followed by independent scientists, the medical profession, the government, environmentalists and, lastly, the media.

The second analysis with perceived response bias as the dependent variable also found a large main effect of source $F(5, 6205) = 1224.10$, $p < 0.001$. Again, *post-hoc* comparisons suggested that the mobile phone industry was thought to have a stricter response bias than any other source (M = 5.48; all $ps < 0.001$). That is to say, although this source was perceived to be most knowledgeable about the risks and thus to have the greatest discrimination ability, it was also the source perceived as being less likely to warn the public about them. The government was thought to have a stricter response bias than any other source except for industry (M = 4.60; all $ps < 0.001$) and thus was also thought to be relatively unwilling to warn the public about any risks. While both medics (M = 3.60) and scientists (M = 3.48) were thought to have stricter response biases than either environmentalists (M = 1.85) or the media (M = 1.73; all ps < 0.001), the amount of bias appears to be moderate and they did not differ significantly in terms of perceived response bias from each other. Finally, environmentalists and the media also did not differ in terms of response bias as both were thought to have a relatively lax criterion for warning the public and to require relatively little evidence before sounding the alarm.

In short, it seems that the six sources can be grouped into three subcategories on the basis of their perceived response bias. The media and environmentalists are not differentiated from each other and both are seen as having a relatively lax response bias, meaning that they are more likely to declare a risk is present. As such we could, following the Social Amplification of Risk Framework (SARF) paradigm, label them risk *amplifiers*. By contrast, industry and the government were seen to have stricter response biases than all other sources, meaning that they are less likely to declare a risk is present. Again, following the SARF approach, we could label these two sources as risk *attenuators*. Finally, independent scientists and the medical profession are perceived to have a more moderate approach, suggesting that we could refer to them (although such a category does not exist in the SARF account) as risk *arbitrators* – that is, they attempt to weigh up and balance competing claims of safety and danger. These distinctions will become important in the analysis examining the influence of individual differences below.

The third analysis with trust as the dependent variable also found a large main effect of source $F(5, 6300) = 1245.13$, $p < 0.001$, with *post-hoc* comparisons suggesting that all sources differed significantly from each other (all ps < 0.001), with the exception of independent scientists (M = 3.89) and the medical profession (M = 3.79). Specifically, these two sources were trusted more than environmentalists (M = 3.19), who were trusted more than the government (M = 1.86), who, in turn, were trusted more than the media (M = 1.53), who were trusted more than industry (M = .88). In short, the patterns of trust in the six sources used in the current study are broadly reflective of surveys and studies looking at trust in these sources for other risk issues (Frewer et al, 1996; Jungermann et al, 1996; Langford et al, 2001).

Relations between perceived discrimination ability, response bias and trust
Figure 4.1 plots trust means for each source against their perceived discrimination ability (knowledge). This kind of approach, using the source as the unit of analysis, reflects the way in which hazards are plotted on the two risk dimensions that emerge in the psychometric approach (see Slovic, 1987). The solid diagonal line shows the regression line produced from regressing the mean trust scores for each of the six sources on their mean perceived discrimination ability scores. It is evident that the fit of

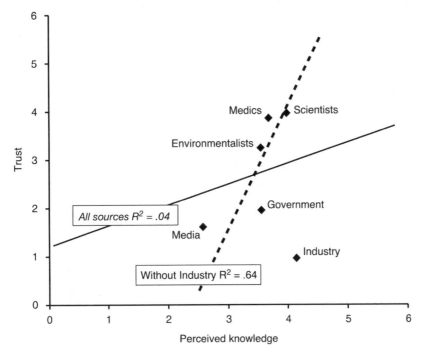

Source: Chapter authors

Figure 4.1 *The aggregate relationship between trust and perceived discrimination ability*

the points to the line is not very good. However, if we ignore the relationship for industry and concentrate on the other five sources, the fit of the points improves remarkably, as shown by the second (dotted) regression line. In short, with the exception of industry, greater perceived discrimination ability appears to be associated with greater trust.

This notion is supported by the correlations between perceived discrimination ability and trust using individual respondents (rather than sources) as the unit of analysis. The correlations between trust and discrimination ability were $r = .49$ for environmentalists, $r = .34$ for the media, $r = .29$ for scientists, $r = .31$ for the medical profession, $r = .13$ for government, but only $r = .04$ for industry. All correlations except the last were significant ($p < 0.001$) and indicate that trust increases as perceived discrimination ability of the source increases.

Figure 4.2 plots the means for perceived response bias and trust for each of the six sources of information. The dotted inverted U-shaped curve represents the best-fit quadratic model using the sources as unit of analysis. It can be seen that the source means are very close to this line, suggesting a good fit. Using this quadratic model, response bias explains some 79 per cent ($R^2 = .79$) of the variance in trust. Clearly, trust is

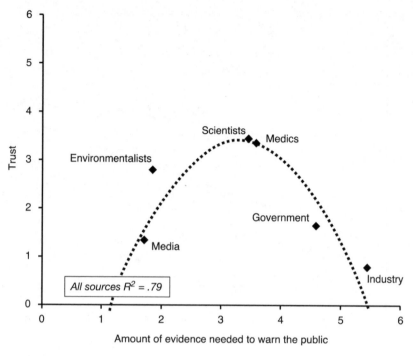

Source: Chapter authors

Figure 4.2 *The aggregate relationship between trust and perceived response bias*

highest for the arbitrators (the scientists and medics), who are thought to have a moderate response bias, and lower for both the attenuators (the government and industry) and the amplifiers (the media and environmentalists), although, on average, the amplifiers appear to be trusted more than the attenuators.

Again, these patterns are (partially) supported when we use individual respondents as the unit of analysis. However, due to the overall quadratic relationship, in order to establish the relationship between trust and perceived response bias *per se* (rather than the direction of response bias), we 'folded' the original response bias variable at the mid-point such that bias would be lowest at the mid-point and largest for either end of the scale. Since the mid-point of a 0 to 6 scale is 3, this was the point at which the scale was folded and a score of 3 was recoded as 0, reflecting *no bias* either way. A source with a score of 0 would require a moderate amount of evidence before warning the public – that is, they would not be alarmist; but at the same time they would not conceal any important signs of danger. One-point differences either side of the original mid-point (scores of 2 or 4) were recoded as 1 to reflect *low bias*. Sources with low bias will either need a little more or a little less evidence before warning the public. Two-point differences from the original mid-point (scores of 1 or 5) were recoded as 2 to reflect *moderate bias*. Finally, three-point differences from the original mid-point (scores of 0 or 6) were recoded as 3 to reflect *high bias*.

Correlating this 'folded' response bias score with trust, we found significant negative correlations (p < 0.001) for the media, $r = -.38$; industry, $r = -.31$; government, $r = -.30$; and environmentalists, $r = -.17$. There was no relationship for either scientists ($r = .02$) or the medical profession ($r = -.01$). In short, for risk amplifiers and attenuators, greater deviation from a moderate response bias (in either direction) was associated with lower trust; but it seems that deviations from a moderate response bias for risk arbitrators was not associated with changes in trust. Although one might therefore conclude that issues of response bias are not important for risk arbitrators, the aggregate level analysis seems to indicate that it is precisely because these sources are perceived to have a moderate response bias that they are trusted. This realization indicates a need for caution in interpreting the results from any given statistical approach and supports calls for multi-method, multilevel analysis (see Langford et al, 1999).

Critically, the overall picture to emerge from these analyses is that in line with predictions from an IDT perspective, trust generally increases as a function of greater perceived discrimination ability and decreases as a function of perceived response bias. In particular, in the case of mobile phone technology, at least, the greater the perceived deviation from a moderate perspective, the lower the trust. However, we also suggested that there may be important individual differences with respect to what people deem to be an 'appropriate' response bias, and this possibility is examined below.

Risk perceptions and trust in different sources
Earlier we noted that as part of the survey we asked people about their perceptions of the perceived likelihood of negative health effects from the use of mobile phones, and that based on their responses we categorized them as having low, moderate or high risk perceptions. According to the IDT perspective, respondents with higher risk perceptions should have greater trust in risk amplifiers (who will make fewer misses) than those with lower risk perceptions, who are likely to be more wary of too many false alarms. By contrast, those with lower risk perceptions are likely to have greater trust in risk attenuators (who will make fewer false alarms) than those with high risk perceptions, who should be wary of too many misses.

To test these predictions we carried out a 3 (source type – amplifiers, arbitrators, attenuators) × 3 (perceived risk – low, moderate, high), mixed factor ANOVA with repeated measures on the first factor. There was a large significant main effect of source, $F (2, 2532) = 1773.79$, $p < 0.001$. Arbitrators were trusted significantly more (3.84) than amplifiers (2.36), who were trusted significantly more than attenuators (1.37, both ps <

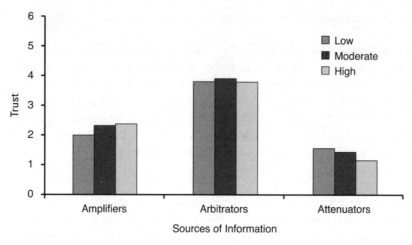

Source: Chapter authors

Figure 4.3 *Trust as a function of source type and perceived risk of mobile phones*

0.001). This pattern is in keeping with our main hypothesis, which suggests that a moderate response bias would be preferred to one that was relatively lax and this, in turn, would be trusted more than one that was highly strict. There was no main effect of perceived risk, $F(2, 1266) = 1.73$, not significant; so perceptions of risk did not influence the overall amount of trust in the three types of source. However, as predicted, there was a significant interaction between source and perceived risk $F(4, 2532) = 34.66$, $p < 0.001$. This interaction can be seen in Figure 4.3.

Although the order of trust for the three types of source is consistent across levels of perceived risk, there are clear intra-source differences as a function of perceived risk. As predicted, trust in amplifiers increased as perceptions of risk increased ($F(2, 1266) = 31.72$; $p < 0.001$), whereas trust in attenuators decreased as risk perceptions increased ($F(2, 1266) = 16.22$; $p < 0.001$). There was no overall difference in trust in arbitrators as a function of perceived risk, $F < 2$. In short, while there is general consistency across respondents about what counts as an 'appropriate' response bias, there are also important individual differences related to personal perceptions of risk that partially moderate these more global perceptions.

To sum up, the results of the mobile phone survey suggest that trust is higher in sources with greater discrimination ability (that is, sources who will make few errors distinguishing safety from danger) and in sources who will balance any errors they make (misses and false alarms) by adopting a moderate response bias. Of the sources which are perceived to have a distinct bias in the amount of evidence they need before warning the public, amplifiers (those with a tendency to make more false alarms) are generally more trusted than attenuators (those with a tendency to make more misses). However, there are also important individual differences for trust in sources as a function of personal perceptions of risk. Those high in perceived risk are generally less tolerant of sources who would tend to make misses and relatively more tolerant of sources who would be prepared to make false alarms, compared to those low in perceived risk.

DISCUSSION

The findings from the mobile phone survey offer preliminary support for the notion that people act as intuitive detection theorists when deciding whether to trust others' messages about the potential risks of a new technology that may be difficult to assess personally. Specifically, it seems that perceptions of both discrimination ability and response bias are important for perceptions of good performance and, in turn, trust. High discrimination ability alone will not engender trust, as witnessed by the fact that despite being perceived as the most knowledgeable source, the industry was trusted least. Moreover, a moderate response bias, as shown by scientists and medics, was also not a unique feature for trust since there were significant positive correlations between knowledge and trust for these sources as well.

In short, trusted sources are those who are perceived to be able to correctly discriminate many instances of safety from danger and when uncertain are likely to make what the public deems to be 'appropriate' errors. In the case of mobile phones, where there are obvious benefits, what is deemed to be appropriate may reflect more tolerance of risk than for a technology whose benefits are less obvious (see, for example, Finucane et al, 2000), and it remains to be seen whether the pattern of trust may differ across different potential risks. Indeed, the moderating effects of individual risk perceptions would support such a contention, so it might be, for example, that for nuclear power, environmentalists may be even more trusted than independent scientists precisely because the former are seen to be adopting a response bias that would result in even fewer misses.

Of course, the two key dimensions proposed by the IDT approach have many similarities with dimensions proposed by other authors. However, while discrimination ability appears highly related to competence, the notion of response bias appears to add significant clarity in the attempt to understand the 'second dimension'. Specifically, a range of proposed dimensions from shared values to care and concern can all be reinterpreted as reflecting the degree to which a source's perceived response bias is deemed 'appropriate' by the observer. Thus, we see the IDT perspective not as an alternative to other perspectives, but as highly compatible to them, and it may prove to be particularly useful as a method for identifying the underlying similarities of these other approaches.

Moreover, we also believe that viewing trust in risk management contexts as a consideration of the costs and benefits of correct and incorrect judgements will also shed additional light on the processes of how trust can be lost as well as gained. For example, while Slovic (1993) has argued that negative information about a source will lead to greater decreases in trust than positive information will lead to trust increases, he did not consider the issue of trust dimensionality. So, based on the discussion of the IDT approach above, we would predict that evidence of a source having made a miss would lead to a greater loss in trust than evidence of a false alarm, despite the fact that both are pieces of negative information. Moreover, a hit would likely increase trust more than a correct rejection because a hit is indicative of a response bias, which is more alert to any sign of danger. In fact, it may even be that someone who makes false alarms (an error) is more trusted than someone who makes correct rejections despite the fact that the former outcome provides negative information and the latter positive information. Indeed, we have already found evidence to support such a possibility (European Commission, 2000).

We also believe there are a number of important overlaps with other areas of the trust and risk perception literature. We have already noted the similarities between perceptions of response bias and the SARF framework, for example; but there are other key connections as well. First, the notion of response bias can be used to understand one of the most well-known reactions to uncertainty in relation to environmental hazards – namely, the precautionary approach (European Commission, 2000). Principle 15 of the Rio Declaration (UNEP, 1992), for example, states that:

> Where there are threats of serious or irreversible damage, lack of full scientific certainty shall not be used as a reason for postponing cost-effective measures to prevent environmental degradation.

The consideration of 'threats of serious or irreversible damage' is focusing on the potential costs of misses – that is, failures to detect and prevent dangers to the environment. Moreover, that the lack of scientific certainty of such dangers should not be 'used as a reason for postponing' preventative measures suggests that the response bias parameter should be set so as to avoid misses. Finally, the mention of 'cost-effective measures' effectively tempers this call for a lax criterion for detecting and preventing environmental risks since the principle is drawing attention to the costs of attempting to avoid misses.

Second, we have also suggested that there may be important individual differences in preferences for response bias based on risk perception. This appears related to research on 'cognitive styles' in relation to risky messages (Janis and Mann, 1977). Specifically, people showing what Janis and Mann call *vigilance* (a term with obvious connections to SDT) appear to be making attempts to improve discrimination ability by finding out more information about risks and by paying greater attention to the fact that risks may be present. By contrast, those exhibiting *defensive avoidance* appear to be setting a very strict response bias. These people are likely to avoid false alarms but will also tend to make more misses. Finally, people showing *hypervigilance* appear to be setting a rather lax criterion for accepting that a risk is present, and will tend to make far more false alarms, but are less likely to make misses (for a similar typology, see Miller, 1995).

Finally, there is also a clear role for an 'affective' component (Finucane et al, 2000; Slovic et al, 2002) in what might otherwise seem a highly cognitive approach to decision-making. Specifically, a decision-maker's wants, hopes, fears and concerns can all influence their preferred response bias. So, for example, a person who has negative 'affect' with respect to nuclear power is perhaps more likely to set a cautious response bias requiring a lot of evidence before accepting that it is safe. Moreover, given limited cognitive resources, it is assumed that more or less heuristic and systematic processing will occur across contexts (Petty and Cacioppo, 1986; Chaiken et al, 1996). For example, when the issue is of particular importance to the individual, the other's assessment of the risks is likely to be scrutinized more carefully than when personal importance of the hazard is low.

Of course, we are also aware that the survey results presented here are not definitive evidence for the IDT approach. In particular, we have said relatively little about the role of transparency and openness, although, as noted earlier, more on this aspect and its relations to discrimination ability and response bias can be found elsewhere. Moreover, there is a need for more investigation into how people collect and process information about these dimensions. This is necessary in order to address issues such as whether there are instances where information about one dimension becomes more important than another, and how information integration can be achieved quickly and efficiently. Experimental research is needed to investigate the relationships between perceived discrimination ability and response bias to establish whether perceptions of one dimension influence the other, or whether, like formal detection theorists, people are able to keep them relatively separate.

To conclude, while there is clearly much work to be carried out within the IDT framework, we believe the early signs are promising. Starting from a theoretical position on decision-making under uncertainty, our approach to understanding trust in risk management contexts suggests dimensions of trust that are broadly reminiscent of those proposed using more descriptive data-driven perspectives. However, our approach is able to offer a possible solution to the current ambiguity that exists in the literature on trust dimensionality and, moreover, is able to propose new insights into trust processes, such as the relationship between negative information and how trust is lost. In short, we believe that a greater understanding of how 'trust is shaped, altered, lost or rebuilt in the processing of risk' can be achieved by viewing people as intuitive detection theorists.

NOTES

1 This research was funded by the UK's regulator for health and safety at work, the Health and Safety Executive, contract no 4104/R71.046.
2 Indeed, even these authors have apparently now accepted the need for a competence dimension in the guise of *past performance* (see, for example, Siegrist et al, 2003).
3 In this study we did not explicitly separate response bias from communication issues such as transparency, although this was achieved in later studies; see European Commission (2000) for more details.
4 Since discrimination ability and response bias are repeated measures variables, it was not possible to carry out ordinary *post-hoc* comparisons. Thus, planned repeated contrasts were carried out and significance was set at $p < 0.001$ to ward against type I errors.

REFERENCES

Chaiken, S., Wood, W. and Eagly, A. H. (1996) 'Principles of persuasion', in E. T. Higgins and W. Kruglanski (eds) *Social Psychology: Handbook of Basic Principles*, Guilford Press, New York, pp702–742

Earle, T. C. and Cvetkovich, G. T. (1995) *Social Trust: Toward a Cosmopolitan Society*, Praeger, Westport, Connecticut

Einhorn, H. J. (1974) 'Expert judgement: Some necessary conditions and an example', *Journal of Applied Psychology*, vol 59, pp562–571

Eiser, J. R. (1990) *Social Judgement*, Open University Press, Milton Keynes

European Commission (2000) *Communication of the Commission on the Precautionary Principle*, www.europa.eu.int/comm/dgs/health_consumer/library/pub/pub07_en.pdf

Finucane, M. L., Alhakami, A., Slovic, P. and Johnson, M. (2000) 'The affect heuristic in judgements of risks and benefits', *Journal of Behavioural Decision Making*, vol 13, pp1–17

Freudenberg, W. R. and Rursch, J. A. (1994) 'The risks of "putting the numbers in context": A cautionary tale', *Risk Analysis*, vol 10, no 3, pp949–958

Frewer, L. (2003) 'Trust, transparency and social context: Implications for social amplification of risk', in N. Pidgeon, R. E. Kasperson and P. Slovic (eds) *The Social Amplification of Risk*, Cambridge University Press, Cambridge, pp123–137

Frewer, L. J., Howard, C., Hedderley, D. and Shepherd, R. (1996) 'What determines trust in information about food related risks?', *Risk Analysis*, vol 16, pp473–486

Gigerenzer, G., Todd, P. M. and ABC Research Group (1999) *Simple Heuristics that Make us Smart*, Oxford University Press, Oxford

Gilovich, T., Griffin, D. and Kahneman, D (2002) *Heuristics and Biases: The Psychology of Intuitive Judgement*, Cambridge University Press, Cambridge

Green, D. A. and Swets, J. A. (1974/1988) *Signal Detection Theory and Psychophysics*, Wiley, New York
Heider, F. (1958) *The Psychology of Interpersonal Relations*, Wiley, New York
IEGMP (Independent Expert Group on Mobile Phones) (2000) *Mobile Phones and Health*, National Radiological Protection Board, Oxford
Janis, I. L. and Mann, L. (1977) *Decision Making*, The Free Press, New York
Johnson, B. B. (1999) 'Exploring dimensionality in the origins of hazard-related trust', *Journal of Risk Research*, vol 2, pp325–354
Jungermann, H., Pfister, H. R. and Fischer, K. (1996) 'Credibility, information preferences, and information interests', *Risk Analysis*, vol 16, no 2, pp251–261
Kasperson, R. E., Renn, O., Slovic, P., Brown, H. S., Emel, J., Goble, R., Kasperson, J. X. and Ratrick, S. (1988) 'The social amplification of risk: A conceptual framework', *Risk Analysis*, vol 8, pp177–187
Kasperson, R. E., Golding, D. and Tuler, S. (1992) 'Social distrust as a factor in siting hazardous facilities and communicating risks', *Journal of Social Issues*, vol 48, pp161–187
Kasperson, J. X., Kasperson, R. E., Pidgeon, N. and Slovic, P. (2003) 'The social amplification of risk: Assessing fifteen years of research and theory', in N. Pidgeon, R. E. Kasperson and P. Slovic (eds) *The Social Amplification of Risk*, Cambridge University Press, Cambridge, pp13–46
Kelley, H. H. (1967) 'Attribution theory in social psychology', in D. Levine (ed) *Nebraska Symposium on Motivation*, vol 15, University of Nebraska Press, Lincoln, pp192–240
Kraus, N., Malmfors, T. and Slovic, P. (1992) 'Intuitive toxicology: Expert and lay judgements of chemical risks', *Risk Analysis*, vol 12, pp215–232
Langford, I. H., Marris, C. and McDonald, A.-L. (1999) 'Simultaneous analysis of individual and aggregate responses in psychometric data using multilevel modeling', *Risk Analysis*, vol 19, pp675–683
Langford, I. H., Marris, C. and O'Riordan, T. (2001) 'Public reactions to risk: Social structures, images of science, and the role of trust', in P. Bennett and K. Calman (eds) *Risk Communication and Public Health*, Oxford University Press, Oxford, pp33–50
Macmillan, N. A. and Creelman, C. D. (1991) *Detection Theory: A Users Guide*, Cambridge University Press, New York
Maeda, Y. and Miyahara, M. (2003) 'Determinants of trust in industry, government, and citizen's groups in Japan', *Risk Analysis*, vol 23, no 2, pp303–310
Metlay, D. (1999) 'Institutional trust and confidence: A journey through a conceptual quagmire', in G. T. Cvetkovich and R. E. Lofstedt (eds) *Social Trust and the Management of Risk*, Earthscan, London, pp100–116
Miller, S. M. (1995) 'Monitoring versus blunting styles of coping influence the information patients want and need about cancer', *Cancer*, vol 76, pp167–177
Mishra, A. (1996) 'Organizational responses to crisis: The centrality of trust', in R. M. Kramer and T. R. Tyler (eds) *Trust in Organizations: Frontiers of Theory and Research*, Sage, London, pp261–287
Peters, R. G., Covello, V. T. and McCallum, D. B. (1997) 'The determinants of trust and credibility in environmental risk communication: An empirical study', *Risk Analysis*, vol 17, no 1, pp43–54
Petty, R. E. and Cacioppo, J. T. (1986) 'The elaboration likelihood model of persuasion', in L. Berkowitz (ed) *Advances in Experimental Social Psychology*, vol

19, Academic Press, New York, pp123–205

Poortinga, W. and Pidgeon, N. (2003) 'Exploring the dimensionality of trust in risk regulation', *Risk Analysis*, vol 23, no 5, pp961–972

Renn, O. and Levine, D. (1991) 'Credibility and trust in risk communication', in R. E. Kasperson and P. J. M. Stallen (eds) *Communicating Risks to the Public*, Kluwer, The Hague, pp175–218

Siegrist, M., Earle, T. and Gutscher, H. (2003) 'Test of a trust and confidence model in the applied context of electromagnetic fields (EMF) risks', *Risk Analysis*, vol 23, pp705–716

Slovic, P. (1987) 'Perception of risk', *Science*, vol 236, pp280–285

Slovic, P. (1993) 'Perceived risk, trust and democracy', *Risk Analysis*, vol 13, no 6, pp675–682

Slovic, P., Finucane, M., Peters, E. and MacGregor, D. G. (2002) 'The affect heuristic', in T. Gilovich, D. Griffin and K. Kahneman (eds) *Heuristics and Biases: The Psychology of Intuitive Judgement*, Cambridge University Press, Cambridge

Starr, C. (1985) 'Risk management, assessment, and acceptability', *Risk Analysis*, vol 5, pp97–102

Swets, J. A. (1996) *Signal Detection Theory and ROC Analysis in Psychology and Diagnostics: Collected Papers*, Erlbaum, Mahwah, NJ

Swets, J. A., Dawes, R. B. and Monahan, J. (2000) 'Psychological science can improve diagnostic decisions', *Psychological Science in the Public Interest*, vol 1, pp1–26

Tetlock, P. E. (2002) 'Intuitive politicians, theologians, and prosecutors: Exploring the empirical implications of deviant functionalist metaphors', in T. Gilovich, D.Griffin and D. Kahneman (eds) *Heuristics and Biases: The Psychology of Intuitive Judgement*, Cambridge University Press, Cambridge, pp582–599

Tetlock, P. E. and Boettger, R. (1994) 'Accountability amplifies the status-quo effect when change creates victims', *Journal of Behavioural Decision Making*, vol 7, no 1, pp1–23

UNEP (United Nations Environment Programme) (1992) *Rio Declaration on Environment and Development*, www.unep.org/Documents/Default.asp?DocumentID=78

White, M. P., Pahl, S., Buehner, M. and Haye, A. (2003) 'Trust in risky messages: The role of prior attitudes', *Risk Analysis*, vol 23, pp717–726

White, M. P., Eiser, J. R. and Harris, P. (2004) 'Risk perceptions of mobile phone use while driving', *Risk Analysis*, vol 24, pp323–334

Wynne, B. (1980) 'Technology, risk, and participation: On the social treatment of uncertainty', in J. Conrad (ed) *Society, Technology and Risk Assessment*, Academic Press, London

5 Scepticism, Reliance and Risk Managing Institutions: Towards a Conceptual Model of 'Critical Trust'[1]

Nick Pidgeon, Wouter Poortinga and John Walls

> I don't generally trust... if the government came out and said 'GM food is safe', I wouldn't even believe it. But if they came out and said GM food wasn't safe, I'd have to sort of question that as well, because I have to question everything, every piece of information that's thrown at me... I wouldn't say I totally distrust everyone, because I make it sound like I do, don't I? But I actually question everything. (Male, Norwich, speaking about genetically modified food, 2003)[2]

INTRODUCTION

In the field of risk research the issue of trust in risk-managing institutions, and its relationship with public risk perception and acceptance, has been a matter of debate for a number of years now. In an early paper, the sociologist Brian Wynne (1980) argued that with technological risks some of the differences between 'expert' and 'lay perspectives' could be traced to differing evaluations of the trustworthiness of risk-managing institutions. If people disagree with experts about the acceptability of nuclear power, this might be because they do not trust the claims of the proponents or regulators of the technology. The relationship between trust judgements and risk perception has since gained widespread attention, particularly since the early 1990s (see, for example, Renn and Levine, 1991; Bord and O'Connor, 1992; Pidgeon et al, 1992; Slovic, 1993; and Cvetkovich and Löfstedt, 1999). In psychometric studies where both have been measured, risk perceptions and trust are often found to be inversely related. For example, Pijawka and Mushkatel (1992) found a strong negative relationship between trust (in government) institutions and the perceived risks of a high-level nuclear waste repository. Since this was accompanied by high levels of opposition, they concluded that both people's risk perceptions and their opposition to the siting of the repository was caused by a lack of trust in the US Department of Energy (DOE). In terms of wider societal impacts, distrust has also been reported to be associated with stigmatization of technologies (Flynn et al, 2001), as well as social amplification effects following major failures of risk regulation (see Pidgeon et al, 2003a). Accordingly, rebuilding trust is seen as a prerequisite for effective risk communication (for example, Kasperson et al, 1992), and has also, in a somewhat separate development, become a central concept within wider theories of late modernity or 'risk society' (Giddens, 1990; Beck, 1992).

In public policy circles, trust (and its presumed absence among some sections of society at large) has also become a lively topic for debate. In the UK and Europe, the events surrounding the mad cow (BSE) crisis of the late 1990s forced decision-makers to reassess the place of traditional science communication efforts alongside that of public trust. For example, the UK House of Lords Science and Technology Committee (2000) *Report on Science and Society* noted an apparent crisis of confidence in the governance of science, and with it a failure of the traditional one-way deficit model of science risk communication. In other words, we can no longer assume that merely educating and informing people about science, technology and risk will *of itself* lead to a resolution of risk controversies (see also Fischhoff, 1995). In many respects, the former diagnosis, that technological controversy was fundamentally down to an 'information deficit' among the lay population at large, has been replaced in many public policy discourses by the assumption that there now exists a 'trust deficit'. As a result, one of the oft-stated objectives of public engagement and participation approaches to technological controversies is that of increasing trust in the decision-making and risk management process (see Beierle and Cayford, 2002).

In this chapter we argue that contemporary approaches to public dialogue and institutional openness by risk-managing institutions need to take account of the complex nature of public 'trust' in organizations. In particular, the notions of 'trust' and 'distrust', when used as all or nothing concepts, tend to simplify what is a much more subtle and complex set of relationships in the discourses and perceptions that people hold about risk-managing organizations, particularly where an individual has no choice but to rely upon the agency in question. Hence, for example, expressions of distrust may signal outright rejection or merely reliance coupled with a healthy scepticism of all organizations. Equally, very few statements of trust, whatever the track record or credentials of a risk manager, are likely to prove unconditional for people. Doubts always remain over motives, resources or other aspects of the truster–trustee relationship.

In order to further explore some of the complexities, we introduce the concept of 'critical trust', an idea which seeks to reflect both people's scepticism about, as well as their reliance upon, risk managing organizations. The chapter also advocates a mixed-methods approach to empirical investigation in that both quantitative and qualitative data are used to illuminate complementary facets of the trust concept. Although debates during the late 1980s and 1990s tended to see qualitative and quantitative research as distinctive paradigms of enquiry (see Henwood and Pidgeon, 1992), more nuanced and complementary facets of qualitative and quantitative methods have more recently come to be discussed in so-called mixed-methods designs and approaches (Bryman, 2001; Todd et al, 2004). Accordingly, this chapter draws upon three separate empirical studies (one qualitative and two quantitative) in order to illustrate its main conceptual themes. In the concluding section, we return to the themes of risk management, risk communication and public policy in attempting to identify lessons from our conceptual and empirical analysis.

CONCEPTUALIZING TRUST: DIMENSIONAL AND CONTEXTUAL APPROACHES

Quantitative approaches to the study of trust have tended to conceptualize it as comprising a core set of dimensions: the so-called 'dimensional' approach. For example, Renn and Levine (1991) identify five core components or attributes – namely, *perceived competence*, which represents the degree of technical expertise of the source; *objectivity*, reflecting the absence of bias in information; *fairness*, or the degree to which the source takes into account all relevant points of view; *consistency*, or the predictability of arguments and behaviour based on past experience and previous communication efforts; and *faith*, which reflects the perception of 'goodwill' of the source. Similarly, Kasperson et al (1992) identify four key dimensions that play an important role in the development and maintenance of trust: *commitment*, *competence*, *caring* and, finally, *predictability*.

Summarizing the empirical evidence from a range of studies, Johnson (1999) identifies *care* and *competence* as the two core dimensions of trust, with a third component – *consensual* values – derived from more recent conceptual and empirical work by Earle and Cvetkovich (1995) on 'social trust' and the salient value similarity hypothesis (Siegrist et al, 2000). Earle and Cvetkovich (1995) have argued that for most people it is far too demanding to base trust on evidence of competence and fiduciary responsibilities. Rather, it is perceived sympathy and agreement around core values that is the principal basis for trust. According to these authors, social trust is particularly critical where complex socio-technical systems generate risks that are remote from everyday experience. Under these circumstances, most people will not have the resources or interest to make a detailed assessment of whether or not it is worthwhile to trust a particular institution.

Dimensional models have a number of conceptual origins. In one respect, they reflect an empirical reality: specifically, the range of common-sense concepts which ordinary people use to talk about trust issues (notions of expertise, care for others, political or other interest-based bias) expressed in ways that can be readily operationalized in quantitative surveys. However, dimensional approaches are also partly based, implicitly or explicitly, upon categories that are predefined by the researchers' conceptual frameworks. Most obviously, as Earle (2004) points out, those working within the dimensional approach often seek to relate judgements of trust to normative matters – specifically, what ethical or other conditions *should* matter for trust to pertain – as found, for example, in philosophical or legal discussions of the issue (see O'Neill, 2002). Hence, if we expect to trust organizations that are 'objective', this can reflect the everyday concept of freedom from political or financial inducement, an epistemological commitment to a 'true' description of the world and its hazards, or both. By contrast, the salient value similarity models are based primarily upon theoretical concerns derived from the social psychology of interpersonal and group behaviour, and, in particular, upon the long-standing empirical observation that we trust and cooperate with people whom we perceive are like us in some fundamental way, or who are members of a social category or group with which we identify.

Contextual studies take a more qualitative and locally grounded approach to issues of public understanding of risk and trust, without making strong prior commitments to normative questions, and tend, accordingly, to be grounded in the traditions of constructivist human geography, social psychology or micro-sociology. Such studies are less concerned with normative issues, preferring to access a range of *in situ* or everyday referents of trust – often also highlighting their relationship with such local contextual factors as economic dependence, place, identity and stigma (see, for example, Irwin et al, 1999; Bickerstaff and Walker, 2003; Horlick-Jones et al, 2003; Walls et al, 2004). In addition, in such research, investigators have tended to move away from a theoretical framing of trust as a simple 'commodity' or attribute that attaches to an organization or individual towards a situation whereby trust is provisionally conceptualized as multifaceted, potentially dynamic and strongly dependent upon social context. Such a perspective also transcends a monolithic view of trust, which tends to portray social agents as essentially passive agents, towards one which opens up the possibility of developing an understanding of trust that recognizes people's capacity for active sense-making.

A striking aspect of much of research within this latter tradition, using as it does various forms of public discussion on institutional trust and risk issues, is that 'trust' often seems an inappropriate category for many participants (see also Wynne et al, 1993). While overt expressions of 'trust' do, indeed, at times draw upon an extremely full range of the conventional dimensions hypothesized in the academic literature, such dimensions always seem lacking (when used as analytic frames for the discourses involved) in full explanatory power[3] – for example, in relation to the absences and hesitations present in participants' discourse about trust, as well as an ultimate reluctance to admit (in the semi-public forum of an interview or focus group, at least) that they fully and absolutely trusted any of the government agencies, environmental groups, professions or individual persons whom researchers might suggest to them. And, ultimately, the closer one questions participants about the bases for their 'trust', the less visible it seems to become in the data! As noted above, Earle and Cvetkovich (1995) argue that although conventional trust concepts (such as competence) may often be used in discourses on trust, this is without regard to whether these really are the critical factors underlying people's judgements. Rather, people may use these concepts to provide an explanation for their general agreement, sympathy or trust in an institution because these underlying beliefs might not be regarded as a socially acceptable justification. It may well be that in the same way that social researchers have developed a sophisticated contemporary view of the epistemological and methodological linkages between qualitative and quantitative research methods, the time is right for a closer integration between the dimensional and contextual approaches to trust. In this chapter we attempt to do this by triangulating data and findings from three separate studies.

TRUST AND SCEPTICISM: THE UK HEALTH AND SAFETY EXECUTIVE

In this section we present qualitative data drawn from a recent research project conducted to identify the factors underlying trust (or not) in the Health and Safety Executive (HSE), the UK regulator of workplace safety (for full details of methodology and data analysis, see Pidgeon et al, 2003b; Walls et al, 2004). A focus group methodology was adopted as the starting point to mapping discourses of trust. Focus groups are informal facilitated group interviews that use simple focused and open-ended questions to get participants discussing a specific topic. Initially developed in commercial market research, such groups are now regarded as an important component of the methodological tool kit used by social researchers (Krueger, 1988). Focus groups allow the generation of relatively unstructured and naturalistic conversations and, through this, the collection of fragments of 'ordinary conversation' in a manageable and cost-effective way. It should be stressed that a focus group-based investigation does not aim to be 'representative' in the manner of quantitative survey research. Rather, focus groups provide a powerful means of exploring group norms and shared ways of conceptualizing, understanding and talking about things: those shared resources that make intelligible conversation and meaningful social interaction possible. In the study reported here, a total of 30 groups (n = 202 participants) were held, each composed of six to eight members of the voting-age public and convened for between 1.5 and 2 hours. The overall sampling strategy was designed to embrace a range of standard socio-demographic variables (age, gender and social class), as well as regional diversity. In total, five groups were held in each of six regions across Great Britain, with data collected between the late autumn of 2000 and the early summer of 2001, and a common elicitation protocol for all of the groups.

The data analysis we report here – drawn from the corpus of qualitative data across the 30 groups – entailed a grounded theory approach (Glaser and Strauss, 1967; Pidgeon and Henwood, 2004). In particular, constant comparison was undertaken between and within transcripts (and linking related categories together in code networks where appropriate) to reveal patterns and categories of reasoning displayed in the conversations between participants. Negative cases were also actively sought to contradict emergent assumptions. In this way, a structured account of the complexities inherent in the data is systematically built up. In grounded theory, the process of abstraction from raw data to higher-order conceptual categories is disciplined, but not determined, by the requirement that categories and codes should throughout 'fit' (or provide a recognizable description of) the data to which they refer. Through analysis of the discourse in this way, the research team developed a number of salient themes that we label here *tentative knowledge*; *inspection, investigation and enforcement*; *altruism*; and *critical sentiments*.

Theme 1: Tentative knowledge

A salient and recurrent theme among our participants – not surprising given people's lack of direct contact with such government agencies, in general – was an almost total lack of detailed knowledge of what HSE did. While most participants were able to demonstrate an awareness of HSE, as we had expected, knowledge of HSE's role and remit was, for many, slight. Few participants reported having had direct experience of or contact with the organization, although many were able to demonstrate an awareness derived from secondary sources – that is, the media, social and/or familial networks:

> I'm not that familiar; I am familiar that they exist, but as to actually what they do, I don't know. (Male, Norwich)

> Are they a government organization? I mean, is it under parliament or something? They draw the lines, examine what's gone wrong if there is a very serious problem or accident perhaps; you know, that's what I'd expect them to do, but I wouldn't say I know. (Female, Norwich)

> The Health and Safety Executive: it's not a spin-off from a trade union is it? It's just the word executive gives that impression. (Female, Monmouth)

Theme 2: Inspection, investigation and enforcement

Because knowledge was tentative, this did not mean participants had no understanding of the functions of the organization. A consistent understanding expressed in the focus groups was that HSE, or an equivalent body, inspected workplaces, investigated accidents, and took action against those who broke health and safety law. References to 'inspectors' were widespread; but linkages were not always made with HSE. For example, participants with little knowledge of HSE often confused their employer's own health and safety advice with HSE guidelines and/or regulation, and the activities of other forms of government inspectors with those of HSE personnel:

> My knowledge of them is not good; I am aware that they are responsible for rules and regulations in public buildings and places like that. And workplaces, whatever. I have never had any dealings ... directly or indirectly with the work that they do. (Male, Reading)

> Well, in places where I have worked they have come in and sorted something, or when I have seen them on the news talking to somebody they have just kind of, usually because it is accidents. (Male, Cardiff)

> At our place maybe every couple of years somebody will just turn up and it is like ... to try and catch you using something that maybe you shouldn't be using, or spraying something in an area you shouldn't be spraying. And like, where I work, they don't take any chances with the health and safety; they just don't want to take any chances because they just close them straightaway. They don't mess about with them at all. (Female, Birmingham)

In the main, expressed knowledge of HSE was limited to functional aspects of HSE activity in the workplace, with the inspection function verbalized as the most recognized role for HSE across all of the focus groups. In the main, the inspection of premises was seen as a necessary activity by most of our participants. Overall, the dominant set of beliefs, or mental model, was anchored in the three interrelated operational aspects of *inspection*, *investigation* and *enforcement*, referenced in particular (although not exclusively) to an industrial safety context. The quotes above illustrate some representative remarks from our groups regarding this.

Theme 3: Altruism

Despite this tentative knowledge about the agency, in a conventional trust-rating task (similar to many available in the published literature) given to all of the focus group members, and as shown in Table 5.1, HSE gained the *highest* trust score of all agencies, including a number of much better-known organizations. The participant in the following quote explicitly articulates the difficulties of providing a rating under conditions of limited knowledge:

> Well, sometimes you are answering these and not really knowing what these people actually do; you know, you don't know whether to trust or not trust what you are told. (Female, Birmingham)

Table 5.1 *Average trust ratings of focus group participants*

Organization	Mean	Standard deviation
Health and Safety Executive	2.50	.78
Environmental groups (e.g. Greenpeace)	2.30	.91
Consumer groups (e.g. Consumers Association)	2.27	.77
Food Standards Agency	2.27	.74
Local environmental health officers	2.27	.70
Department of Health	2.23	.79
Environment Agency	2.13	.63
National Farmers Union	2.00	.73
Ministry of Agriculture, Fisheries and Food	1.99	.72
Local council	1.82	.61
Confederation of British Industry	1.80	.60
Railways Inspectorate	1.74	.70
Government ministers	1.68	.59

Note: 'For each of the following bodies, please indicate how much you would trust them to protect people's health, safety and well-being: scale values of 1 (trust very little); 2 (trust to some extent); 3 (trust a great deal); 4 (trust almost entirely).'

So, what underlay people's conceptions and high ratings of trust? Although these findings may be, in part, due to the fact that the agency's name matches the question wording ('How much do you trust them to protect health, safety and well-being?'), this was rarely articulated as a reason (and certainly not the sole reason) for the high ratings given. Rather, the dominant discourse underlying expressed levels of trust in HSE was a perception that the agency is *motivated to act in the public interest*. In particular, the agency is perceived as performing a fundamentally *altruistic role* (in particular, raising awareness of safety issues, in activities demonstrating care for ordinary people in the workplace, and as a fundamental orientation relatively free from other pressures such as political or financial interests). The following quotes illustrate the typical sentiments offered:

Because you feel that they are just advising you from a health and safety point, they are not trying to score points politically... they are just more or less saying, well, you know you have to watch out for this or that or the other in the workplace because it can, you know, damage your health in some way. (Male, Reading)

Well, because they should be, their name suggests that they should be there looking after our interests and should have the public's welfare at heart; whether they actually do or not, I'm not 100 per cent sure and I doubt if anybody is really sure. (Male, Norwich)

Well, my personal experience with HSE is that they are effective. I don't believe they are answerable to any particular government and I think the HSE continues; the personnel doesn't change when the government changes, and from my personal experience, when they say something has to be done, to improve a situation, then it gets done. (Male, Norwich)

So a lot of it is looking at their title and say they are looking after that for me, and just hope that they are doing their job. (Male, Birmingham)

In essence, within a dimensional model, we would label this altruistic aspect of perceptions of HSE primarily as one of care. However, a closer reading of the phrasings used in these particular quotes might also be interpreted as demonstrating a degree of implicit agreement around shared goals or values ('our interests', 'advising you', 'for me'), as might be predicted from the salient value similarity model. Indeed, Earle (2004) demonstrates by using verbal protocol analysis that in a situation similar to this one, where direct knowledge of an agency is low or non-existent (specifically, regarding the fictional Institute for Climate Change Policy), trust judgements will be driven primarily by agreement or sympathy with perceived institutional goals rather than by more normative (dimensional) considerations. In the case of the Health and Safety Executive, it is reasonable to assume that the protection of personal well-being at work is a goal to which most people would subscribe, providing a powerful basis for inferring value similarity directly from the agency's name and, with this trust, judgements, in the absence of other more definitive information. This is supported by evidence in our transcripts of a range of inferences, made in the face of limited knowledge, directly from the agency's name.[4]

Note, also, in the third quote above, that there is a judgement about the effectiveness of the organization ('my personal experience with HSE is that they are effective'). This latter aspect we would label *competence*. But perhaps more significant for our purposes here is the fact that in several of the quotes (and these were by no means atypical), the discourse displays a hesitancy, a tentativeness indexed by phrases such as 'you feel that', 'they are just more or less', 'I doubt', or 'just hope'. These phrases point to the need to take a more nuanced view of the statements of trust being offered.

Theme 4: Critical sentiments

Further analysis of the data indicated that while the agency (in marked contrast to some of the other agencies discussed in the groups) was, indeed, seen primarily as mission oriented, independent from politics, and, above all, performing a function of care for people's health and safety in the workplace, critical sentiments nevertheless existed. For example, a small number of participants suggested that the organization was seen to act only retrospectively – that is, after a major incident had occurred:

> But it tends to be after the event, doesn't it, rather than before?
> Yes, but you only see them after; you don't see them before, do you? (Females, Cardiff)

Above all, though, critical sentiments were directed towards the practical effectiveness of the organization. As the following quote and dialogue illustrate, our participants could discuss HSE as a necessary institution that they rated as trustworthy, but at the same time not necessarily as effective an organization as it could be, given a potential lack of funds or the operation of subtle political influences:

> They are grossly understaffed; you just have to look at the one at Norwich, the one at St Giles, etc., and, quite frankly, they have too big a caseload and they can't follow through. (Male, Norwich)

> *Moderator*: So you have seen their literature. Do you think they do a good job in what they are supposed to do?
> They do get it over to people, I suppose, yeah.
> The only thing is they are fighting against the manufacturers, as well, like, so they don't go around as much as they should do; well, maybe because they don't have enough agents to do it. There are lots of places in this country where the working conditions are still very poor for people. (Males, Reading)

Participants who verbalized these opinions did reconcile these critical sentiments with the overwhelmingly positive perceptions outlined above, suggesting that stated trust can be the outcome of an active reconciliation of potentially contradictory sentiments and ideas about a risk regulator. Participants' sense of such critical sentiments was not sufficient to overwhelm the positive impressions of HSE based on its perceived altruistic role, but functioned to counteract a naive trust. Moreover, these critical comments almost never became the central focus of group discussions about HSE, but were often the result of individual experience or *ad hoc* reasoning. As Walls et al (2004) point out, social trust emerges in these accounts as multidimensional and fragmented, as a product of a reconciliation of ideas and knowledges and impressions, so that the perceived limited effectiveness of HSE can coexist with a high degree of relative trust, where it appears that it is based on an organization's perceived intention to act in the public interest as effectively as possible, tempered by a pragmatic and 'common-sense' observation about the role that 'government', for instance, plays:

> I trust a great deal because they are aware of what they are doing, you know, and they are respected; I feel that within the limits that they are given, they have created their own competence as far as we can tell. But I mean they are all subject to whims and fashions of government.
> *Moderator*: So, they are there to protect the health and safety of people?
> Yes, the profits and things like that totally are of no interest to them. Their thing is just health and safety, isn't it, as the name implies. (Male, Chesterfield)

Our overall analysis of the data led us to propose the concept of *critical trust* as an organizing principle (see Figure 5.1). Critical trust lies on a continuum between outright scepticism (rejection) and uncritical emotional acceptance. Such a concept attempts to reconcile the actual *reliance* by the public on institutions, while simultaneously possessing a *critical attitude* towards the effectiveness, 'motivations' or independence of the agency in question. In a study of chemical industry hazards within Greater Manchester, Irwin et al (1996) similarly report high levels of local scepticism in industry (as trustworthy sources of information), coupled with a pragmatic reliance upon the same source (as the only party who can do anything to reduce hazards). They also note that 'this critical evaluation does not solely apply to local industry – all other sources are likely to (and, indeed, do) undergo similar scrutiny' (Irwin et al, 1996, p57).

The argument is that verbalizations of trust (or, conversely, distrust) are rarely, if ever, made at the extremes or discussed in a one-dimensional manner. Rather, as we have seen, people can negotiate an array of impressions and experiences to arrive at a temporary 'fix' for the context they find themselves in, and it is the eventual weight which people give to these factors that accounts for their statements about trust. In our analysis, the concept of critical trust reflects the way in which participants reason and talk about risk regulators. Throughout the data there was little discussion of naive or blind trust. Rather, lay discourses tended to contain elements of critical reflection as part of a negotiated balance of scepticism and reliance, and in which a range of contextual variables impacted upon a participant's stated attitude. Our coding schema shows that discussion of limiting factors

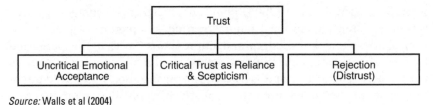

Source: Walls et al (2004)

Figure 5.1 *Critical trust*

about HSE were evident in 23 out of a total of 30 focus groups,[5] while discussions of HSE's altruistic role occurred in 26 out of the 30 groups (Walls et al, 2004). We have characterized the outcome of the active negotiation of these opposing sentiments as reflecting critical trust.

TRUST, SCEPTICISM AND RELIANCE

Further evidence for the critical trust hypothesis, which sees reliance and scepticism as part of a negotiated tension in people's beliefs about institutions, comes from a separate quantitative survey of public beliefs about government risk regulation in five technological domains. This study (see Poortinga and Pidgeon, 2003a, 2003b) had the objective of examining people's perceptions of the following risk cases – namely, climate change, radiation from mobile phones, radioactive waste, genetically modified food (GM food) and human genetic testing – on a wide range of trust-related aspects. These cases were chosen because they are all prominent societal risk issues and have complementary, as well as contrasting, facets of risk, benefits and uncertainty. We also wished to empirically investigate the relationship between a range of concepts comprising distinctive dimensions of trust, particularly whether people's perception of government's risk policies can be described by a limited number of underlying dimensions. The design was developed primarily to provide a test of the empirical model suggested by Metlay (1999), who in his study of trust in the US Department of Energy found that a range of theoretically distinct dimensions factored into two simple concepts: first, a predominantly affective general *trustworthiness dimension* (involving, among other things, items indicating care) and, second, a *competence* dimension.

Data for our study were collected in the summer of 2002 across the UK (England, Scotland and Wales) by the market research company MORI. Fully trained and supervised market research interviewers carried out face-to-face interviews with people in their own homes, which lasted, on average, about 30 minutes. The total sample of 1547 people aged 15 years and older comprised five *separate* quota samples of about 300 respondents, each covering one of the five risk cases.

Participants were presented with various standardized statements on the government and its policy on one of the five risk cases, depending upon the version of the questionnaire. The statements were selected from a review of previous work on trust (for example, Renn and Levine, 1991; Frewer et al, 1996; Peters et al, 1997; Johnson, 1999; Metlay, 1999), with 11 items designed to measure the trust-related aspects of *competence, credibility, reliability, integrity* (vested interests), *care, fairness* and *openness* (see Poortinga and Pidgeon, 2003a).

Table 5.2 *Items and factor loadings after varimax rotation*

	Climate change		Mobile phones		Radioactive waste		GM food		Genetic testing	
	1	2	1	2	1	2	1	2	1	2
The government is doing a good job.	**.80**	−.11	**.75**	−.27	**.73**	−.18	**.77**	−.39	**.79**	−.30
The government is competent enough.	**.72**	−.26	**.71**	−.27	**.76**	−.16	**.74**	−.42	**.81**	−.29
The government has the necessary skilled people.	**.57**	−.20	**.47**	−.35	**.70**	.13	**.66**	−.31	**.71**	−.07
The government distorts facts in its favour.	−.23	**.80**	−.23	**.86**	−.22	**.73**	−.28	**.82**	−.19	**.83**
The government changes policies without good reasons.	−.23	**.81**	−.21	**.88**	−.09	**.86**	−.22	**.86**	−.27	**.83**
The government is too influenced by industry.	−.11	**.73**	−.36	**.60**	−.24	**.76**	−.35	**.73**	−.19	**.74**
The government is acting in the public interest.	**.73**	−.18	**.65**	−.30	**.70**	−.28	**.73**	−.30	**.75**	−.25
The government listens to concerns raised by the public.	**.72**	−.06	**.72**	−.21	**.71**	−.24	**.77**	−.15	**.79**	−.09
The government listens to what ordinary people think.	**.68**	−.16	**.75**	−.25	**.70**	−.28	**.82**	−.26	**.76**	−.26
I feel that the way in which the government makes decisions is fair.	**.79**	−.29	**.75**	−.20	**.75**	−.29	**.77**	−.37	**.79**	−.28
The government provides all relevant information to the public.	**.71**	−.23	**.76**	−.20	**.56**	−.26	**.76**	−.19	**.70**	−.30
Eigenvalue	4.25	2.13	4.14	2.42	4.08	2.29	4.77	2.73	4.83	2.40
Explained variance	38.6	19.4	37.6	22.0	37.1	20.8	43.4	24.8	43.9	21.9
Average agreement	2.53	3.66	2.60	3.49	2.61	3.55	2.48	3.64	2.65	3.51
Cronbach's α	.87	.74	.88	.78	.87	.75	.92	.82	.92	.78

Notes: The scale ranged from 1 (totally disagree) to 5 (totally agree). Factor loadings that are higher than .40 are in bold. Factor interpretations: 1 (general trust); 2 (scepticism).

In order to examine whether the 11 statements evaluating government could be described by a number of underlying dimensions, separate principal components analyses (PCAs) with varimax rotation were conducted for each of the five risk cases. Since the results were similar for all risk cases, one single PCA was conducted across all risk cases. Table 5.2 shows that the PCAs in each of the five risk cases produced very similar factor solutions. The initial 11 items were described by two main components, which in each case jointly accounted for about 60 per cent of the original variance in the responses. As noted above, our initial expectation had been that trust in government with regard to the five risk cases would be in accordance with the findings of Metlay (1999). In other words, the evaluation of government could best be described by a component that represents technical competence, and one that represents a general trustworthiness dimension encompassing care for the public interest (among other things). The first component in the current results (Table 5.2), reproduced across five different issues and samples of respondents, was concerned with a wide range of trust-relevant aspects, such as competence, care, fairness and openness: that is, a general 'affective' trust dimension similar to that found by Metlay (1999), but one that conflates the competence and care dimensions which, in his study, were separate factors. The second factor obtained here resembled the vested interest factor of Frewer et al (1996), including the aspects of credibility and reliability. This factor seems to reflect a sceptical view regarding how risk policies are brought about and enacted (similar to the 'critical sentiments' observed by Walls et al, 2004) and was therefore labelled as a *scepticism* dimension. What is important in this discussion is that at a meta-level of analysis, and as also occurred in the qualitative data associated with the HSE study, our data suggest that sentiments of overall trust can coexist independently with different degrees of scepticism about the organization or agency in question.

As noted above, the concept of critical trust incorporates both scepticism in *and* reliance upon an agency or organization. The converse of trust being mediated by scepticism is for distrust to be mediated by reliance: that is, even where an institution is distrusted, people may still make a pragmatic evaluation of its role in the decision-making process about hazards or risk. The final data that we use to illustrate our argument derives from a study conducted as one part of a wider evaluation of a national public participation process on agricultural biotechnology that took place across the UK in the summer of 2003. Data for the study were collected in the UK by the market research company MORI (see Poortinga and Pidgeon, 2004). A nationally representative quota sample of 1363 people aged 15 years and older were interviewed face to face in their own homes.[6] In the study, and among a large number of other questions, respondents were asked to indicate to what extent they trusted various sources to tell them the truth about GM food (see Table 5.3). Consistent with previous work on trust (see, for example, Frewer et al, 1996; Corrado, 2001; Worcester, 2001; Poortinga et al, 2004), Table 5.3 shows that doctors, consumer rights organizations, environmental organizations and scientists working for universities were the most trusted information sources. More than three-quarters of the general population indicated that they trusted these information sources (a little or a lot) 'to tell the truth about GM food'. More than half of the sample said that they trust scientists working for environmental groups, the Food Standards Agency (FSA), the Department for Environment, Food and Rural Affairs (DEFRA) and farmers to tell the truth about GM food. Perhaps not surprisingly, the least trusted sources were scientists working for government, scientists working for the biotechnology industry, local authorities, the biotechnology industry, food manufacturers, the European Union (EU) and the national government.

Table 5.3 *'To what extent do you trust the following organizations and people to tell the truth about GM food?' (percentage)*

	Distrust a lot	Distrust a little	Neither/nor	Trust a little	Trust a lot	No opinion
Doctors	1	2	14	42	39	2
Consumer rights organizations (e.g. Consumers Association)	2	5	13	43	33	4
Environmental organizations	2	6	14	45	31	3
Scientists working for universities	2	4	17	46	29	2
Scientists working for environmental groups	2	7	15	48	25	2
Food Standards Agency (FSA)	3	6	14	45	26	5
Department for Environment, Food and Rural Affairs (DEFRA)	5	7	18	44	20	6
Farmers	5	10	26	36	19	3
Scientists working for government	15	21	19	34	8	3
Scientists working for the biotech industry	14	21	23	28	9	5
Local authorities	10	17	33	31	5	3
Biotechnology industry	15	20	25	27	8	5
Food manufacturers	17	26	21	29	5	3
European Union (EU)	20	19	25	25	7	4
National government	23	25	19	25	5	3

Note: Weighted dataset: n = 1363.
Source: Poortinga and Pidgeon (2004)

Table 5.4 *Factor loadings after varimax rotation*

	Factor 1: Government	Factor 2: Production and regulation of biotechnology	Factor 3: Watchdogs
Doctors	−.04	.26	.43
Consumer rights organizations (e.g. Consumers Association)	.09	.06	**.61**
Environmental organizations	.09	.02	**.65**
Scientists working for universities	.15	.17	**.79**
Scientists working for environmental groups	.12	.08	**.82**
Food Standards Industry (FSA)	.15	**.58**	.47
Department for Environment, Food and Rural Affairs (DEFRA)	.25	**.56**	.39
Farmers	−.02	**.59**	.15
Scientists working for government	**.75**	.35	.12
Scientists working for the biotech industry	.41	**.76**	.06
Local authorities	**.69**	.26	.14
Biotechnology industry	.40	**.76**	.04
Food manufacturers	.41	**.56**	−.09
European Union (EU)	**.79**	.09	.22
National government	**.85**	.25	.08
Eigenvalue	3.05	2.86	2.84
Explained variance	17.9	16.8	16.7
Average trust	2.84	3.28	3.95
Cronbach's α	.85	.82	.77

Notes: The scales were coded to range from 1 (totally disagree) to 5 (totally agree). Factor loadings that are higher than .40 are in bold.
Source: Poortinga and Pidgeon (2004)

A PCA with varimax rotation was then conducted in order to examine whether there was an underlying pattern in respondents' judgements about these sources. Table 5.4 shows the factor loadings after varimax rotation. The first factor comprised scientists working for the government, local authorities, the national government and the EU, and accounted for 17.9 per cent of the original variance of trust in information sources. This factor can be interpreted as *trust in government institutions*. The sources that loaded highly on the second factor, and accounted for 16.8 per cent of the original variance, were food manufacturers, the biotechnology industry, scientists working for the biotechnology industry, the FSA and DEFRA. This suggests that these institutions and groups may be seen as being part of a wider system of *production and governance of food biotechnology*. The third factor accounted for 16.7 per cent of the original variance and included

Table 5.5 *'How much do you agree or disagree that the following should be involved in making decisions about genetically modified food?'* (percentage)

	Strongly disagree	Tend to disagree	Neither/ nor	Tend to agree	Strongly agree	No opinion
Environmental organizations	1	3	8	43	41	4
Food Standards Agency (FSA)	1	2	7	42	42	5
Scientists working for environmental groups	1	3	10	45	36	3
Consumer rights organizations (e.g. Consumers' Association)	1	4	10	43	37	5
Doctors	1	4	11	43	37	3
Department for Environment, Food and Rural Affairs (DEFRA)	2	3	9	42	37	6
Scientists working for universities	1	3	13	46	31	4
Farmers	4	7	13	38	35	4
Food manufacturers	8	15	11	40	22	4
Scientists working for government	7	11	15	40	22	4
National government	9	10	15	41	21	4
Local authorities	6	14	20	40	16	5
Biotechnology industry	8	12	18	39	16	7
Scientists working for the biotechnology industry	8	13	18	37	18	7
European Union (EU)	13	12	16	34	19	6

Note: Weighted dataset: n = 1363.
Source: Poortinga and Pidgeon (2004)

consumer organizations, environmental organizations, scientists working for environmental groups and scientists working for universities. The latter set of trust judgements represents *trust in watchdogs* – that is, independent organizations and experts that keep a critical eye on developments in genetically modified food and those who inform the public about the possible consequences of GM food.[7] The solution found in this study is largely comparable to the one found by Poortinga et al (2004) regarding trust in information sources to tell the truth about the UK foot and mouth disease outbreak in 2001.

Table 5.4 shows (the average) trust in watchdog organizations as the highest: 3.95 on a five-point scale, represented here as ranging from 1 (totally disagree) to 5 (totally agree), with 3 (neither agree nor disagree) as the scale mid-point. Government institutions were trusted the least to tell the truth about GM food (2.84). Although industry seems to be moderately trusted (average 3.28), a closer examination of this factor reveals that the two regulators loading on this factor, the FSA (3.88) and DEFRA (3.71), as well as farmers (3.55), are far more trusted than food manufacturers (2.79), the biotechnology industry (2.91) and scientists working for the biotechnology industry (2.96).

People were then asked to what extent they agreed that the same people or organizations *should* be involved in making decisions about GM food (see Table 5.5). It appeared that a large majority (more than 80 per cent) agreed that watchdog organizations, such as environmental organizations, scientists working for environmental groups, consumer organizations and doctors, should be involved in making decisions about GM food. In addition, there was strong agreement that the FSA, which happens to be the independent food safety regulatory agency set up in the wake of the BSE crisis in the UK, should also be involved in decisions. Slightly fewer people (but still more than 60 per cent) felt that the Department for Environment, Food and Rural Affairs (the ministry responsible for regulation), scientists working for universities, farmers, food manufacturers, scientists working for government and the national government should be involved in making decisions about GM food. The lowest agreement levels were found for local authorities, the biotechnology industry, scientists working for the biotechnology industry and the EU. Figure 5.2 plots the percentage who agree or strongly agree that the group or agency is trusted to tell the truth alongside (for each institution, respectively) the judgements about the extent to which they should be involved in decision-making. The pattern of judgements here is clear, with the least trusted organizations having much higher involvement ratings – so much so that with the relatively distrusted institutions, more than half of the sample feels that these groups *should* be involved in making decisions about GM food. This data demonstrates that (despite overt statements of distrust) among the general public, a wide variety of sources for decision-making is desired. Again, this finding can be interpreted within the framework provided by the critical trust idea that scepticism goes hand in hand with a pragmatic reliance.

CONCLUSIONS: RISK MANAGEMENT, COMMUNICATION AND PUBLIC DIALOGUE

We have argued that what is frequently called 'trust' or 'distrust' exists along a continuum, ranging from uncritical emotional acceptance to (downright) rejection. Somewhere between these extremes a healthy type of distrust can be found called *critical trust*. Critical trust can be conceptualized as a practical form of reliance on a person or institution, combined with a degree of scepticism. The combined findings of our

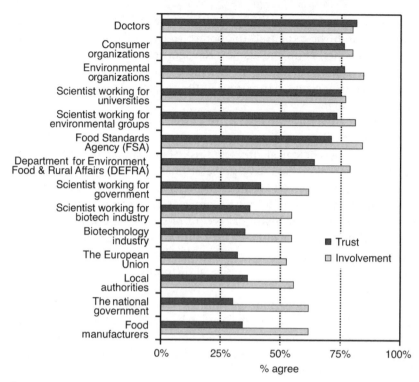

Source: Poortinga and Pidgeon (2004)

Figure 5.2 *Trust in various information sources and agreement about involvement in decision-making about genetically modified (GM) food*

studies hold a number of important implications, both for theories of trust, and for risk communication and management practice, more generally.

In theoretical terms, it may well be the case, as both Metlay (1999) and Poortinga and Pidgeon (2003b) posit, that we have to revisit the dimensional approach as a truly *descriptive* or psychological model of how trust judgements are formed. The salient values similarity approach advocated by Earle and Cvetkovich (1995) argues from a similar premise, too. In one sense, lack of descriptive validity is hardly a surprise, given the normative origins (albeit partial) of many of the conventional 'dimensions' to trust. Our results reinforce the need to seek more psychologically plausible (as well as sociologically and locally grounded) mechanisms that can explain the contingent operation of trust as a combination of both reliance and scepticism. Equally, our results also suggest that shared values do not necessarily provide an entirely sufficient explanation for the complex pattern of trust in institutions or organizations that we have observed. Even in the case of the Health and Safety Executive, where knowledge and awareness of the agency was very low and a valued and shared goal was at least implicitly available (the conditions under which social trust might be expected to operate to the full), few respondents vested full acceptance in the agency. Recognition of the practicalities of politics, or of lack of resources, forced a more critical stance to be taken in people's discourse, even when ratings of the trustworthiness of the source were relatively high.

A promising avenue for further research, suggested by the current results, concerns the way in which institutions are perceived in affective terms. Work on psychological affect and risk (Finucane et al, 2000) and its relationship with more cognitive processes has, until now, focused upon how *hazards* are perceived in affective terms. Further research might explore whether, and how, affect becomes ascribed to institutions, and how this relates to their historical record of competence, whether this varies with different levels of familiarity with what the agency does, or the 'closeness' to (shared objectives/values with) the perceiver, and how these then all feed through into evaluations of different levels of both general trust and scepticism.

The results also hold implications for risk communication and its management. When controversies occur, risk communication is often seen as a panacea for risk management and policy (see Fischhoff, 1995). A number of comments are in order here. First, in policy terms, decision-makers should not confuse 'critical trust' with outright 'distrust' or rejection. Of course, these are not the same thing. Nor do they necessarily demand similar policy responses. Where deep distrust exists, perhaps based upon poor past performance or a fundamental disagreement over goals and values, then it makes sense (instead of focusing on how to increase trust through risk communication) to give attention to the inter-action between institutional structures and goals, agency behaviour and the properties of perceptions of trust. In other words, what kind of relationship between people and risk management institutions is achievable and desirable, taking in account (above all) the need to manage the hazards appropriately? On the issue of trust and risk communication, Bier (2001) argues that, especially in a situation of distrust, one must begin with listening to the concerns of the public before giving them new information. Trying to increase trust by simply providing information may well be interpreted as not taking concerns seriously, and is more likely to destroy than to create trust. However, even more participatory or two-way risk communication and public engagement exercises – generally thought to hold advantages over conventional one-way provision of information – may fail to meet their objectives of inclusive and fair dialogue where stakeholders are deeply divided over an issue, and mistrustful of those who manage it (for example, as occurred during the 2003 public debate on agricultural biotechnology in the UK; see Pidgeon et al, 2005). In addition, recent empirical work has begun to suggest that where strong attitudes exist about an issue (for example, over nuclear power or GM food), trust may be, in part, a consequence rather than a cause of the acceptability of hazardous technologies (see Eiser et al, 2002; Poortinga and Pidgeon, 2005). Under such circumstances, it also makes far greater sense for regulators and managers to address the factors that have led to the underlying concerns about the hazard in the first place (perhaps the distribution of risk and benefits, or catastrophic potential, or a record of poor management in the past), rather than to merely attempt to influence acceptability through enhancing trust.

On the other hand, where the relationship is one of critical trust, particularly where people are, indeed, prepared to place reliance upon an institution despite background doubts, it could well be more suitable to have critical but involved citizens in many situations. In such situations, the absence of full trust may even be normative for decisions. Trettin and Musham (2000) have recently questioned the importance of enhancing trust in institutions. That is, the public does not necessarily expect or see trust as an achievable goal in their relation with institutions. And as Barber (1983, p134) has observed:

> The public is less passively deferential in its relations with experts and others in authority and is more likely to take an active part in monitoring the fulfilment of professionals' claims to absolute trustworthiness.

Barber (1983) notes that the importance of full trust tends to be exaggerated (see also O'Neill, 2002). In his view, the public has become *more* competent and knowledgeable to have 'effective' distrust. Questioning of institutions in this sense is not destructive, but can be seen as an essential component of political accountability in a participatory democracy. Under such circumstances, risk communication, alongside risk management and stakeholder engagement processes, plays an important role in the societal debates about hazards and new technology, and may also feed useful information into the risk management process itself. Freudenburg (2003) notes that rather than responding to the substance of citizen concerns, policy-makers of the past have sometimes employed 'diversionary reframing' – attempts to divert attention away from citizens' real concerns by 'reframing' the debate as being 'about' something else – and that such efforts have often failed to counteract credibility problems, frequently backfiring. For the future, according to Freudenburg, there would appear to be significant potential benefit in exploring an entirely different approach – not of trying to divert policy attention *away* from citizen concerns, but, instead, taking them seriously as one part of the risk assessment and management process. Not only have a growing number of analysts of organizational failures in the risk management domain (see, for example, Vaughan, 1996; Pidgeon and O'Leary, 2000) come to focus on the problems associated with what Barry Turner (1978) originally called the 'disregard of non-members', but some have also come to suggest that increased 'institutional permeability' may deserve much greater attention as an antidote to the blind spots created or reinforced by forms of structural insularity. Agencies, in particular, rather than dismissing outside critics as necessarily ill informed or biased, might do well to examine their *own* recent history, roles, assumptions and frames of reference for understanding risk, and the extent to which risk management will fail if key assumptions (especially when challenged from critics outside) are subsequently proved wrong.

This chapter has developed a conceptual argument, also making connections between contextual and dimensional explanations of trust. It makes no more than a start in the process of exploring the concept of 'critical trust'. Further conceptual and empirical work is clearly necessary at a number of levels: for example, on the bases of scepticism; on the relevance of the concept of 'critical trust' (as value position) to the 'salient value similarity' thesis; and on the better operationalization of measures to indicate varying categories of 'scepticism', 'distrust' and 'trust'. Our chapter also illustrates the productive synthesis that can be achieved through combining both qualitative and quantitative research methods in contemporary trust research. A final observation to arise from the discussion concerns the relative *invisibility* of many risk-managing organizations in people's everyday lives, something which tends to be overlooked by risk managers themselves, as well as academic studies that take a purely quantitative approach to the topic. As our qualitative data shows, trust in risk management is likely to be grounded in a process of inference, rather than direct personal experience, from such things as the name of an agency, from second-hand anecdotes, accounts and assumptions about shared values, and from beliefs about the legitimacy of the organization's intervention in relation to the issue at hand. A key policy lesson, then, is that risk managers should *not* take a naive view of 'trust', or the ways in which their organization is constructed and perceived by ordinary people, in their attempts to design risk communication, public engagement or future risk management programmes.

NOTES

1   The research studies in this chapter were supported by grants from the UK Health and Safety Executive (R64.062), the Economic and Social Research Council (ESRC) Science in Society Programme (L144250037) and the Leverhulme Trust under a programme grant (RSK990021) to the University of East Anglia and partners. We wish to thank Tim Earle, Lynn Frewer and Michael Siegrist for helpful comments on an earlier version, along with Dick Eiser, Tom Horlick-Jones, Branden Johnson, Andrew Weyman and Mathew White for illuminating discussions of trust issues over the years. As ever, the content is the responsibility of the authors alone.
2   This quotation is from Poortinga (2004).
3   This is somewhat in the same way that the concept of 'risk' often stands in for other concerns.

4   And, accordingly, two of the major policy conclusions that we recommended to the sponsors of the study (the HSE) were that:
    1   their name and its associations were a significant current asset; but that
    2   current trust was relatively high, while also potentially fragile (Pidgeon et al, 2003b).
5   Included within the NVivo code for 'critical trust' are the following critical/limiting factors: petty regulation; too reactive/prior notice; underfunded. For a more thorough discussion see Pidgeon et al (2003b).
6   The overall sample was made up of a core British sample of 1017 interviews, a booster survey in Scotland of 151 interviews and a booster survey in Wales of 195 interviews.
7   A fourth factor, not reported here, included items that were interpreted as *trust in personal sources* – specifically, trust in 'friends and family', plus trust in 'people from your own community'.

REFERENCES

Barber, B. (1983) *The Logic and Limits of Trust*, Rutgers University Press, New Brunswick

Beck, U. (1992) *Risk Society: Toward a New Modernity*, translated by M. A. Ritter, Sage Publications, London

Beierle, T. C. and Cayford, J. (2002) *Democracy in Practice: Public Participation in Environmental Decisions*, Resources for the Future, Washington, DC

Bickerstaff, K. and Walker, G (2003) 'The place(s) of matter: Matter out of place – public understandings of air pollution', *Progress in Human Geography*, vol 27, pp45–67

Bier, V. M. (2001) 'On the state of the art: Risk communication to the public', *Reliability Engineering and System Safety*, vol 71, pp139–150

Bord, R. J. and O'Connor, R. E. (1992) 'Determinants of risk perceptions of a hazardous waste site', *Risk Analysis*, vol 12, no 3, pp411–416

Bryman, A. (2001) *Social Research Methods*, Oxford University Press, Oxford

Corrado, M. (2001) 'No-one likes us – or do they?', *Science and Public Affairs*, vol 4, pp14–15

Cvetkovich, G. T., and Löfstedt, R. E. (eds) (1999) *Social Trust and the Management of Risk*, Earthscan, London

Earle, T. C. (2004) 'Thinking aloud about trust: A protocol analysis of trust in risk management', *Risk Analysis*, vol 24, pp169–184

Earle, T. C. and Cvetkovich, G. T. (1995) *Social Trust: Towards a Cosmopolitan Society*, Praeger, London

Eiser, J. R., Miles, S. and Frewer, L. J. (2002) 'Trust, perceived risk and attitudes towards food technologies', *Journal of Applied Social Psychology*, vol 32, no 11, pp2423–2433

Finucane, M. L., Alhakami, A. S., Slovic, P. and Johnson, S. M. (2000) 'The affect heuristic in the judgement of risks and benefits', *Journal of Behavioral Decision Making*, vol 13, pp1–17

Fischhoff, B. (1995) 'Risk perception and communication unplugged – 20 years of process', *Risk Analysis*, vol 15, no 2, pp137–145

Flynn, J., Slovic, P. and Kunreuther, H. (2001) *Risk Media and Stigma: Understanding*

Public Challenges to New Technologies, Earthscan, London

Freudenburg, W. R. (2003) 'Institutional failures and the organizational amplification of risks: The need for a closer look', in N. F. Pidgeon, R. K. Kasperson and P. Slovic (eds) *The Social Amplification of Risk*, Cambridge University Press, Cambridge, pp102–122

Frewer, L. J., Howard, C., Hedderley, D. and Shepherd, R. (1996) 'What determines trust in information about food-related risks? Underlying psychological constructs', *Risk Analysis*, vol 16, no 4, pp473–485

Giddens, A. (1990) *The Consequences of Modernity*, Polity Press, London

Glaser, B. and Strauss, A. (1967) *The Discovery of Grounded Theory*, Aldine, New York

Grove-White, R., Macnaghten, P., Mayer, S. and Wynne, B. (1997) *Uncertain World: Genetically Modified Organisms, Food and Public Attitudes in Britain*, Lancaster University, Centre for the Study of Environmental Change, Lancaster

Henwood, K. L. and Pidgeon, N. F. (1992) 'Qualitative research and psychological theorizing', *The British Journal of Psychology*, vol 83, pp97–111

Horlick-Jones, T., Sime, J. and Pidgeon, N. F. (2003) 'The social dynamics of environmental risk perception', in N. F. Pidgeon, R. K. Kasperson and P. Slovic (eds) *The Social Amplification of Risk*, Cambridge University Press, Cambridge, pp262–285

House of Lords Committee on Science and Technology (2000) *Report on Science and Society*, HL Paper 38, February, London, Westminster

Irwin, A., Dale, A. and Smith, D. (1996) 'Science and Hell's kitchen: The local understanding of hazard issues', in A. Irwin, and B. Wynne (eds) *Misunderstanding Science? The Public Reconstruction of Science and Technology*, Cambridge University Press, Cambridge

Irwin, A., Simmons, P. and Walker, G. (1999) 'Faulty environments and risk reasoning: the local understanding of industrial hazards', *Environment and Planning A*, vol 31, pp1311–1326

Johnson, B. B. (1999) 'Exploring dimensionality in the origins of hazard related trust', *Journal of Risk Research*, vol 2, no 4, pp325–354

Kasperson, R. E., Golding, D. and Tuler, S. (1992) 'Social distrust as a factor in siting hazardous facilities and communicating risk', *Journal of Social Issues*, vol 48, no 4, pp161–187

Krueger, R. A. (1988) *Focus Groups: A Practical Guide*, Sage, Newbury Park

Metlay, D. (1999) 'Institutional trust and confidence: a journey into a conceptual quagmire', in G. T. Cvetkovich and R. E. Löfstedt (eds) *Social Trust and the Management of Risk*, Earthscan, London

O'Neill, O. (2002) *A Question of Trust*, Cambridge University Press, Cambridge

Peters, R. G., Covello, V. T. and McCallum, D. B. (1997) 'The determinants of trust and credibility in environmental risk communication', *Risk Analysis*, vol 17, no 1, pp43–54

Pidgeon, N. F. and Henwood, K. L. (2004) 'Grounded theory', in M. Hardy and A. Bryman (eds) *Handbook of Data Analysis*, Sage, London, pp625–648

Pidgeon, N. F. and O'Leary, M. (2000) 'Man-made disasters: Why technology and organizations (sometimes) fail', *Safety Science*, vol 34, pp15–30

Pidgeon, N. F., Hood, C., Jones, D., Turner, B. and Gibson, R. (1992) 'Risk perception', in *Risk Analysis, Perception and Management: Report of a Royal Society Study Group*, The Royal Society, London, Chapter 5, pp89–134

Pidgeon, N. F., Kasperson, R. K. and Slovic, P. (2003a) *The Social Amplification of*

Risk, Cambridge University Press, Cambridge

Pidgeon, N. F., Walls, J., Weyman, A. and Horlick-Jones, T. (2003b) *Perceptions of and Trust in the Health and Safety Executive as a Risk Regulator*, Research Report 2003/100, HSE Books, Norwich

Pidgeon, N. F., Poortinga, W., Rowe, G., Horlick-Jones, T., Walls, J. and O'Riordan, T. (2005) 'Using surveys in public participation processes for risk decision-making: the case of the 2003 British GM Nation? Public debate', *Risk Analysis*, vol 25, pp467–479

Pijawka, K. D. and Mushkatel, A. H. (1992) 'Public opposition to the siting of the high-level nuclear waste repository: The importance of trust', *Policy Studies Review*, vol 10, no 4, pp180–194

Poortinga, W. (2004) *Public Perceptions of and Trust in the Regulation of Genetically Modified Food*, PhD thesis, School of Environmental Sciences, University of East Anglia, Norwich

Poortinga, W. and Pidgeon, N. F. (2003a) *Public Perceptions of Risk, Science and Governance: Main findings of a British Survey on Five Risk Cases*, Technical report, Centre for Environmental Risk, Norwich

Poortinga, W. and Pidgeon, N. F. (2003b) 'Exploring the dimensionality of trust in risk regulation', *Risk Analysis*, vol 23, pp961–972

Poortinga W. and Pidgeon, N. F. (2004) *Public Perceptions of Genetically Modified Food and Crops, and the GM Nation? Public Debate on the Commercialisation of Agricultural Biotechnology in the UK*, Understanding Risk Working Paper 04–01, Centre for Environmental Risk, Norwich, pp1–54

Poortinga, W. and Pidgeon, N. F. (2005) 'Trust in risk regulation: Cause or consequence of the acceptability of GM food?', *Risk Analysis*, vol 25, pp197–207

Poortinga, W., Bickerstaff, K., Langford, I., Niewöhner, J. and Pidgeon, N. F. (2004) 'The British 2001 foot and mouth crisis: A comparative study of public risk perceptions, trust and beliefs about government policy in two communities', *Journal of Risk Research*, vol 7, no 1, pp73–90

Renn, O. and Levine, D. (1991) 'Credibility and trust in risk communication', in R. E. Kasperson, and P. J. M. Stallen (eds) *Communicating Risks to the Public*, Kluwer, The Hague

Siegrist, M., Cvetkovich, G. T. and Roth, C. (2000) 'Salient value similarity, social trust, and risk/benefit perception', *Risk Analysis*, vol 20, no 3, pp353–362

Slovic, P. (1993) 'Perceived risk, trust and democracy', *Risk Analysis*, vol 13, no 6, pp675–682

Todd, Z., Nerlich, B., McKeown, S. and Clarke, D. (2004) (eds) *Mixing Methods in Psychology*, Psychology Press, Hove and New York

Trettin, L. and Musham, C. (2000) 'Is trust a realistic goal of environmental risk communication?', *Environment and Behavior*, vol 32, no 3, pp410–426

Turner, B. A. (1978) *Man-Made Disasters*, Wykeham Science Press, London

Vaughan, D. (1996) *The Challenger Launch Decision: Risky Technology, Culture, and Deviance at NASA*, Chicago University Press, Chicago

Walls, J., Pidgeon, N. F., Weyman, A. and Horlick-Jones, T. (2004) 'Critical trust: Understanding lay perceptions of health and safety risk regulation', *Health, Risk and Society*, vol 6, no 2, pp133–150

Worcester, R. M. (2001) 'Science and society: What scientists and the public can learn from each other', *Proceedings of the Royal Institution*, vol 71, Oxford University Press, Oxford

Wynne, B. (1980) 'Technology, risk and participation: On the social treatment of uncertainty', in J. Conrad (ed) *Society, Technology and Risk Assessment*, Academic Press, New York

Wynne, B., Waterton, C. and Grove-White, R. (1993) *Public Perceptions and the Nuclear Industry in West Cumbria*, Centre for the Study of Environmental Change, Lancaster

6 Societal Trust in Risk Analysis: Implications for the Interface of Risk Assessment and Risk Management

Lynn Frewer and Brian Salter

INTRODUCTION

It is frequently speculated that risk assessment is isolated from societal values, and that values only influence risk management and risk communication. From this, it has been argued that creating a functional separation between risk assessment and risk management will result in increased public trust in risk management since values or political considerations will not influence risk assessments, which exist purely in the 'natural' world of science. This assumption has been somewhat speculative and has not been subject to systematic empirical test, but nonetheless has resulted in the creation of institutional structures, which reflect structural, as well as functional, separation of risk management and risk analysis.

Traditional models of risk analysis have assumed that risk communication follows on from risk management, which, in turn, is the outcome of risk assessment. More recent frameworks have assumed some level of integration between these three elements of risk analysis (for example, FAO/WHO, 1998). However, the decline in public confidence in those institutions and industries with responsibility for risk analysis continues unabated, indicating that, in itself, trust is not created through the functional separation of risk management and risk assessment. This may raise further issues relating to public recognition that values permeate the whole process of risk analysis. As a result, it is possible that public confidence in risk analysis may only be developed and maintained under circumstances where they are taken into consideration through the whole of the risk analysis process.

Much public negativity associated with the way in which risks are managed and regulated has been the result of risk managers, assessors and other key actors in the process of risk analysis failing to take account of the concerns of the public when assessing, managing and communicating about risks. One consequence has been increased public distrust in the motives of regulators, science and industry in taking certain decisions or actions in relation to potential hazards and risk mitigation activities (Slovic, 1993; Cvetkovich and Löfstedt, 1999; Frewer, 1999). An illustrative example is found in the area of food safety – the focus of much public controversy in recent years. Jensen and Sandøe (2002) have observed that despite the creation of new food safety institutions, such as the European Food Safety Authority (EFSA), the decline in public confidence in food safety matters continues. This may, it is argued, be partly the result of communication about food safety issues (including genetic modification applied to food production) being based on scientific risk assessments alone, and failing to incorporate public concerns and fears into a broader societal debate (Levidow and Marris, 2002). Communication that does not explicitly address public concerns is likely to have a limited role in reassuring the public (Frewer et al, 2004). Jensen and Sandøe (2002) argue that this is because risk assessment is *presented to the public* by interested institutions as a purely objective scientific event located in the natural world, which is not influenced by societal values or subjective judgements. For this reason, risk assessment has been frequently assumed to be '*functionally separate*' from the other components of risk analysis, risk management and communication. In contrast, risk management is typically described as political in orientation. Risk communication is the process by which the results of both risk management and risk assessment are communicated to the general public and, thus, is prone to influence by risk managers.

Jensen and Sandøe (2002, p247) note that:

> Risk assessments are determined by the exact choice of putative hazard ... to be assessed for possible unwanted consequences, and by the exact demarcation in time and space of the possible consequences to be addressed... These choices clearly affect the outcome of the risk assessment; but they are not themselves the results of a scientific process.

Other values and preferences may be brought to bear on the risk assessment – for example, how to handle probability and variability (Thompson and Bloom, 2000), and, of course, *how*, *when* and *if* this information is communicated to risk managers and the general public. Values may also be brought to bear in the determination of how resources are allocated in the identification of emerging hazards, and the conditions under which new assessment and risk mitigation approaches are adopted.

It is of interest to note that, in the past, science has been more closely engaged with the social problems of risk than is seen today. For example, Groenewegen (2002), in his analysis of Dutch toxicology, observed that scientists who are also advisers were, historically, frequently involved in defining the social problem, as well as the research agenda put into place to counter it. Groenewegen concludes that, in the early days of toxicological science (for example, activity which arose during the early 19th century as a response to the public health problems associated with industrialization), research, advice and policy developments were closely linked rather than 'functionally separated'. More latterly, it was concluded, advice, policy and research activities have become more clearly demarcated, which has resulted in differentiation of agendas of different actors within the risk management process.

Indeed, the claimed universalization of scientific culture has been challenged, not only at the level of risk management (where socio-political influences may predominate regulation), but also in risk assessment – for example, in the adoption of preferred methodologies and assessment techniques. A case in point is provided by Rothstein et al (1999), who note that in the area of agrochemical regulation there is tension between international and national harmonization of assessments and regulatory practices. This lack of harmonization implies that the assessment process itself is not founded on a 'purely natural' objective reality uncontaminated by social judgements and values. There is, of course, substantive evidence for local divergences away from universalization in assessment, despite claims that policy initiatives in the area of risk are becoming more global – for example, in the area of consumer and environmental policy (Löfstedt and Vogel, 2001).

The aim of this chapter is, first, to examine the historical context surrounding risk analysis and public trust. The importance (or otherwise) of citizen/consumer trust will then be examined, and discussion made of whether the functional separation of the different components of risk analysis is a causative factor in developing this distrust. Conclusions will be drawn arguing that a more integrated risk analysis framework may be needed if citizen/consumer confidence in the risk analysis process is to be developed and maintained.

SETTING THE CONTEXT: PUBLIC TRUST AND THE RISK ANALYSIS PROCESS

Discussion of public trust and the process of risk analysis arose following the seminal research by Paul Slovic and colleagues regarding the public perception of risk (see, for example, Slovic, 1987, 1993, 2000). The finding that lay people incorporated psychological factors of relevance other than technical risk estimates within their personal assessment of the acceptability of risks was of great interest to the risk management community. This was because the research provided an empirical basis for explaining why risk management decisions acceptable to expert communities were not acceptable to the general public or, at least, sub-groups within the population.

As a result, risk communication activities during the 1970s focused on changing public views on risk, with emphasis on communication directed towards technology acceptance. Hilgartner (1990) has described this process as the 'deficit model', whereby expert and elite organizations and institutions assumed that the publics are, in some way, deficient in their understanding of risk and, indeed, in other areas of science and technology. In other words, communicators adopted the perspective that the public was ignorant of the scientific 'truth' about risk and probability. For this reason, the goal of risk communication was to 'rectify the knowledge gap' between the originators of scientific information and those receiving the information. The resulting public 'enlightenment' regarding risk issues would resolve all problems of technology implementation and commercialization under circumstances where public acceptance had been problematic before. Despite these efforts, the lay public remained deeply sceptical of the motives of scientists, regulators and industrialists.

Thus, despite different risk communication activities on the part of technocratic and policy elites, the sought-for alignment between expert and lay visualization of risk acceptability or otherwise did not occur. The explanation for this disparity appeared to lie in research findings that indicated that there was a decline in public confidence in the motives of scientific institutions, regulatory bodies with responsibility for consumer protection, and industry (Bauer, 1995). Indeed, the acceptance of emerging technologies was thought to be contingent upon public trust in institutions with responsibility for regulating them. For example, Siegrist (1999) has reported that for citizens with high levels of trust in institutions with responsibility for regulating gene technology and its products, perceived risk in the technology is decreased, and perceived benefit resulting from it is increased. Citizen distrust in these same institutions has the converse effect on perceptions of risk and benefit.

Public distrust in the motives of institutional actors appeared to reinforce public cynicism about the process of risk analysis. Thus, it was reasoned by risk assessors and managers, differences between experts and the public were thought to be unrelated to differences in the way that risks were assessed, and more to do with public distrust in those institutions with responsibility for risk assessment, management and communication. One consequence was the development of scientific activity, which focused on restoring public trust in risk management. In association, there was at the same time emphasis on increased transparency in the risk analysis process, particularly in risk assessment and risk management, as it was assumed this would increase public trust in institutional activities. In fact, there is limited evidence to suggest that increased transparency is a precondition for trust in institutional activities to develop, although in itself it is not a trust-increasing event (Frewer et al, 1996). Specifically, lack of transparency may result in increased distrust; but trust *per se* is a result of citizens' perceptions of honesty, concern for public welfare and competence. A second approach to developing trust focused on greater public inclusion in the process of policy development.

In particular, it was thought that more extensive public consultation and participation in risk management and other science and technology issues would restore public confidence in institutions with responsibility for public protection (see, for example, Renn et al, 1995; Rowe et al, 2004). Thus, it appeared that institutions began to acknowledge that citizens' attitudes towards different hazards are not only dependent upon an analytical assessment of risk and benefit. Other factors, such as ethical and moral considerations, were recognized as potentially influential in establishing the acceptability or otherwise of a particular hazard, or societal approval of the measures put into place to contain the associated risks (Select Committee on Science and Technology, 1999).

One result of this institutional concern about public distrust in their activities was increased emphasis on the 'functional separation' between risk assessment and risk management. Risk assessment was, as before, implicitly assumed to exist in the 'natural world' and, thus, was portrayed as immune from the influence of social values and societal priorities. Political influences and societal concerns were also assumed by decision-makers to be confined to risk management through the process of functional separation of the three components of the risk analysis framework. Risk communication was, arguably, promoted as a tool to reinforce the principle of institutional transparency rather than a mechanism for developing a direct dialogue with citizens and/or consumers.

In fact, the functional separation of the three components, and isolation of risk as an immutable truth, would naturally follow on from the assumption that science was isolated from the influence of society and its predominant values. The ways in which science and the social world are often viewed as independent of each other has been the focus of systematic study. For example, Woolgar (1996) has noted that it is useful to examine what underlines the received view regarding the assumptions of scientific 'neutrality' and its portrayal. The first assumption relates to how objects in the natural world are 'objective' and 'real', and have an existence *independent* of human beings. The actions and beliefs of human beings are incidental to nature (and these must be isolated from scientific 'truth' in order to maintain its purity). From this, scientific knowledge (for example, risk assessment) can only be determined by understanding those objective realities found and observed in the natural world. As a consequence, investigation of scientific knowledge is via a unitary set of methods, the application of which is the result of consensus agreement by scientific experts and elite bodies. Finally, science is an activity that is individualistic (rather than amenable to collectivist or consensus approaches to decision-making) and 'cognitive' or 'logical', rather than 'subjective' or 'affective'. Risk assessment, it is assumed, is founded entirely in the 'objective' reality provided by science, and must be portrayed as immune from values, preferences or investigation by different methodological variants. In contrast, risk management is prone to influence by political judgements. The question of societal trust in risk assessment is therefore not open to question. Public trust can only vary in risk management, where value judgements can influence how decisions are made. As a consequence, it follows on that it therefore becomes necessary to separate risk assessment from risk management in order to gain public trust in the risk analysis process.

The increase in public distrust in science and scientific institutions has been of concern primarily to regulators and politicians. It has been less of a concern to the research community. Scientists and technologists have, apparently, been societally the most unprepared for the public reaction against technology innovations (for example, agricultural applications of genetic modification, human cloning, nuclear technology or, potentially, nano-technology; see Frewer and Salter, 2002). One defence against societal criticism of expert scientific communities and their judgements has been a retreat into reliance on the purely 'objective' risk assessments in order to deflect external criticism of this component of the risk analysis process (Law and Hassard, 1999). As a (somewhat circular) consequence, claims about the legitimacy of a particular risk can only be underpinned by the 'scientific' risk assessment or possible economic cost-benefit analysis, and other risk factors, such as beliefs, aesthetics and quality of life, have little or no significance (Heller, 2002).

Garvin (2001) has argued that there are important issues of epistemology inherent in the process and practice of risk analysis. She specifies that there are three groups of key players in understanding risk: scientists, policy-makers and the public. Arguably, 'each group employs different, although equally legitimate, forms of rationality'. For scientists, rationality is constructed in scientific terms. Members of the scientific community are primarily interested in risk assessment. Within the policy community, the basis of rationality is political and expedient. Members of the policy group are primarily interested in risk management. Members of the public utilize a social form of rationality, which takes account of risk context, cultural factors and variations in local conditions (and, implicitly, societal values in the broadest sense). It is perhaps the different focus of interested actors (assessors, regulators and society) that has not only increased the imperative towards the increased functional separation of the different components of risk analysis, but may also have fuelled the mutual distrust between the different groups of actors.

The inclusion of analysis of broader issues in technology implementation is intended to increase public trust in regulatory practices (Irwin, 2001). New organizations, which take account of both technical and societal issues, appear to be the beginnings of 'boundary organizations' proposed by Guston (1999) that intersect politics, policy and science (and, by implication, public opinion). The extent to which these are influential in terms of risk assessment is open to debate – their primary sphere of influence appears to be on risk communication, and their secondary sphere of influence on risk management. Risk assessment appears to be outside of their sphere of influence, specifically because of the application of 'functional separation' between risk assessment and risk management.

The discounting of societal considerations as having any influence on risk assessment by institutions has become more problematic with increased transparency. This is because the (unacknowledged) application of values to risk assessment practices and policies become more obvious to citizens in general, but is systematically denied by other influential actors involved with risk analysis. The response of the external onlooker may well be to question the motives of those institutional actors involved in this process.

TRUST AND RISK ANALYSIS

The issue of the decline in public trust merits further consideration at this point. Frewer and Salter (2002) have observed that public distrust in risk analysis (and, by implication, risk assessment, management and communication) may be attributable to various changes in society. As a consequence, public reliance on the decisions of expert or elite groups is no longer a tenable way to conduct risk analyses, as has been the case in the past. The rise of the *'consumer citizen'*, for example, means that societal disquiet with risk management and risk assessment may be expressed through consumer preference and choice in the market-place. This may include, for example, increased sales of consumer goods that involve less processing and production technology. Such shifts in consumer preference may occur under circumstances where there is public concern about the development and commercialization of emerging technology, as has been the case with genetically modified food (Bruhn, 2002). Alternatively, the occurrence of a potential hazard may result in consumers switching to a different type of product (Verbeke, 2001; Pennings et al, 2002) or to highly trusted brands, which are perceived to be safe (Chaudhuri and Holbrook, 2001). In any case, there is plenty of potential for individuals to express concern through consumption choices, which may have profound economic consequences for product brands, industries or national economies.

Increased transparency in regulation is often presented as a 'trust-increasing' factor (Lang, 1998; HM Government, 2000). Intuitively, it may be supposed that retrenching transparency may be a trust-reducing activity. Increased transparency is unlikely in itself to increase public trust. But increased transparency has provided society with information about the potential for societal or institutional values to influence all components of risk analysis. However, public distrust in risk assessment is likely to arise under circumstances where uncertainties and variabilities in risk assessment become open to public scrutiny through increased transparency, but are not explained explicitly as part of the risk communication process (Frewer et al, 2002; Miles and Frewer, 2003). This may possibly be the result of institutional concerns that the public is not tolerant of, or able to understand: uncertainty and variability.

As previously noted, other attempts to increase public trust in risk analysis have included attempts to involve the public in decision-making. Jasanoff (1990) has concluded that both increased transparency and public consultation may further undermine the credibility of government authorities if the process of opening up legislative processes and increased public consultation is not considered carefully. For example, the process of developing and conducting participatory exercises should be externally audited, as should the impact of participatory processes on the policy process itself (Rowe and Frewer, 2000, 2004). There is a further need to examine how to take account of the diversity of public opinion in policy decisions. Failure to accommodate these issues within the new transparent risk analysis systems may result in increased distrust, rather than trust, in the activities of risk assessors and risk managers.

A second issue relates to the inclusivity of public consultation. The greater the openness and the transparency of a risk analysis system, the more questions may be raised regarding decision points relevant to the public. As a consequence, this will reduce the segment of the public who is likely to be able and willing to participate in the consultation. For pragmatic reasons alone, the approach to consultation that may be most effective is that of representative, as opposed to direct, democracy (Till and Meyer, 2001; Beierle, 2002).

Trust in risk assessment and risk management is likely to be a particularly important determinant of public confidence under circumstances where people feel that they have very little personal control over their exposure to potential hazards. For example, consider the case of genetically modified foods, which have been judged by consumers to be threatening partly because they judge the consumption of modified foods and ingredients to be outside the personal control of the consumer (Miles and Frewer, 2001). This is because of a public perception that adequate traceability mechanisms through the food chain have not been operationalized, and, as a result, appropriate labelling strategies cannot be implemented (Miles et al, 2005). As a consequence, it is more important that consumers are able to trust institutions with responsibility for consumer protection compared to circumstances where consumers perceive that they are able to make an informed choice about consuming products or not.

In controversial fields such as nuclear science and biotechnology, public trust may be taken as a statement about the legitimacy of the activities that are encompassed by technological development and subsequent commercialization of technology applications. One measure of the success of any technology innovation is the degree of public trust in analytical and regulatory procedures associated with the technology. If public confidence in a particular arena declines, there will be a negative impact in terms of the political exposure of the regulatory institutions, the economic vulnerability of the industrial sector concerned and the likely amplification of critical media interest in the sector overall. Scientific and regulatory authority has lost much of the credibility conferred upon it in the past. In other words, people no longer trust science to legitimize itself by *reference to the technical estimates about risk which science itself produces* (Cabinet Office and Office of Science and Technology, 1999; Frewer and Salter, 2003). This is, in part, again because of the societal recognition that technological risk estimates are prone to influence by values and political factors, as well as hitherto undisclosed uncertainties in assessment (for example, in the extent of the uncertainty or variability associated with risk assessment measurement).

Being able to make judgements about the social impact of issues such as emerging risks (particularly those associated with emerging technologies where there may be a tangible societal benefit associated with technology implementation) requires the setting of standards, or benchmark criteria, regarding acceptable risk. Such judgements are contingent upon the acceptability of the process by which associated risks are traded against potential benefits resulting from the technology application in order to determine risk acceptability. Social values as well as scientific assessment of risk and benefit that claim to exclude normative judgements must inform these standards. A major problem of the latter approach is that despite the frequent use in the scientific discourse of phrases such as 'risk-based regulation', in practice, there has never been any official definition of what that might consist of in precise procedural terms (Levidow and Carr, 1996; van den Daele, 1999). In the absence of such a definition, scientific authority is diminished because it is dependent upon an implicit trust in the process by the public, rather than upon the visible process of scientific reasoning, which in any case is contingent upon values and therefore cannot be described as 'purely objective'.

An example is again provided by the case of the emerging biosciences. Jensen et al (2003) argue that it will be difficult to restore public trust in the approval procedure for genetically modified organisms unless all of the relevant assumptions involved in the procedure are made transparent and become open to public scrutiny and debate. In particular, the authors argue that it is necessary to make explicit the value judgements that are inherent in risk assessments (for example, judgements that imply that, under certain circumstances, a scientific assumption holds true or that a particular assumption takes priority over a viable alternative). Unless such judgements are made explicit, it is difficult to operationalize public debate about the broader ethical issues involved in the approval of genetically modified organisms. Value judgements inherent in the assessment process include, in this example, the demarcation in space and time of the possible adverse effects to be assessed, and which potential hazards are to be assessed for possible adverse effects. Using European directives on the risk assessment of genetically modified crops as case in point (European Commission, 1990, 2001), Jensen et al (2003) demonstrate that the lack of transparency in the risk assessment process is related to the failure to identify the points where value judgements influence the outcome of the assessment process. This failure increases public distrust in the assessment process itself since value judgements are not explicitly acknowledged. The only way to restore public confidence is if the:

> Value judgements underlying the regulation and particularly delimiting the risk assessment ... are ... made explicit; and the limitations of available data, and any uncertainties they give rise to ... are ... made clear. (Jensen et al, 2003, p170)

If the public perceives that there is an attempt by the authorities or the scientific community to conceal the limitations of scientific knowledge regarding the risk assessment process, public distrust in the actors involved and in the process itself will result. In the same spirit, these authors note that effective communication about associated uncertainties is also essential if the precautionary principal is to be applied since its application in itself implicitly provides external observers with the signal that there is some uncertainty associated with the risk assessment.

COMMUNICATING UNCERTAINTY AND VARIABILITY: A PRECONDITION FOR TRANSPARENT RISK ASSESSMENT PROCESSES?

Of course, increasingly transparent risk assessment and management activities mean that the effective communication of uncertainty and variability associated with risk assessment is increasingly important, since information about uncertainty and variability is available in the public domain, but hitherto may not have been explicitly communicated as such. In itself, such communication may not increase trust. However, it behoves risk communicators to develop effective ways of communicating about these issues, since they are in the public domain as a result of increased public distrust: failure to so do may be trust destroying. Communication flow must be established between all key players (assessors, risk managers and the general public) if a fully transparent process is to be developed, and the issues must be debated by all of these key players. Communication should also differentiate between different types of uncertainty – for example, the distinction between outcome uncertainty: 'What might actually happen and with what probability?', and assessment uncertainty: 'To what extent are the results of the analysis likely to change with additional information?' (Brown and Ulvilla, 1987).

In the past, expert communities have assumed that providing lay people with information about uncertainty would result in public negativity towards the whole process of risk analysis. 'In addition, public distrust in science and scientific institutions would increase, and panic and confusion regarding the impact of a given hazard on human health and the environment result' (Wynne, 1992). Expert groups have tended to assume that lay people cannot conceptualize uncertainty in risk assessment or risk management (Frewer et al, 2002). However, scientific and policy communities appear to have underestimated the ability of non-experts to understand uncertainty. In fact, there is evidence that it is the failure of institutional actors to communicate uncertainty that increases public distrust in institutional activities designed to manage risk (Frewer et al, 2002).

Johnson and Slovic (1995, 1998) report that 80 per cent of lay people, at least in the US, appear able to distinguish between point estimates and interval estimates associated with particular risks. For these individuals, discussion of risk uncertainties (as opposed to not communicating about uncertainties at all) signals greater institutional honesty, but lower competence associated with the institutions responsible for risk assessment. There is further evidence that acknowledgement of uncertainty associated with risk assessment does, in fact, act to increase public confidence in regulatory processes. For example, Kuhn (2000) reports that environmental attitudes held by respondents proved to be accurate predictors of environmental risk perceptions if uncertainties associated with assessment were not mentioned. Communicating uncertainty information associated with the assessment *decreased* risk perceptions for those expressing high initial levels of environmental concern. It is of note, however, that a *converse effect* was found for those initially unconcerned about the potential for environmental impact. Moreover, lay people do distinguish between different *kinds* of uncertainty (Frewer et al, 2003). For example, there is a lay differentiation between uncertainty associated with lack of knowledge (for example, lack of scientific information regarding a specific risk, or conflicting scientific information or opinion), and uncertainty about the potential impact or extent of a particular hazard. Lay people also recognize that further research may be needed in order to reduce the uncertainty, and acknowledgement of this need may, in turn, be trust inducing. Indeed, the public appears to be more accepting of uncertainty resulting from inadequacies in scientific process than to uncertainty associated with the failure of institutions to reduce scientific uncertainty through conducting appropriate empirical investigation. This all serves to confirm the National Research Council recommendation (1994) that risk communication should focus on the sources of uncertainty as well as the magnitude of uncertainty associated with a particular hazard.

Of course, other factors associated with risk assessments also influence risk management decisions (for example, the severity and immediacy of the potential risk, the cost and side effects of mitigation options, and the cost and time required for research). Uncertainty associated with risk assessments, risk management and the link between risk assessment and risk management should be communicated to the public and other key stakeholders, as well as to decision-makers, if there is to be an informed public debate about how risks should be handled.

Another risk assessment issue that must be disseminated to all interested parties, including the public, is that of risk variability when the risk varies across a population but the distribution is well known. Vulnerable groups may be identified from this information, which merit targeted communication about both assessment and management processes. Understanding variability may also have implications for the allocation of resources to risk mitigation activities (Morgan and Henrion, 1990), which may also be a focus of public debate. Discussion of how such resources are allocated is important in developing public confidence in risk management and, ultimately, risk assessment.

At present, however, there is insufficient knowledge about how to develop best practice in risk communication about uncertainty and variability. The former is contingent upon developing ways of discussing different kinds of uncertainty; the latter may entail methodological development in targeting information to 'at risk' populations. Both risk uncertainty and risk variability have profound implications for decisions associated with resource allocation (for example, how research funds are distributed across hazards in order to reduce uncertainties, or how risk mitigation activities are prioritized for risks that differentially affect different sub-populations). Public trust in these processes is likely to be low unless there is informed public debate regarding both risk management and risk assessment procedures, which permits the inclusion of wider societal values and priorities within decision-making processes.

CONCLUSIONS

Societal responses to emerging and/or potentially transformative technologies (for example, nano-technology) may reflect increased public distrust unless institutions and organizations act to develop and maintain public confidence in their risk assessment and risk management practices. While public trust is, to some extent, contingent upon institutional transparency, other factors, such as institutional responses to public concerns, are also important. To this end, there has recently been increased emphasis on involving the public in risk management.

One conclusion that may be reached is that the basic premise supporting the functional (and possible structural) separation of risk assessment from the other components of risk management is no longer tenable. This is because increased transparency in risk decision-making has made it apparent to all stakeholders that risk assessment is not a purely objective process as it has been previously portrayed, but is subject to value judgements in the same way as risk management is. Increased transparency has also resulted in risk uncertainties inherent in risk assessment becoming open to public scrutiny, which will result in further challenges to the concept of 'purity' of the objective reality in which risk assessment is currently confined. Participatory processes, therefore, should not only contribute to decisions about how risks are managed, but also how, and which, risks are assessed, in what time scale and geographical location, taking full account of risk variability across populations and landscapes. The latter will become more salient as knowledge about risk variability increases (for example, as more is known about individual susceptibilities to risks through advances in genomic research). There will be greater competition between vulnerable groups of the population, or inhabitants of different local or national environments, for resources to mitigate the risks of specific hazards. Both the selection of risk assessment processes and risk management practices will be increasingly open to public involvement and consultation. The effective implementation and impact evaluation of such participatory processes may be a necessary precondition for them to be the basis of developing public confidence, and further discussion of this issue is made elsewhere (see, for example, Rowe and Frewer, 2004).

In conclusion, therefore, there is evidence that the continued 'functional separation' of risk assessment from risk management is likely to increase public distrust in risk analysis in the future. Without public trust, however, it is unlikely that risk analysis will be effective, and public opposition to many institutional risk-related activities will continue.

REFERENCES

Bauer, M. (ed) (1995) *Resistance to New Technology*, Cambridge University Press, Cambridge

Beierle, T. C. and Cayford, J. (2002) *Democracy in Practice: Public Participation in Environmental Decisions*, Resources for the Future, Washington DC

Brown, R. and Ulvilla, J. W. (1987) 'Communicating uncertainty for regulatory decisions', in V. T. Covello, L. B. Lave, A. Moghiss, and V. R. R. Uppuluri (eds) *Uncertainty in Risk Assessment and Risk Management and Decision Making*, Plenum Press, New York

Bruhn, M. (2002) *Verbrauchereinstellungen zu Bioprodukten – Der Einfluss der BSE-Krise*, Report Number 201, Institut für Agrarökonomie der Universitat Kiel, Lehrstuhl für Agrarmarketing, Kiel

Cabinet Office and Office of Science and Technology (1999) *The Advisory and Regulatory Framework for Biotechnology: Report from the Government's Review*, The Stationery Office, London

Chaudhuri, A. and Holbrook, M. B. (2001) 'The chain of effects from brand trust and brand affect to brand performance: The role of brand loyalty', *Journal of Marketing*, vol 65, pp81–93

Cvetkovich, G. and Löfstedt, R. E. (eds) (1999) *Social Trust and the Management of Risk*, Earthscan, London

European Commission (1990) Council Directive 90/220/EEC of 23 April 1990 on the deliberate release into the environment of genetically modified organisms

European Commission (2001) Directive 2001/18/EC of the European Parliament and of the Council of 12 March 2001 on the deliberate release into the environment of genetically modified organisms and repealing Council Directive 90/220/EEC – Commission Declaration

FAO/WHO (Food and Agriculture Organization/World Health Organization) (1998) *The Application of Risk Communication to Food Standards and Safety Matters*, FAO Food and Nutrition Paper 70, Rome

Frewer, L. J. (1999) 'Risk perception, social trust, and public participation into strategic decision-making – implications for emerging technologies', *Ambio*, vol 28, pp569–574

Frewer, L. J. and Salter, B. (2002) 'Public attitudes, scientific advice and the politics of regulatory policy: The case of BSE', *Science and Public Policy*, vol 29, pp137–145

Frewer, L. J. and Salter, B. (2003) 'The changing governance of biotechnology: The politics of public trust in the agri-food sector', *Applied Biotechnology, Food Science and Policy*, vol 1, pp199–211

Frewer, L. J., Howard, C., Hedderley, D. and Shepherd, R. (1996) 'What determines trust in information about food-related risks? Underlying psychological constructs', *Risk Analysis*, vol 16, pp473–486

Frewer, L. J., Miles, S., Brennan, M., Kusenof, S., Ness, M. and Ritson, C. (2002) 'Public preferences for informed choice under conditions of risk uncertainty: The need for effective risk communication', *Public Understanding of Science*, vol 11, pp1–10

Frewer, L. J., Hunt, S., Kuznesof, S., Brennon, M., Ness, M. and Ritson, R. (2003) 'The views of scientific experts on how the public conceptualise uncertainty', *Journal of Risk Research*, vol 6, pp75–85

Frewer, L. J., Lassen, J., Kettlitz, B., Scholderer, J., Beekman, V. and Berdal, K. G. (2004) 'Societal aspects of genetically modified foods', *Journal of Toxicology and Food Safety*, vol 42, pp1181–1193

Garvin, T. (2001) 'Analytical paradigms: The epistemological distances between scientists, policy makers and the public', *Risk Analysis*, vol 21, pp443–455

Groenewegen, P. (2002) 'Accommodating science to external demands: The emergence of Dutch toxicology', *Science, Technology and Human Values*, vol 27, pp479–498

Guston, D. H. (1999) *Between Politics and Science*, Cambridge University Press, Cambridge

Heller, C. (2002) 'From scientific risk to *paysan savoir-faire*: Peasant expertise in the French and global debate over GM crops', *Science as Culture* vol 11, pp–37

Hilgartner, S. (1990) 'The dominant view of popularization: Conceptual problems, political uses', *Social Studies of Science*, vol 20, pp519–539

HM Government (2000) *The Government Response to the BSE Inquiry*, The Stationary Office, London

Irwin, A. (2001) 'Constructing the scientific citizen: Science and democracy in the biosciences', *Public Understanding of Science*, vol 10, pp1–18

Jasanoff, S. (1990) *The Fifth Branch: Science Advisors as Policy Makers*, Harvard University Press, Cambridge

Jensen, K. K. and Sandøe, P. (2002) 'Food safety and ethics: The interplay between science and values', *Journal of Agricultural and Environmental Ethics*, vol 15, pp245–253

Jensen, K. K., Gamborg, C., Madsen, K. H., Jørgensen, R. B., von Krauss, M. C., Folker, A. P. and Sandøe, P. (2003) 'Making the "risk window" transparent: The normative foundations of the environmental risk assessment of GMOs', *Environmental Biosafety Research*, vol 2, no 3, pp161–171

Johnson, B. B. and Slovic, P. (1995) 'Presenting uncertainty in health risk assessment: Initial studies of its effects on risk perception and trust', *Risk Analysis*, vol 15, pp485–494

Johnson, B. B. and Slovic, P. (1998) 'Lay views on uncertainty in environmental health risk assessments', *Journal of Risk Research*, vol 1, pp261–279

Kuhn, K. M. (2000) 'Message format and audience values: Interactive effects of uncertainty information and environmental attitudes on perceived risk', *Journal of Environmental Psychology*, vol 20, pp41–57

Lang, T. (1998) 'BSE and CJD: Recent developments', in S. Ratzan (ed) *The Mad Cow Crisis: Health and the Public Good*, University College London Press, London, pp65–85

Law, J. and Hassard, J. (1999) *Actor Network Theory and After*, Blackwell, Oxford/Malden

Levidow, L. and Carr S. (1996) 'UK: Disputing boundaries of biotechnology regulation', *Science and Public Policy*, vol 23, pp164–170

Levidow, L. and Marris, C. (2001) 'Science and governance in Europe: Lessons from the case of agricultural biotechnology', *Science and Public Policy*, vol 28, no 5, pp345–360

Löfstedt, R. and Vogel, D. (2001) 'The changing climate of regulation: A comparison of Europe and the United States', *Risk Analysis*, vol 21, pp399–416

Miles, S. and Frewer, L. J. (2001) 'Investigating specific concerns about different food hazards – higher and lower order attributes', *Food Quality and Preference*, vol 12, pp47–61

Miles, S. and Frewer, L. J. (2003) 'Public perception of scientific uncertainty in relation to food hazards', *Journal of Risk Research*, vol 6, no 3, pp267–283

Miles, S., Ueland, O. and Frewer, L. J. (2005) 'Public attitudes towards genetically modified food and its regulation: The impact of traceability information', *British Food Journal*, vol 107, pp246–262

Morgan, M. G. and Henrion, M. (1990) *Uncertainty: A Guide to Dealing with Uncertainty in Quantitative Risk and Policy Analysis*, Cambridge University Press, Cambridge

National Research Council (1994) *Science and Judgement in Risk Analysis*, National Academy Press, Washington, DC

Pennings, J. M. E., Wasink, B. and Meulenberg, M. T. G. (2002) 'A note on modelling consumer reactions to a crisis: The case of the mad cow disease', *International Journal of Research in Marketing*, vol 29, pp91–100

Renn, O., Webler, T. and Wiedemann, P. (1995) *Fairness and Competence in Citizen Participation*, Kluwer, Dordrecht

Rothstein, H., Irwin, A., Yearly, S. and McCarthy, E. (1999) 'Regulatory science and the control of agrochemicals', *Science, Technology and Human Values*, vol 24, pp241–264

Rowe, G. and Frewer, L. J. (2000) 'Public participation methods: An evaluative review of the literature', *Science, Technology and Human Values*, vol 25, pp3–29

Rowe, G. and Frewer, L. J. (2004) 'Evaluating public participation exercises: A research agenda', *Science, Technology and Human Values*, vol 29, pp512–556

Rowe, G., Marsh, R. and Frewer, L. J. (2004) 'Evaluation of a deliberative conference using validated criteria', *Science, Technology and Human Values*, vol 29, no 1, pp88–121

Select Committee on Science and Technology (1999) *The Scientific Advisory System: Genetically Modified Foods Session 1998–99*, HC286-I, The Stationery Office, London

Siegrist, M. (1999) 'A causal model explaining the perception and acceptance of gene technology', *Journal of Applied Social Psychology*, vol 29, no 10, pp2093–2106

Slovic, P. (1987) 'Perception of risk', *Science*, vol 236, pp280–285

Slovic, P. (1993) 'Perceived risk, trust and democracy', *Risk Analysis*, vol 13, pp675–682

Slovic, P. (2000) *The Perception of Risk*, Earthscan, London.

Thompson, K. M. and Bloom, D. L. (2000) 'Communication of risk assessment information to risk managers', *Journal of Risk Research*, vol 3, pp333–352

Till, J. E. and Meyer, K. R. (2001) 'Public involvement in science and decision making', *Health Physics*, vol 80, pp370–378

van den Daele, W. (1999) 'Dealing with the risks of genetic engineering as an example of "reflexive modernization?"', *New Genetics and Society*, vol 18, pp65–77

Verbeke, W. (2001) 'Beliefs, attitude and behaviour towards fresh meat revisited after the Belgian dioxin crisis', *Food Quality and Preference*, vol 12, pp489–498

Woolgar, S. (1996) 'Psychology, qualitative methods and the ideas of science', in S. T. E. Richardson (ed) *Handbook of Qualitative Research Methods*, British Psychological Society, Leicester, pp11–24

Wynne, B. (1992) 'Uncertainty and environmental learning: Reconceiving science and policy in the preventive paradigm', *Global Environmental Change*, June, pp111–127

7 Rebuilding Consumer Trust in the Context of a Food Crisis[1]

Lucia Savadori, Michele Graffeo, Nicolao Bonini, Luigi Lombardi, Katya Tentori and Rino Rumiati

INTRODUCTION

In the recent history of food production, numerous scares have occurred that have severely threatened consumers' health. These scares are usually followed by crises of consumption. A food crisis can be defined as a sudden decrease of the aggregate demand for a product following a food scare. Certainly, the biggest food crisis that ever happened was mad cow disease, or bovine spongiform encephalopathy (BSE). On this occasion, European Union (EU) beef consumption dropped from 21.5kg per person in 1990 to 19.7kg in 1998, reaching a low of 18.6kg per person in 1996 when the UK suggested a link between BSE and the new variant of Creutzfeldt-Jakob disease (Roosen et al, 2003). An additional 27 per cent fall was registered during the most recent crisis in 2000 (*The Economist*, 2001).

The factors causing a food scare can be diverse. In the history of food crises, we have observed food contamination scares due to salmonella, listeria, *Escherichia coli*, illegal hormones, dioxin, abuse of agrochemicals, and, of course, the BSE agent.

A food crisis might be the result of agricultural terrorism; fortunately, until now, no such act has been reported. Agricultural terrorism is defined as the deliberate introduction of a disease agent against livestock, crops or into the food chain for the purposes of undermining socio-economic stability and/or generating fear. In contrast with other types of food scares, which act specifically at one level of the food production chain, such as production or storage, agricultural terrorism can affect the whole food supply chain.

Another major concern for consumers that seemingly has never caused serious scares is the use of genetically modified food. This chapter, though, will limit its analysis to trust in the context of food crises, even if we do not exclude the possibility that the same trust-building mechanisms that are effective in re-establishing consumer demand could work to increase public acceptance of genetically modified (GM) food.

Each crisis affects the market to a different degree. The effect may be felt by the individual company, the entire industry or even a group of similar industries. Due to the social amplification of risk (for a recent review, see Pidgeon et al, 2003), consumers can, as well, decide to avoid similar products from similar industries that are not influenced by the scare, but are in some way associated to the target product. An example would be to avoid all dairy products because one of them has been contaminated with salmonella.

There is also a notable difference among countries in reaction to a food scare. In Italy and Germany, due to the BSE scare, beef consumption declined by as much as 30 to 50 per cent in 1990 and 1998 (Verbeke and Viaene, 1999). Other studies showed that BSE produced a stronger reaction in Germans than in Americans and Dutch consumers (self-reported measures in Pennings et al, 2002), and German consumers reported they were more risk averse ('For me, eating beef is not worth the risk') than American or Dutch consumers. It is important to consider, however, that BSE was never a problem in the US, so respondents in this study had to deal with a scare that never directly concerned them. Nevertheless, Dutch consumers were involved and notable differences were reported in their reactions to BSE in comparison with German consumers. One explanation for these different levels of concern may be because American and Dutch consumers are more trusting of the information they receive from their respective governments. In fact, in the US, 83 per cent of the population trust the US Food and Drug Administration (FDA) (Wansink and Kim, 2001).

When a food crisis occurs and the aggregate demand for a product suddenly decreases, the problem that the company or the industry has to face is how to regain consumer trust when the scare no longer exists. Learning how consumers respond to a crisis is the first step in this direction.

CONSUMER RESPONSE TO A FOOD CRISIS

Consumers use 'short cuts', 'heuristics' or what are sometimes called 'rules of thumb' to complete their purchasing decisions. Since they have incomplete information on food and limited processing capabilities, they normally apply what has been called a 'routine response behaviour' based on their experience (Howard, 1977; Kaas, 1982). Consumers apply decision rules that give them a satisfactory result until they receive a strong-enough signal to make them revise their prior beliefs or decision rules. The question is, on what dimensions are these decision rules based?

Food is both a functional and an expressive product in the sense that consumer preferences are based on dimensions such as the healthiness, naturalness, freshness and price, but, of course, also on dimensions such as pleasure and personal taste, as well as environmental and animal welfare concerns. Usually, safety is not among the top-level dimensions when consumers choose a food (Green et al, 2003). In fact, if a line has to be drawn between what is negative and what is positive, food would certainly be placed on the positive side. In other words, food normally has more benefits than risks. It is only when a crisis occurs that food safety becomes a primary dimension by which food choices are made. Consequently, the set of consumer preferences is adapted to also include safety evaluations. What type of evaluations are these?

Food safety is, in essence, an experiential attribute in the sense that it cannot be perfectly observed prior to consumption. No one, however, would use experience to determine the safety of a food. There is also a strong credence component from the point of view that people rely on the message associated with the food.

When a crisis occurs, consumers show an immediate and sharp drop in demand that slowly recovers (although never to the original level) as safety is established (Liu et al, 1998). Usually, this pattern of behaviour is attributed to media coverage, and the impact of the message is modelled by the level of trust that the receiver has in the information source (Slovic, 1992; Frewer et al, 1996).

Research on economic behaviour has nevertheless shown that media coverage is not the only factor shaping consumer response to a food crisis. In particular, trust in the supplier is taken into account when making a purchasing decision in the context of a food crisis (Bocker and Hanf, 2000). Trust in the supplier is seemingly used to simplify the process of considering the whole range of information related to the hazard. For a fully informed choice, consumers would need an incredible amount of information about the product contamination level, the health effects of exposure and many other factors that are difficult to discover and nearly impossible to compute into a decision rule (Ravenswaay and Hoehn, 1996). Trust is a way of reducing uncertainty and making the decision-making process easier. For example, during the BSE crisis in Germany, the share of fresh beef sold by local butcheries rose from 13 to 20 per cent between March and May 1996 (Loy, 1999), presumably because local butchers were more personally trusted to provide safe meat products.

All of these data seem to point to the fact that when a food accident occurs, the best way to re-establish product demand to its original level is to restore consumer trust. In the following section we will look at a study that explores the role of trust in the context of a food crisis.

CONSUMER TRUST IN THE CONTEXT OF A FOOD CRISIS

Here, we report on a study using structural equation modelling intended to examine the relative influence of trust and attitude on consumption intentions in the context of a dioxin food scare (Graffeo et al, 2004). The study had three interrelated aims. First, it intended to increase our understanding of the actual influence of trust on consumption intentions in the context of a food scare. Second, it intended to identify which, among a series of antecedents of trust, was the most important in this respect. Last, it studied the relationship between trust and attitude and their relative importance in determining consumption. In relation to this last point, it is commonly assumed that after the outbreak of a food crisis the restoration of consumer trust is necessary to recover from the drop in consumption; but this assumption might not be entirely true, or might be true only partially.

Usually, when a researcher states that his/her study is about trust, often the definitions used do not really cover such a concept or they cover it only partially. Indeed, many definitions of trust have been given by researchers from different perspectives – psychological, sociological and economic. Within each perspective itself, many definitions exist; but very few have defined trust in the domain of food safety.

Some insight is provided by the sociological perspective where trust in food safety has been described as consisting of three levels (Kjærnes and Dulsrud, 1998). At a first level, we have an *individual trust*, which is simply an attitude towards a certain product. This type of trust is essentially a coping strategy, that is, it is translated into an action. For example, a person might suspect that a chicken is infected with salmonella and has two options: to eat or not to eat the chicken. At a second level, trust can be *system oriented* or *structural*. This kind of trust concerns systems: the ability of food producers or government institutions to maintain adequate levels of food safety. System distrust may result in:

- a decision to avoid the product;
- political activism; or
- the establishment of alternative markets.

Lastly, trust may be *relational*. This trust emerges from the interaction of individuals and is based on direct experience with a person. For example, purchasing food directly from farms or from a farmers' market is a choice based on this type of trust.

System trust is very similar to what has been called *social trust* in psychological studies (see Renn and Levine, 1991; Johnson, 1999; Cvetkovich and Löfstedt, 2000). Social trust is the willingness to rely on experts and institutions in the management of risks and technologies. Siegrist (2000), in his study, found that social trust was positively related to perceived benefits of gene technology and negatively related to the perceived risk.

In the study by Graffeo et al (2004), trust was defined as 'the willingness for a party to be vulnerable to the actions of another party based on the expectation that the other will perform a particular action important to the truster, irrespective of the ability to monitor or control that other party' (Mayer et al, 1995). This definition has been chosen because it highlights the interdependency between two parties: a truster and the trustee. In the context of a food scare, the truster is the consumer and the trustee refers to any of the actors in the food supply chain, such as the breeders, the sellers or the authorities in charge of food safety. As used in this study, the definition captures the system trust and the relational trust of the sociological model and is very similar to the psychological notion of social trust.

Nevertheless, the aim of the study is not just to discover the effect of trust, but also to learn what antecedents of trust are most effective in determining food consumption in the context of a food scare. In truth, one of the difficulties that has hindered previous research on trust has been a lack of clear differentiation among factors that contribute to trust, trust itself and the outcomes of trust (Cook and Wall, 1980).

For this purpose, the starting model considers three antecedents of trust, as suggested by Mayer et al (1995). In the model, trust in another person or party is supposed to be based on:

1   competence (ability);
2   benevolence; and
3   shared values (integrity).

Competence is defined as the perception that the trustee is capable and expert in the specific domain. Benevolence refers to the extent to which a trustee is believed to want to do good to the truster, aside from any egocentric profit motive. Benevolence suggests that the trustee has some specific attachment to the truster. Finally, 'shared values' refers to the truster's perception that the trustee adheres to a set of principles that the truster finds acceptable For example, a party who is committed solely to the principle of profit-seeking at all costs would be judged high in shared values, but probably low in trust.

In this model, risk is part of the context and enters into the dynamics of trust as a covariate factor. In other words, it is assumed that risk, or any investment, is a requisite for trust. The resulting behaviour is what has been called risk-taking in relationship, which means that one must take a risk in order to engage in a trusting action.

The model suggested by Mayer et al (1995) was originally applied to the description of developing a trust relationship between two people who work together and are in a hierarchical relationship. In particular, Mayer and colleagues concentrate on this aspect: what makes a person in a lower position of power – the subordinate (truster) – trust their superior (trustee)? This model can be easily adapted to describe the behaviour of consumers in their purchasing decisions since these two contexts share the same dynamics. Compared to their superior, the employee has less negotiating power, less relevant information upon which to base their decisions and, in general, has little chance of influencing the decisions taken at a higher level. In the same way, the consumer has much less information regarding a product than, for example, the seller, and has decidedly less negotiating power. With these limitations, the consumer does not have the means to fully appreciate the data that are provided by the seller. Nonetheless, the consumer can attach a value to these data. The consumer has an opinion regarding the seller, and decides whether they can or cannot trust him or her.

In the final model (see Figures 7.1 and 7.2), a further antecedent – the perceived truthfulness of information provided by the actor – was added. This element was introduced because, in the context of a food scare, consumers might be very confident in the competence, benevolence and shared values of the actors; nevertheless, they might feel that these actors do not provide them with true information, perhaps for reasons of public order.

Finally, the factors influencing the choice of food include the individual's *attitude* towards it, which is one of the most widely studied areas. Attitude towards poultry, for instance, is essentially determined by the beliefs relating to health, eating enjoyment and, to a lesser extent, safety. Alone, these factors predict 64 per cent of the variance in behavioural intentions of eating poultry (McCarthy et al, 2004). Similar findings have emerged regarding beef (McCarthy et al, 2003). For this reason, when studying consumer behavioural intentions, we cannot exclude attitude from the analysis without severely affecting the validity of the results.

As can be seen in Figure 7.1, attitude has been included in the model. An initial set of 18 bipolar semantic differential-type scales (for example, positive versus negative; good versus bad) was reduced to two factors: an *evaluative*, or *experiential*, factor and an *instrumental* factor. The evaluative–instrumental attitude was defined by such attributes as 'positive versus negative', 'bad versus good', 'pleasant versus unpleasant', 'harmful versus beneficial' and 'risky versus safe'. The instrumental attitude was defined by attributes such as 'convenient versus inconvenient', 'disadvantageous versus advantageous' and 'opportune versus inopportune'. A third *moral* factor was excluded from the model because the pattern of correlations showed that it had no relationship with any behavioural intentions.

In the study, 104 consumers were exposed to a threatening scenario regarding dioxin contamination[2] and then asked to answer a questionnaire measuring trust, attitude and consumption intentions. The threatening scenario was as follows:

> *Dioxin: A real problem for health.* A considerable threat to our health, disappointingly very seldom detected, is the risk posed by the consumption of food contaminated by dioxin. Dioxin is extremely

toxic and is especially used as an additive in oils for motors and condensers. Getting rid of old machinery that used dioxin is difficult and costly. For this reason, in the absence of effective controls by the authorities, thoughtless individuals will continue to dump old machinery in the environment. Once it has been left in the environment, dioxin will make its way into the surrounding vegetation. The vegetation then becomes fodder for a large range of breeding animals. The major risk posed by dioxin is due to its tendency to accumulate in animal fat. As a result, lower initial concentrations in the fodder increase at every processing phase, ultimately reaching high levels of risk in the breeding animals. Researchers have demonstrated a large variety of effects on the human body. The organs most at risk include the liver, reproductive and neurological organs, as well as the immune system. The Environmental Protection Agency (EPA) has classified dioxin as a potential cancerous substance.

The largest dioxin scandal was the 'chicken scandal' in 1999 in Belgium. During this food scare, large quantities of chicken meat were seized and destroyed because they were heavily contaminated with dioxin.

The current risk connected with dioxin is now considerably lower, even though recently the authorities that check food safety have found cases of chicken and salmon contaminated with dioxin in the Triveneto region, where the participants of the study lived.

The final models resulting from structural equations are described in Figures 7.1 and 7.2. The first model refers to salmon and the second to chicken.

As shown by the model in Figure 7.1, the intention to consume salmon in the context of a food scare was affected both by the extent to which the consumer believed that the breeder, fishmonger or the authorities shared their own values (.37) and by the extent to which the consumer liked salmon (.32). An interesting result was that trust in the actor was predicted by competence and shared values, as well as truthfulness of information, but had apparently no relevance in orienting consumer decisions – whereas one of the trust antecedents (shared values) was the best predictor.

A very similar model was found for chicken (Figure 7.2). Here, relationships were somewhat stronger than those observed for salmon. The intention to consume chicken was significantly affected by the extent to which the consumer believed that the actor shared their own values (.45) and by the extent to which the consumer liked chicken (.28), as well as by the conviction that consuming chicken was cheap (.22). As for the previous model, no significant relationship was found between trust and the intention to consume chicken.

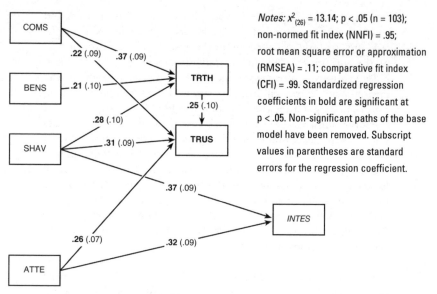

Notes: COMS: to what extent do you think that Italian fishmongers (or breeders/authorities in charge of food safety) are competent in their work? BENS: to what extent do you think that Italian fishmongers (or breeders/authorities in charge of food safety) are concerned about your health? SHAV: to what extent do you think that Italian fishmongers (or breeders/authorities in charge of food safety) share your values? ATTE: the experiential attitude. TRTH: to what extent do you trust Italian fishmongers (or breeders/authorities in charge of food safety) to tell the truth about salmon meat? TRUS: to what extent do you trust Italian fishmongers (or breeders/authorities in charge of food safety)? INTES: to what extent do you trust eating salmon?

Source: for a detailed description of the procedure and statistical analyses, see Graffeo et al (2004)

Figure 7.1 *Final path model (salmon) with standardized regression weights*

These results are also in line with those found by Frewer et al (2003), who investigated the role of trust, attitude and risk perception in the field of GM food. Their study showed that trust is a consequence rather than a cause of attitudes towards emerging technologies. In particular, the extent to which people trusted the information sources was determined by people's attitudes to genetically modified foods, rather than trust influencing the way in which people reacted to the information.

These results confirm the importance of attitude in affecting food choices, but also show some novel findings: trust did not feature among the most important predictors of choice. If these results are confirmed in future studies, the central role of trust in recovering consumption after a food crisis needs to be revised. Moreover, the most important finding of this study was that shared values were the best predictors of consumption in the event of a food scare, even more important than having a positive attitude. This result is crucial because it shows that an effective strategy for regaining the original consumption levels should be based on convinc-

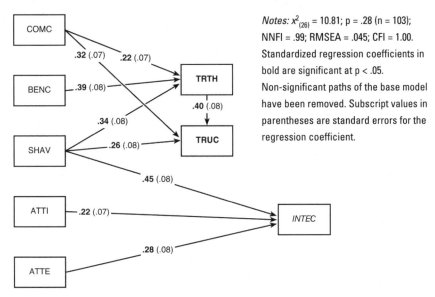

Notes: COMC: to what extent do you think that Italian butchers (or breeders/authorities in charge of food safety) are competent in their work? BENC: to what extent do you think that Italian butchers (or breeders/authorities in charge of food safety) are concerned about your health? SHAV: to what extent do you think that Italian butchers (or breeders/authorities in charge of food safety) share your values? ATTI: instrumental attitude. ATTE: experiential attitude. TRTH: to what extent do you trust Italian butchers (or breeders/authorities in charge of food safety) to tell the truth about chicken meat? TRUC: to what extent do you trust Italian butchers (or breeders/authorities in charge of food safety)? INTEC: to what extent do you trust eating chicken?
Source: for a detailed description of the procedure and statistical analyses, see Graffeo et al (2004)

Figure 7.2 *Final path model (chicken) with standardized regression weights*

ing the consumers that the supplier – or whoever else is affected by lack of trust – shares their own values.

An extraordinary resemblance of this model to Siegrist and Earle's model of cooperation (Earle et al, 2002; Siegrist et al, 2003) can be noted. In their model, consumer acceptability was, indeed, determined by shared values (see Chapter 1 of this book).

SOLUTION TO THE CRISIS AND SUGGESTIONS FOR RISK MANAGEMENT

The solution to a food crisis lies essentially in two types of strategy:

1   a more effective communication; and
2   a drastic measure with respect to product supplies, such as recall or discontinuation of a product.

As shown by the diverse reactions of different countries to mad cow disease, consistent communication is sometimes effective, whereas more extreme measures in product supplies are most effective at other times (Pennings et al, 2002). It has to be remembered that trust in a food supplier or an institution is largely built *before* the crisis occurs. *After* a crisis has occurred, though, a supplier or an institution can *maintain* or even *build* new trust by means of specific trust-building behaviours focused on shared values. In an extreme case, when a company or a government is solely responsible for the food crisis, we observe a sudden drop in consumer trust. Nevertheless, even in these cases, accurate communication of shared values can be of paramount importance.

As noted in several studies, during a food crisis consumers tend to react differently to positive and negative information: bad news has a more immediate impact on consumption than good news (Siegrist and Cvetkovich, 2001). For example, this has been noted in the milk contamination crisis in Hawaii, in 1982 (Liu et al, 1998). As noted by Bocker and Hanf (2000), this asymmetric impact of negative and positive information can be easily understood if we consider that a product failure or a supplier's misbehaviour is clear evidence of misconduct falling under the supplier's responsibility, whereas information regarding the supplier's adherence to food safety regulations alone cannot be regarded as a proof of the supplier's benevolence and trustworthiness.

Some researchers also point out that in the attempt to create trust, suppliers should not discriminate against competitors, showing that they are not so careful about food safety, but should clearly highlight their own reliability, taking consumer protection seriously and doing their best to avoid putting the consumer at risk (Bocker and Hanf, 2000).

The model in Figure 7.1 has a further implication. It shows that consumer choices are not only dependent upon the education of the general public in terms of safe food-handling practices and healthy diet, which is the main prerogative in the interventions based on the knowledge deficit model – experts are right and people are wrong (for a discussion, see Hansen et al, 2003). Consumer choices, indeed, are also dependent upon the extent to which consumers believe that the actors in the food supply chain share their own values.

Central to a crisis solution is the understanding of why and how consumers react to it. Many crises are seen as catastrophic and with irrevocable consequences. Crises, such as mad cow disease, have a strong emotional impact upon consumers and they increase susceptibility and reaction to future similar scares. The first step towards understanding consumer reaction is to map the structure of preferences for food safety, which means detecting which dimensions are relevant in food choice when consumers are evaluating safety.

A second step is to understand what it means for a consumer to share similar values with a producer, a supplier or an institution. At a purely speculative level, we can reason that shared values may relate to a set of moral rules of behaviour. Such values such as honesty and environmental and animal welfare protection are general moral principles, and, certainly, count; but when a consumer builds their evaluation of similarity with the values of a supplier, what else are they looking at? Furthermore, what type of message can convey these values?

In this regard, an interesting analysis has been carried out on the effectiveness of different strategies of meat labelling after the BSE crisis (Roosen et al, 2003). Following the BSE crisis, EU beef demand fell dramatically and in order to counteract the trend, beef producers and retailers attempted to signal the quality of their products. Origin labels were rated by consumers as the most important among the strategies studied, even if differences among countries were noticed. Origin labels presumably convey a message relating to geographic origins and the physical production environment, but mostly to traditions of agricultural and product transformation practices. Likely enough, these traditions share principles or values similar to those accepted by the consumers.

Another example of research that can increase our understanding of consumer values is one by Rozin et al (2004). In this study, consumer preferences for natural food compared to the food produced with human intervention were studied. Explicitly asking the reasons for these preferences revealed that healthfulness was the major factor conveying this judgement. Nevertheless, when healthfulness of the natural and artificial exemplars was indicated as equivalent, the great majority of consumers who have shown a preference for natural food continued to prefer it. This clearly suggests that 'natural' is an attribute that cannot be easily broken down into its components, and probably is a basic 'value' that can be important in food safety.

The findings collected in this chapter emphasize that more research is needed to understand what values are important in determining food choice in the context of a food scare. The knowledge of these values places a company or an industry a step ahead in the process of restoring consumer trust in their products after a food crisis.

NOTES

1   Supported by the European Commission's Quality of Life Programme, Key Action 1 Food, Nutrition and Health, Research Project: Food Risk Communication and Consumers' Trust in the Food Supply Chain – TRUST (contract no QLK1-CT-2002-02343).
2   A comparative analysis of 50 consumers who were not presented with the threatening scenario showed that it did, significantly, increase perceived risk, but also decreased trust and the consumption intention of those consumers to whom it was presented.

References

Bocker, A. and Hanf, C. H. (2000) 'Confidence lost and – partially – regained: Consumer response to food scares', *Journal of Economic Behaviour and Organization*, vol 43, pp471–485

Cook, J. and Wall, T. (1980) 'New work attitude measures of trust, organizational commitment, and personal need nonfulfillment', *Journal of Occupational Psychology*, vol 53, pp39–52

Cvetkovich, G. and Löfstedt, R. (2000) *Social Trust and the Management of Risk: Advances in Social Science Theory and Research*, Earthscan, London

Earle, T. C., Siegrist, M. and Gutscher, H. (2002) 'Trust and confidence: A dual-mode model of cooperation', unpublished manuscript, Western Washington University, Bellingham

The Economist (2001) 'A new type of farming?', *The Economist*, 3 February, pp32–33

Frewer, L. J., Howard, C., Hedderley, D. and Shepherd, R. (1996) 'What determines trust in information about food-related risks? Underlying psychological constructs', *Risk Analysis*, vol 16, pp473–486

Frewer, L. J., Scholderer, J. and Bredahl, L. (2003) 'Communicating about the risks and benefits of genetically modified foods: The mediating role of trust', *Risk Analysis*, vol 23, pp1117–1133

Graffeo, M., Savadori, L., Lombardi, L., Tentori, K., Bonini, N. and Rumiati, R. (2004) 'Trust and attitude in consumer food choices under risk', *AGRARWIRTSCHAFT. Zeitschrift für Betriebswirtschaft, Marktforschung und Agrarpolitik*, vol 53, pp319–327

Green, J. M., Draper, A. K. and Dowler, E. A. (2003) 'Short cuts to safety: Risk and "rules of thumb" in accounts of food choice', *Health, Risk and Society*, vol 5, pp33–52

Hansen, J., Holm, L., Frewer, L., Robinson, P. and Sandoe, P. (2003) 'Beyond the knowledge deficit: Recent research into lay and expert attitudes to food risks', *Appetite*, vol 41, pp111–121

Howard, J. A. (1977). *Consumer Behaviour: Application of Theory*, McGraw-Hill, New York

Johnson, B. B. (1999) 'Exploring dimensionality in the origins of hazard-related trust', *Journal of Risk Research*, vol 2, pp325–354

Kaas, K. P. (1982) 'Consumer habit forming, information acquisition and buying behaviour', *Journal of Business Research*, vol 10, pp3–15

Kjærnes, U. and Dulsrud, A. (1998) 'Consumption and mechanisms of trust', Paper presented to the ESA sub-group at the conference on the Sociology of Consumption, Milan, 16–17 September

Liu, S., Huang, J. C. and Brown, G. L. (1998) 'Information and risk perception: A dynamic adjustment process', *Risk Analysis*, vol 18, pp689–699

Loy, J. P. (1999) 'Auswirkungen der BSE-Krise auf die Verbraucherpreise für Rindfleisch in Deutschland' ['Impacts of the BSE crisis on German consumer prices for beef'], *Schriften der Gesellschaft für Wirtschafts- und Sozialwissenschaften des Landbaues* e.V., vol 35, pp249–256

Mayer, R. C., Davis J. H. and Schoorman, F. D. (1995) 'An integrative model of organizational trust', *Academy of Management Review*, vol 20, pp709–734

McCarthy, M., de Boer, M., O'Reilly, S. and Cotter, L. (2003) 'Factors influencing intention to purchase beef in the Irish market', *Meat Science*, vol 65,

pp1071–1083
McCarthy, M., O'Reilly, S., Cotter, L. and de Boer, M. (2004) 'Factors influencing consumption of pork and poultry in the Irish market', *Appetite*, vol 43, no 1, pp19–28
Pennings, J. M. E., Wansink, B. and Meulenberg, M. T. G. (2002) 'A note on modeling consumer reactions to a crisis: The case of the mad cow disease', *International Journal of Research in Marketing*, vol 19, pp91–100
Pidgeon, N., Kasperson, R. E. and Slovic, P. (2003) *The Social Amplification of Risk*, Cambridge University Press, Cambridge
Ravenswaay, E. O. and Hoehn, J. P. (1996) 'The theoretical benefits of food safety policies: A total economic value framework', *American Journal of Agricultural Economics*, vol 78, pp1291–1296
Renn, O. and Levine, D. (1991) 'Credibility and trust in risk communication', in R. E. Kasperson and P. J. M. Stallen (eds) *Communicating Risks to the Public*, Kluwer, Dordrecht, pp175–218
Roosen, J., Lusk, J. L. and Fox, J. A. (2003) 'Consumer demand for and attitudes toward alternative beef labeling strategies in France, Germany and the UK', *Agribusiness*, vol 19, pp77–90
Rozin, P., Spranca M., Krieger, Z., Neuhaus R., Surillo, D., Swerdlin, A. and Wood, K. (2004) 'Preference for natural: Instrumental and ideational/moral motivations, and the contrast between foods and medicines', *Appetite*, vol 43, pp147–154
Siegrist, M. (2000) 'The influence of trust and perceptions of risks and benefits on the acceptance of gene technology', *Risk Analysis*, vol 20, pp195–204
Siegrist, M. and Cvetkovich, G. (2001) 'Better negative than positive? Evidence of a bias for negative information about possible health dangers', *Risk Analysis*, vol 21, pp199–206
Siegrist, M., Earle, T. C. and Gutscher, H. (2003) 'Test of a trust and confidence model in the applied context of electromagnetic field (EMF) risks', *Risk Analysis*, vol 23, pp705–716
Slovic, P. (1992) 'Perceptions of risk: Reflections on the psychometric paradigm', in S. Krimsky and D. Golding (eds) *Social Theories of Risk*, Praeger, Westport
Verbeke, W. and Viaene, J. (1999) 'Consumer attitude to beef quality labeling and associations with beef quality labels', *Journal of International Food and Agribusiness Marketing*, vol 10, pp45–65
Wansink, B. and Kim, J. (2001) 'The marketing battle over genetically modified foods: Mistaken assumptions about consumer behaviour', *American Behavioural Scientist*, vol 44, pp1405–1417

8 Trust and Risk in Smallpox Vaccination

Ann Bostrom and Emily Atkinson

INTRODUCTION

Ubiquitous vaccination is widely considered one of the top public health successes of the last century; but some fear that the public is becoming increasingly averse to the risks of vaccines. Although vaccine policies have changed to reduce vaccine risks (for example, the replacement of oral polio vaccine with injected killed polio vaccine), public health experts tend to regard avoiding vaccines as an extremely risky strategy. Recent focus on smallpox as a potential weapon for bioterrorism increased the importance of understanding how people think about vaccines, vaccination policies and related risks, and complicated the kinds of risk trade-offs that a vaccination choice might entail.

Smallpox vaccine was the first to enable worldwide elimination of a disease (last case in 1977), and the first to be eliminated from the otherwise increasing battery of childhood vaccines routinely administered in the US. In the wake of the 11 September 2001 terrorist attack and the anthrax attacks in the nation's capital, the Bush administration's new smallpox policy was widely awaited. However, its implementation did not go according to plan as fewer than 50,000 emergency responders and medical personnel had been vaccinated against smallpox (military personnel were required to be vaccinated) by September 2003, compared to the publicized goal of vaccinating 500,000 people in the initial round of vaccination (IOM, 2003). An obvious question is whether trust played a role in the lack of enthusiasm for smallpox vaccination, or whether it could be explained by other beliefs and perceptions, particularly risk perceptions. To begin to address this question, this study examines trust in information sources about smallpox vaccine or disease, behavioural intentions, and mental models of both smallpox disease and vaccine. The study is an extension of previous research on vaccination and trust (Bostrom, 1999), mental models of hazardous processes (for example, Bostrom et al, 1992; Morgan et al, 2001), and smallpox vaccine and disease-related survey research (Blendon et al, 2003).

METHODOLOGY

Research on mental models of hazardous processes has adapted research methodologies from decision research, cognitive anthropology and cognitive science, including, especially, cognitive aspects of survey methodology. The approach assumes that a verbal protocol (see Ericsson and Simon, 1993) provides some valid insights into thought processes, and that risk analysis can improve risk mitigation decisions. The framework underlying the latter was adapted from early work by Hohenemser et al (1985), and is laid out nicely by Kates (2001). A mental models study begins with two projects:

1   an expert analysis of the risk mitigation decision; and
2   an open-ended interview designed to reveal the decision-makers' mental model(s) of the underlying hazard(s), followed by increasingly structured questions designed to probe their risk-related beliefs and perceptions (this can be called a mental models interview).

Subsequently, the interview results are compared to the expert decision model to design a survey instrument in order to allow collection of a larger sample and to permit more robust conclusions. Further description of mental models research on risk perceptions for risk communication design is provided elsewhere (see, for example, Morgan et al, 2001; Bostrom et al, 1992).

In this study we used a qualitative expert decision model derived from the Centers for Disease Control and Prevention's (CDC's) *Vaccine Information Statement for Smallpox Vaccine*. Vaccine information statements (VISs) are written by staff at the CDC, generally in the National Immunization Program, and peer reviewed within the agency before release to the public. The VISs are designed to meet a federal mandate requiring distribution of information to the public about vaccines in order to support informed consent for vaccination.

The interview protocol starts with a few very general, open-ended questions; it then becomes more structured in order to explore what mental model(s) the interviewee may be using to make decisions about smallpox vaccine and disease risks. For purposes of comparison and a reliability check, this study also included several of the questions from a study led by Robert J. Blendon and the Harvard School of Public Health during 8 October to 8 December 2002 (Blendon et al, 2003, hereafter referred to as the Blendon study).

DATA COLLECTION AND ANALYSIS

We interviewed 24 Georgia Tech students enlisted in the psychology experiment pool in the spring of 2003. The ages of our participants ranged from 17 to 26, with the average age being 20.6. Of our 24 respondents, 11 were male and 13 were female; all were current students; none of the 24 respondents had any children; 20 had never had any training in public health; 5 respondents had family members who were in the medical profession; 22 respondents were American citizens, and those who were not had lived in the US for an average of six years; and the most common ethnicity was Caucasian, followed by: Asian; other (Burmese/Myanmar, Caucasian and Middle Eastern, and mixed); and one respondent each in the categories African-American, Asian-American and Hispanic.

Blendon et al (2003) conducted their interviews via phone to a nationally representative sample of 1006 respondents who were 18 years of age and older.

The opening two questions of the interviews were transcribed verbatim and coded for these analyses by two coders independently. Discrepancies were discussed and resolved; almost all discrepancies were due to omitted codes by one or the other coder. Only the first open-ended section of the interview protocol is analysed here, along with the responses to three trust questions from near the end of the interview protocol from 21 respondents. Other interview data reported here are from all 24 respondents, from questions with closed-ended response scales.

RESULTS

Mental models: Qualitative results

The most common opening remark, in response to 'Tell me about smallpox' was along the lines of: 'Really, I don't know about it.'

Two other responses illustrate the kinds of basic knowledge other respondents had:

> Well, I know it's a very deadly disease, at least from what I've heard. I believe it's extinct. I think. But I know probably a lot of people died of it and probably we have some sort of vaccination. My knowledge is quite limited as to the subject. But that's just my general, from growing up, and school and whatever.

> I really know very little about the disease ... in pictures in my history book and [it] looks quite gruesome. ['What looks gruesome about it?'] People ... post-smallpox: I guess the survivors, and I don't know ... it looks like lumps on the skin. Their skin is darker in certain areas; they look disfigured.

For smallpox disease, respondents volunteered the following:

- About half mentioned that smallpox is deadly.
- Over one third mentioned that it is highly contagious.

- About one quarter mentioned either terrorism or that smallpox could be a bioweapon, as a source of exposure: 'A lot of people think that smallpox could be used as a BT agent, bioterrorism agent.'
- About one quarter mentioned that it causes a rash ('spots', 'corpuscles').
- A few mentioned that it causes high fevers.
- A few mentioned that there might be a new genetically engineered vaccine-resistant strain of smallpox out there: 'From what I['ve] heard on the news lately, there's a new strain of the smallpox virus and ... it's resistant to the original vaccine.'
- One or two mentioned that smallpox is an old disease (one participant mentioned the 1600s; another said, when asked to elaborate, that it had appeared in the 1950s) and a few mentioned that smallpox has been eradicated.

For smallpox vaccine, respondents volunteered the following:

- About one quarter mentioned, generally, adverse effects from the vaccine.
- About one quarter thought (erroneously) that the vaccine contains killed or altered smallpox virus.
- Approximately the same one quarter stated that the vaccine builds immunity (by giving a person a tiny case of the disease).
- A few mentioned that smallpox vaccination at some point in history came from cowpox.
- A few mentioned that they thought one could get the disease from the vaccine.
- Many stated that smallpox was a childhood vaccination commonly administered, and almost half (10 out of 24) thought that they had been vaccinated for smallpox – either as a child, or in the process of moving to the US from another land (note that the average age of this group is 21, and that smallpox vaccine has not been administered regularly to children in the US for the last two decades).
- Several volunteered information about the administration of the vaccine, either mentioning pricks or, in some cases, one or a series of shots.

Overall, respondents' mental models of the vaccine as expressed in their responses to the first open-ended question ranged from confounds with other childhood vaccines to general beliefs regarding its current development to counteract smallpox as a bioweapon:

> Um, all that I can really think about ... is it's something we had when we were children for elementary school, I think before we can start kindergarten.

It's mostly, it's being developed right now to counter any kind of terrorism threat: the use of smallpox. Some people have recently gotten access to it by getting laboratory samples and they can produce some quantity of smallpox.

Related responses to closed-ended questions (see, for example, Table 8.1 and Figure 8.1) are consistent with the open-ended responses. There is, however, a systematic tendency for the Georgia Tech respondents to think smallpox vaccine is more effective and that side effects are less likely than do respondents from the national sample in the Blendon study.

Table 8.1 *Beliefs About the effectiveness of smallpox vaccine*

If a person has never been exposed to smallpox, how effective do you think the smallpox vaccine is in preventing people from coming down with the disease?

	Very effective	Somewhat effective	Not very effective	Not at all effective	No response
Georgia Tech	67%	29%	4%	0	0
Blendon study	54%	39%	3%	2%	2%

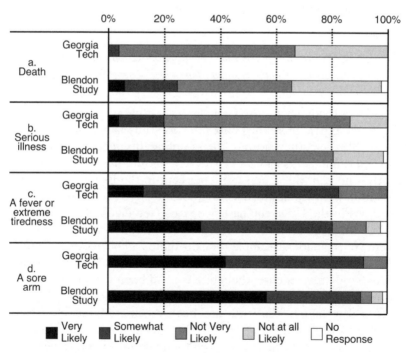

Note: At the time of the study, there had been recent deaths from smallpox vaccination in the US, but no recent publicly known use of smallpox as a bioweapon.

Source: Chapter authors

Figure 8.1 *Beliefs regarding the likelihood of side effects from smallpox vaccination in the Blendon et al (2003) study compared to this study*

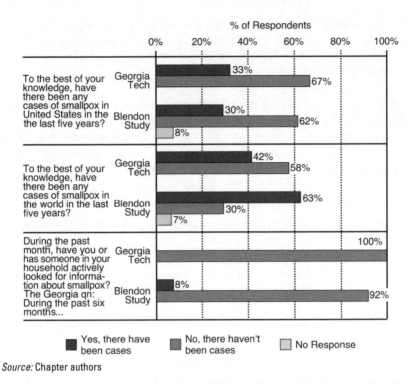

Figure 8.2 *Search for information about smallpox, and beliefs regarding the incidence of smallpox in the US and in the world during the past five years*

Aside from a tendency for the Georgia Tech students to trust technology and experts a little more, responses in this study largely resemble those from the Blendon study, despite the sampling differences between them. Almost no respondents had actively looked for information about smallpox recently – none in this study and 8 per cent in the Blendon study; about one third in both studies mistakenly thought there had been recent cases of smallpox in the US in the past five years (see Figure 8.2). Slightly less than half (42 per cent) of the Georgia Tech students and a majority (63 per cent) of the national sample thought there had been cases of smallpox in the world in the last five years (see Figure 8.2). Two-thirds of the Blendon study respondents thought they had been vaccinated for smallpox, and some might have been. However, 42 per cent of the Georgia Tech students also thought that they had been vaccinated, putting smallpox vaccine into the category of routine childhood immunizations. These beliefs are presumably mistaken, given their age and circumstances.

About half of the respondents in both studies thought that the news media have been reporting the risks of smallpox vaccination accurately; 25 per cent of the Blendon study respondents thought that the media were exaggerating the risks, compared to 42 per cent of the Georgia Tech students.

Trust in information sources

While none of our respondents had searched for information about the disease or the vaccine in the last six months, most had heard something at some time about either the disease or the vaccine. When asked what sources this information had come from, in an open-ended question, about one third of respondents (38 per cent) volunteered the news media. As one put it: 'Just the media. Just TV.' Almost as many mentioned school or classes. A few mentioned the internet or websites. Many of these mentioned multiple sources – for example: 'I don't know if I saw it on TV or radio, or in history or something like that', or 'What I know I've learned from various health classes. I read about it on the internet back in middle school.' One named his family's physician, one her (Korean) parents, one the encyclopaedia, and one just laughed – not having heard about smallpox vaccine or disease as best he recollected (this question came up more than two-thirds of the way through the interview).

When asked who they would trust most for information about smallpox vaccine or disease, respondents were most likely to mention the medical or scientific studies, publications or authorities, including the government and CDC, in particular, (n = 6) and professional health publications and/or associations (for example, the American Medical Association). Media were mentioned by a few (3), including CNN specifically, with personal physicians and online medical sources coming close behind, although these were not always specified:

> Doctors, reliable news articles and magazines, or news shows. Anything that looks, at least anything that looks like it's from a, you know, scientific medical source.

> Probably like – I don't trust the newspaper much. Like if you go to, like, a medical institution or something, the government, grabs something off the internet: I mean, like a viable source.

> Websites and government agencies.

One person said he simply trusted what others told him, that he generally trusted people: 'Probably just what I've heard from other people.'

Asked why they trusted these sources, our respondents often referred to the sources' scientific or medical expertise, with specific reference to experts' personal experience with smallpox issues:

> Because they're experts in the field; they're, like, epidemiologists. They actually get personal experience researching this stuff (refers to the American Medical Association and the Centers for Disease Control and Prevention).

> Because I know that they have to be extremely intelligent to be able to be in the position they're in, and that it would be essential to them to be correct; they're in the top position. They don't want to be false to you and ruin their reputation (refers to surgeon general, other government health officials and 'well-known schools').
>
> Just because of their education. And they're around all sorts of people all the time [who] know a lot about these things, and personal experience and maybe even contact with smallpox issues (refers to personal physician and 'people in the medical field').
>
> They've done their research and they have, like, approval from all these other boards and doctors and stuff (refers to medical institution or the government, off the internet).
>
> Just cause they're people [who], people [who] would know the virus: educated people (refers to the CDC and the American Health Association).
>
> Doctors, you kind of have to [trust them], and the internet, you know, has all the information that you need.
>
> ... they should know what they're talking about (refers to the Georgia Tech Health Center and 'other healthcare people').

One respondent stated that he trusted the physician and the internet. Asked why, he volunteered: because 'the internet has all the information there is'.

As the above quotes illustrate, the most common threads regarding why these sources are trusted are expertise or knowledge, and experience with smallpox.[1]

Economists refer to expressed and revealed preferences. One might analogously construe the above as expressed trust and compare it to revealed trust in terms of real or hypothetical actions that people might take depending upon what others say or do. Questions regarding vaccination choices illustrate this, although the responses are hypothetical and, hence, technically expressed rather than revealed.

At the outset of the interview, respondents were asked if they would receive a smallpox vaccination now or have their children vaccinated. Because the mental models interviews systematically explore how the respondent characterizes both the disease and the vaccine, we also asked these questions after the mental models questions to see how thinking through the issues may have changed their responses. Personal vaccination preferences flipped (54 per cent said yes at the outset; 43 per cent said yes after the mental models questions). As one might expect, at the outset of the interview, all of those (3) who thought that someone could not die from a smallpox vaccination were among the 54 per cent who said that

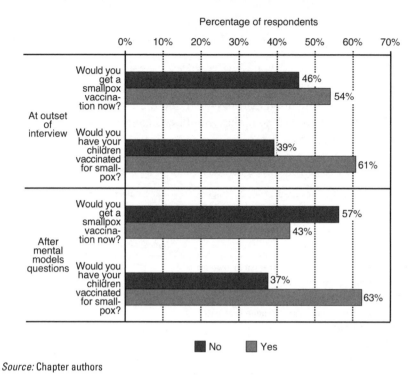

Figure 8.3 *Hypothetical vaccination decisions before and immediately after mental models questions*

they would receive a smallpox vaccination, demonstrating the consistency of confidence in the vaccine with expressed trust. But both before and after the mental models questions, a majority of respondents said that they would vaccinate their children for smallpox now (see Figure 8.3). This lack of 'sensitivity' could be related to the fact that none of our respondents had children at the time of the interview, so that this question was, in a sense, a double hypothetical for them.

In the closed-ended section of the interview, after the mental models questions, interviewees were asked the following: 'Would you receive a vaccination or have your child vaccinated as a precaution against terrorist attacks using the smallpox disease?' The Blendon survey asked the children's vaccination question separately. A large majority (79 per cent) of our respondents said yes. In the Blendon study, 61 per cent answered yes, and 64 per cent said that they would have their child vaccinated as a precaution against terrorist attacks using the smallpox disease. It is worth noting that the 79 per cent of respondents who said yes to this more specific vaccination question are more numerous than the percentage who said that they would be vaccinated against smallpox in response to the earlier, less specific, question. This appears to be a result of availability bias, illustrating the conjunction fallacy (Tversky and Kahneman, 1983):

respondents changed their minds because the specific mention of terrorism as a source of exposure to smallpox made the risk from smallpox more imaginable.

The interview branched at this point to ask those (n = 19 for this study) who said yes if they would change their mind and *not* want to be vaccinated depending upon events (see Table 8.2), and to ask those (n = 5 for this study) who said no if they would change their mind and *want* to be vaccinated (see Table 8.3).

Table 8.2 *Hypothetical questions to those in favour of vaccination*

If the following situations occurred, would it change your mind and make you not want to be vaccinated?

	No: would not change mind	Yes: would change mind
Your doctor and many other doctors refuse to become vaccinated against smallpox because they believe the vaccine to be too risky.	11% (21%)	89% (79%)
You heard that some people who had been vaccinated had died from the vaccine.	53% (33%)	47% (67%)
You would be forced to stay out of work for two weeks (one non-response).	84% (44%)	11% (56%)
You heard that 1 in 1 million people who are vaccinated die from the vaccine.	100% (51%)	(49%)

Note: A subsample of 19 respondents who responded that they would be vaccinated as a precaution against a terrorist attack using smallpox disease was examined. The corresponding percentage of respondents from the Blendon study is given in parentheses.

With regard to trust, it appears from Table 8.2 that people are more likely to change their minds based on the actions of their and others' doctors than they are for any of the other reasons presented, which confirms the expression of trust in physicians. It is also evident that the prospect of missing work did not bother students as much as it did respondents from the national sample (hardly a surprise). People were much more likely to change their minds based on specific known deaths from vaccines than based on an abstract statistic of a 1 in 1 million chance of death from the vaccine. This general pattern is possibly reflective of availability, as discussed above, or attributable to the identifiable victim effect, in which people react much more strongly to an identifiable victim than to statistics or general information, which Jenni and Loewenstein (1997) find is attributable to the size of the reference group. However, it is difficult to distinguish these potential effects from subjective increases in exposure estimates due, for example, to source proximity.

While only a few students said they would not become vaccinated, Table 8.3 suggests that they resemble the national sample to some extent,

Table 8.3 *Hypothetical questions to those against vaccination*

If the following situations occurred, would it change your mind and make you *want* to be vaccinated?

	No: would not change mind	Yes: would change mind
Cases of smallpox are reported somewhere in the world.	100% (35%)	(65%)
Your doctor and most other doctors are vaccinated against smallpox.	80% (27%)	20% (73%)
President Bush and his family are vaccinated against smallpox.	60% (34%)	40% (66%)
Cases of smallpox are reported in the US.	40% (25%)	60% (75%)
Cases of smallpox are reported in your community.	(12%)	100% (88%)

Note: A subsample of five respondents who responded that they would *not* become vaccinated as a precaution against a terrorist attack using smallpox disease was examined. The corresponding percentage of respondents from the Blendon study is given in parentheses.

although they do not respond to their doctor becoming vaccinated – and so, perhaps, do not trust physicians as much, do not trust President Bush and his family quite as much, and are somewhat more likely to respond to proximate risk than is the comparable national sample.

As this quote from one of our respondents shows, smallpox vaccination decisions are 'live' risk–risk trade-offs, although this trading includes heeding social norms ('what the majority does') and trusted experts ('what top-level people think'):

> Well, I guess I'd sort of do a thorough search of, you know, what the side effects were and, you know, weigh the – I know this is beating around the bush – but I'd weigh the benefits with the... Well, I'm not crazy about ... I'm not going to get every vaccine that's out there. So, you know, I want to do what the majority does, and that, you know, what top-level people think, what they advise the public. Then, you know, I'll probably... But definitely to take all cases necessary to prevent my kids from, you know, getting the disease.

When asked to whom they would first go for treatment if they developed smallpox symptoms or the smallpox disease, 88 per cent of Georgia Tech students volunteered that they would go to a hospital emergency room, compared to only 40 per cent of the national sample. A mere 8 per cent of Georgia Tech students said that they would go to their personal doctor, compared to 52 per cent of the national sample. This may be because some students are away from home and do not have a personal doctor, whereas respondents in the national sample do, although other explanations are plausible as well.

CONCLUSIONS AND IMPLICATIONS

Smallpox creates contentious analytic and policy challenges because the risk considered most likely is from the deliberate spread of the disease: bioterrorism. This also means that the risk could be catastrophic in nature – very low probability with potentially very high consequences. Determining what responses to this risk are rational or acceptable is not for the faint-hearted.

The students we interviewed were not thinking much about (or looking for information about) smallpox vaccines or diseases, and held opinions fairly similar to those of the general public in the US. Some differences appeared. The Georgia Tech students interviewed appear to trust medical technology and expertise more than the general public, and, perhaps as a consequence, were less concerned than the general public about smallpox vaccine and underestimated its risks (note that an inactivated smallpox vaccine was in use at the time of the interviews outside the US).

While Georgia Tech students' mental models of smallpox vaccine and disease ranged from simple analogies between smallpox and chickenpox to quite sophisticated sets of beliefs, including some history and microbiology, both lack of information (gaps in their mental models) and misconceptions, as well as trust, appear to have influenced their behavioural intentions. Despite what has been portrayed in the media as the catastrophic potential of a smallpox bioweapon, in this context trust appeared to be driven less by any conscious comparison of specific values and more by general, previously determined, attitudes towards technology, expertise and government. This supports recent findings on the importance, in trust, of prior attitudes (see, for example, Cvetkovich et al, 2002), and the importance of trust when knowledge is lacking (see Siegrist and Cvetkovich, 2000) and the tendency of those who respect expertise to trust experts more (see Siegrist et al, 2001), as illustrated by the greater willingness of the students to be vaccinated. However, in contrast with the findings of Siegrist et al (2001), trust in public health authorities appeared high in this study. The high credibility of government and public health authorities in this regard may explain the rather surprisingly low perceptions of risk from smallpox vaccine (compare Trumbo and McComas, 2003).[2]

With regard to trust, the study suggests that CDC has high name recognition and is a trusted source of information about smallpox, but that any personal or identifiable (as opposed to abstract statistical) experiences with the vaccine or the disease are likely to swamp that trust in short order. Given that all of the hypothetical mind-changers occurred (medical personnel refusing to be vaccinated; a few people who were vaccinated for the disease died from the vaccine; people were forced to stay out of work for a week or two; and the risk of the vaccine was discussed widely as being at least a 1 in 1 million risk of death), it is not surprising that the voluntary vaccination programme sputtered.

This comparison of the CDC vaccine information statement (VIS) for smallpox with respondents' mental models suggests that the VIS CDC dwelled disproportionately little on characterizing exposure to and effects of the disease, and should have addressed some common confusions between smallpox vaccine and other vaccines, as well as perceptions of possibly genetically altered and vaccine-resistant smallpox disease. An inherent advantage of this kind of in-depth mental models study is that it can suggest substantive changes in risk communications that might not be discernible in more structured studies. Additional research with appropriately representative samples, however, would be required to validate this and related findings.

It is important to recognize that people's knowledge and beliefs reflect the realities of their context, and that they may have little opportunity or reason to learn about extremely low probability risks. It was not the authors' expectation that those interviewed would know or have focused much effort on learning a lot about the smallpox disease or vaccine, given their relative lack of relevance to people's daily lives. Many of those interviewed commented in the debriefing that they felt they did not know enough or just felt 'dumb'. At least equally striking to the authors was the thoughtfulness of many responses, as well as how much some of the respondents did know.

Notes

1   Weber (2003) has commented that the evidence from lay and medical experts' experiences with vaccines can conflict, since medical experts 'experience' large numbers of vaccinations, and the patient presumably bases her or his decision on a description, which can lead to an overweighting of information about rare adverse events. However, patients may be influenced most by their own and close friends' and family's negative experiences, where such exist – that is, risk as feelings (compare Slovic and Weber, 2002), which would exacerbate the emphasis on adverse events.
2   Interestingly, vaccination was perceived as least or next-to-least risky of 30 activities and technologies by a league of women voters, students and club members in Slovic (1987), whereas experts rated it 25th out of 30.

References

Blendon, R. J., DesRoches, C. M., Benson, J. M., Herrmann, M. J., Taylor-Clark, K. and Weldon, K. J. (2003) 'The public and the smallpox threat', *The New England Journal of Medicine*, vol 348, no 5, pp426–432 (full survey and data available from www.hsph.harvard.edu/press/releases/survey_results.pdf, accessed 13 February 2006)

Bostrom, A. (1999) 'Who calls the shots? Credible vaccine risk communication', in G. Cvetkovich and R. E. Lofstedt (eds) *Social Trust and the Management of Risk*, Earthscan, London

Bostrom, A., Fischhoff, B. and Morgan, M. G. (1992) 'Characterizing mental models of hazardous processes', *Journal of Social Issues*, vol 48, pp85–100

Cvetkovich, G., Siegrist, M., Murray, R. and Tragesser, S. (2002) 'New information and social trust: Asymmetry and perseverance of attributions about hazard managers', *Risk Analysis*, vol 22, no 2, pp359–367

Ericsson, K. A. and Simon, H. A. (1993) *Protocol Analysis: Verbal Reports as Data* (revised edition), MIT Press, Cambridge

Hohenemser C., Kasperson R. E. and Kates, R. W. (1985) 'Causal structure', in R. W. Kates, C. Hohenemser and J. X. Kasperson (eds) *Perilous Progress: Managing the Hazards of Technology*, Westview, Boulder, pp25–42

IOM (Institute of Medicine of the National Academies) (2003) *Review of the Centers for Disease Control and Prevention's Smallpox Vaccination Program Implementation*, Letter Report no 4, Committee on Smallpox Vaccination Program Implementation Board on Health Promotion and Disease Prevention, National Academy Press, Washington, DC

Jenni, K. and Loewenstein, G. (1997) 'Explaining the identifiable victim effect', *Journal of Risk and Uncertainty*, vol 14, pp235–257

Kates, R. W. (2001) 'Queries on the human use of the Earth', *Annual Review of Energy and Environment*, vol 26, pp1–26

Morgan, M. B., Fischhoff, B., Bostrom, A. and Atman, C. J. (2001) *Risk Communication: A Mental Models Approach*, Cambridge University Press, Cambridge

Siegrist, M. and Cvetkovich, G. T. (2000) 'Perception of hazards: The role of social trust and knowledge', *Risk Analysis*, vol 20, no 5, pp713–720

Siegrist, M., Cvetkovich, G. T. and Gutscher, H. (2001) 'Shared values, social trust and the perception of geographic cancer clusters', *Risk Analysis*, vol 21, no 6, pp1047–1054

Slovic, P. (1987) 'Perception of risk', *Science*, vol 236, pp280–285

Slovic, P. and Weber, E. (2002) 'Perception of risk posed by extreme events', Paper presented at the conference on Risk Management Strategies in an Uncertain World, Palisades, New York

Trumbo, C. and McComas, K. A. (2003) 'The function of credibility in information processing for risk perception', *Risk Analysis*, vol 23, no 2, pp343–353

Tversky, A. and Kahneman, D. (1983) 'Extensional versus intuitive reasoning: The conjunction fallacy in probability judgment', *Psychological Review*, vol 90, no 4, pp293–315

Weber, E. (2003) 'Origins and functions of the perception of risk', Paper presented at the National Cancer Institute Workshop on Conceptualizing and Measuring Risk Perceptions, 13–14 February 2003, Bethesda, Maryland, available at www.cancercontrol.cancer.gov/brp/presentations/weber.pdf, accessed 13 February 2006

9 The What, How and When of Social Reliance and Cooperative Risk Management[1]

George Cvetkovich and Patricia L. Winter

One reaction to an earlier book in this series (Cvetkovich and Löfstedt, 1999) recognized a lack of consensus on definitions of key concepts regarding social reliance and trust (Fischhoff, 1999). Agreeing that this conceptual jumble hinders scientific advancement and handicaps the ability to offer effective practical advice, we cautiously make some suggestions. We offer these suggestions based on our joint research on cooperative management of US national forests, trust research within other cooperative risk management domains, and on the first author's examinations of trust in other domains of human functioning. There are many domain-specific aspects to relying on others. A child's trust of a parent or trust in a romantic partner, obviously, are different in many ways from trust in the individuals whom you have never met and who are in charge of an environmental protection or natural resource management agency. But there are also some important similarities. Recognition of the similarities as well as differences can aid the development of a consensus on basic definitions and directions for further research.

Our discussion addresses three questions:

1. The question of 'what?' deals with the definitions of three key terms – social reliance, trust and relational assurances. We deal in this section with questions about the nature of trust. Is trust a social emotion, a rational, objective judgement or something else?
2. The question of 'how?' deals with the social psychological processes underlying trust – how does trust, or distrust, come about?
3. The question of 'when?' deals with the identification of the circumstance determining the importance of trust to judgements about cooperative risk management. We conclude with a discussion of the conceptual, theoretical and practical implications of these distinctions.

WHAT IS TRUST?

Discussions about trust are in agreement on some general points. For example, trust involves risking betrayal, but provides certain potential benefits. Beyond some of these agreements about trust's basic properties and outcomes, there exists an amazing array of conceptualizations and definitions. Our view is that this babble will remain indeterminate and that new putative distinctions will continue to be creatively spawned unless conceptualizations are grounded in broader understandings of psychological functioning that are empirically tested.

We suggest that two aspects of functioning are important, even basic, to defining what trust is. The first aspect of functioning has to do with the mode of information processing. We use the labels 'implicit mode' and 'explicit mode' of information processing to mark these differences. The second aspect of functioning has to do with working out distinctions related to the kinds of representations that humans use when relying on others. We use the labels 'relational assurances' and 'trust' to mark these differences.

Implicit and explicit modes of information processing, trust and social reliance

Information processing is concerned with the manner by which sensations, perceptions, impressions and other 'inputs' are transformed or 'computed' into judgements, choices and actions. The human mind engages in two modes of information processing: an implicit, rapid, automatic mode and an explicit, slower, controlled mode. These differences are sometimes referred to as affect/emotion and cognition. Table 9.1 shows other commonly used labels and additional characteristics of the two modes. The distinction between the implicit and explicit modes has a long history[2] and has been supported by brain imaging studies (Damasio, 1994; Winston et al, 2002). Research on risk perception has recently begun to take these human dual modes of information processing seriously (Slovic, 1999; Trumbo, 1999; Finucane et al, 2000; Slovic et al, 2002, 2004).

Having recognized the mixed-mode nature of information processing, we hasten to add that it seems fair to conclude that with trust, the balance is clearly towards implicit mode processing.[3] If this is so, we can then conclude that trust is primarily a social emotion. Social emotions such as fear, happiness, sadness and trust are 'systems of coordinated changes in physiology, cognition and behaviour'. They are 'specialized modes of operation shaped by evolution that increase the capacity and tendency to respond adaptively to threats and opportunities' (Nesse, 1990).

As a social emotion, trust reflects the characteristics of implicit information processing shown in Table 9.1. This table shows that the computational rules used in implicit processing are simple unconscious associations of similarity in concrete or generic representations or tempo-

Table 9.1 *Characteristics of implicit and explicit modes of information processing*

	Implicit	Explicit
Common names	Emotion/affect Experiential mode Habit Heuristic processing Intuition Instinct 'Hearts' in the phrase 'hearts and minds'	Decision-making Logic Analysis Problem-solving Rational thought 'Minds' in the phrase 'hearts and minds'
Source of knowledge	Personal experience Evolved mechanisms	Language, culture and formal systems Symbol mediated experience
Level of awareness	Unconscious – unaware of process (aware of result)	Conscious – aware of process and result Used to explain process
Voluntary control	Involuntary Reflexive	Controlled Deliberate
Effort	Effortless	Effortful
Computation rules	Associations – similarities of structure and temporal contiguity (frequencies and correlations)	Rule based and symbol based Algorithms
Speed of processing	Rapid	Slow
Nature of representations	Concrete and generic concepts, images, stereotypes and feature sets Gist representations	Concrete and generic, as well as abstract concepts Abstracted features Compositional symbols Verbatim representations
Focus	Holistic/pattern	Constituent parts
Basis	Elicitation of evolved mechanism or well-learned expertise following identification of problem	Iterative interaction with the environment
Ability to communicate	Difficult or impossible	Possible
Criteria of truth/validity	Coherence	Correspondence

Source: adapted from Sloman (1996)

ral continuity. In line with this characterization, trust is an example of peripheral processing (Petty and Cacioppo, 1986; Verplanken, 1991; Petty et al, 1994). Peripheral processing involves the influence of factors outside of the main content of a message. One peripheral factor is the judged trustworthiness of the communicator. The association of the trusted communicator with the message leads to acceptance of the message. Credibility of the communicator has been found to have more influence on individuals who are using implicit rather than explicit processing. These are individuals for whom an issue is not personally relevant, who are less educated or who lack relevant technical expertise (Chaiken, 1980; Petty et al, 1981; Earle et al, 1990; Cacioppo et al, 1996). In contrast, explicit processing involves symbol-based rules applied to the content of the message. The message's content is evaluated as right or wrong on its own merits, regardless of the peripheral matter of who said it.

The standard of truth for the social emotion of trust and other forms of implicit information processing (see Table 9.1) is the internal standard of coherence (Hammond, 1996). A judgement is true if it is internally consistent, feels right and, in the colloquial phrase, 'makes sense'. The holistic comprehensive nature of trust, a characteristic of coherence, has often been noted. The standard of truth for the explicit mode processing is correspondence. Correspondence asks 'external' questions: does the judgement or conclusion fit available evidence? Was the judgement or conclusion reached through a logical, defensible process that can be explained to other people?

While trust primarily has the characteristics of a social emotion, trust-related judgements, like those related to other emotions, often involve some amount of explicit processing. Explicit processing is likely to be activated when people are called upon to communicate about or explain their trust. This includes when people are questioned about their trust by researchers. The two modes operate together, their relative contribution being weighted in different proportions at different times (Gray, 2004). People who are incapable of using one or the other mode due to traumatic or organic damage to parts of their brains make bad decisions (Damasio, 1994; Bechara et al, 2000).

Social reliance occurs when an individual risks allowing another person to control his health, safety or another aspect of well-being. The putative benefits motivating reliance have been extensively discussed. By relying on another person, one can reduce the time and effort required from information seeking, evaluation of evidence and decision-making. Relying on another person may also produce social capital by reducing transaction costs (Fukuyama, 1996; Putnam, 2000).

In addition to the distinction between implicit and explicit modes of information processing, we make a second distinction based on two routes of information processing. In agreement with other theorists, we have concluded that it is useful to conceptualize a route of relational assurances and a route of trust (Luhmann, 1979; Yamagishi et al, 1998a; Hardin, 1993; Earle et al, 2000; Siegrist et al, 2002).

The route of relational assurances

Toshio Yamagishi and his colleagues have conducted studies confounding common beliefs about national differences in trust (Yamagishi et al, 1998a;

Yamagishi et al, 1998b; Yamagishi and Yamagishi, 1994). Contrary to widely held stereotypes, this important research shows that Americans are more trusting than are the Japanese. Yamagishi and colleagues define trust in a similar fashion as we do here. It is reliance on another based on an assessment of the other's morality, personality, goals, motives or other personal characteristics. Americans base reliance on trust more often than do Japanese, who prefer basing reliance on the assurance of an established interpersonal social relationship.

Willingness to rely on another person might be based on another assurance besides an established interpersonal social relationship. Confidence has been identified in other chapters in this book (see Chapter 1) as reliance based on a record of past performance. Confidence based on past performance, in our view, is an example of relational assurance. In addition to established social relationships and past performance, other relational assurances that possibly induce reliance are systematically enforced laws, procedures attempting to ensure fair and just decisions, institutionalized accountability, and opportunities to voice one's view. Fukuyama (1996) identifies a number of innovations in trading and banking, such as letters of credit, which induced the level of reliance needed for the development of international trade in Europe during the Renaissance.

The evolution of eBay, the largest online auction site, is a recent history example of implementing new relational assurances to induce potential customers to rely on each other and on the internet trading network (Weidenbacher and Cvetkovich, 2003). Included among these assurances are Pay Pal, a system of payment that ensures anonymity of the buyer's credit card information, the exclusion of traders found to have used deception, the identification and elimination of shill betting, and the ability to use escrow accounts that release payment to a seller only if the buyer is satisfied with the quality of the purchase. An important relational assurance provided by eBay is making available purchasers' ratings of sellers' past performance.

The route of trust

Most of us, unless plagued by a phobic fear of flying, rely on the commercial airline industry because we are assured by safety regulations, crew training, anti-hijacking security measures, and other safety measures. Should something seem amiss, should we not feel assured, we may engage in efforts to characterize the individuals who should be protecting us or who may jeopardize us. The airline hijackings on 11 September 2001 resulted, in part, from failed airline security. Airline passengers no longer felt that their safety was ensured. Many began scrutinizing other passengers following the terrorist events. In one widely reported case, this scrutiny led passengers on one flight to conclude that a passenger could not be trusted, and there was a collective refusal to allow the plane to fly until this individual was made to leave the plane.

Reliance based on the route of trust involves the making of attributions about other individuals. Attributions about others reflect social representations of who they are – what their motives, goals, personal characteristics and, especially, their morality are. Social representations are shared knowledge: sets of organized attitudes, values and beliefs about something (Breakwell, 2001; Moscovici, 2001). Social representations serve two functions:

1 They provide understanding and allow us to derive meaning.
2 They allow us to communicate with others.

Trust involves knowledge of how the human mind works and the meaning of particular actions. These representations might be specific to a particular individual or they might be about people who are members of a particular group (for example, bureaucrats, police and elected officials), or they might be about how all human minds work. The information involved in trust couples relevant representations about the human mind with representations of what a particular person did, said or even looks like to arrive at a generalization about the characteristics of the person, including their trustworthiness. The colourful term 'mind reading' has been applied to this process.

To summarize, we have made a distinction between two routes to relying on other people – trust and relational assurances. We rely on another person when we have concluded that there are conditions assuring the nature of our relationship to another. We trust when we have concluded that an individual's mind operates in ways that will result in reliable judgements and behaviours.

Those of us interested in cooperative risk management have not given much attention to relational assurances. Our focus has been mostly on trust (at least in name). In keeping with the theme of this book, we will focus primarily on trust in much of the remainder of the chapter.

THE 'HOW' OF TRUST: SOCIAL PSYCHOLOGICAL PROCESSES INVOLVED IN TRUSTING

Our discussion of the 'what?' of reliance indicates that trust is based on attributions about the other person's psychological characteristics. It is a conclusion that the other person has the characteristic of being trustworthy. On what grounds is the conclusion of trustworthiness reached? The salient values similarity (SVS) model offers an answer to this question on the basis of relevant social representations. Trust is an in-group phenomena; it is an emotional reaction to other group members. Distrust is an out-group phenomenon; it is an emotional reaction to members of other groups. Social identity is an important aspect of trust, risk perception and judgements about risk management (see the Chapter 2 in this book;

Breakwell, 1986; Satterfield, 2002; Clayton and Opotow, 2003). Trusting someone occurs when there is a recognition that the person is similar to one's self and is, therefore, 'one of us'. Distrusting someone occurs when the person is identified as being dissimilar to one's self and is, therefore, 'one of them'.

In his classic best-selling book *The Nature of Prejudice*, social psychologist Gordon Allport (1954, p433) has described this issue thus:

> ... human beings are characterized by *obligatory interdependence*... For long-term survival, we must be willing to rely on others for information, aid and shared resources, and we must be willing to give information and aid and to share resources with others. At the individual level, potential benefits (receiving resources from others) and costs (giving resources to others) of mutual cooperation go hand in hand and set natural limits on cooperative interdependence. The decision to cooperate ... is a dilemma of trust since the ultimate benefits depend on everyone else's willingness to do the same. A cooperative system requires that trust dominate over distrust. But indiscriminate trust (or indiscriminate altruism) is not an effective individual strategy; altruism must be contingent on the probability that others will cooperate as well.
>
> Social differentiation and clear group boundaries provide one mechanism for achieving the benefits of cooperative interdependence without the risk of excessive costs... in-groups (are) bounded communities of mutual trust and obligation that delimit mutual interdependence and cooperation.

The SVS model identifies two important sets of trust-relevant social representations: *salient values* are the individual's representations of the goals and means that should be followed in responding to a problem. Salient value representations include implicit and explicit meanings, such as an understanding of what problem is being faced, what options are available and the likely consequences of options. *Value similarity* representations consist of comparing one's own salient values to those that are concluded to be salient for the person whose trustworthiness is being judged. If the other person's represented salient values are similar to one's own salient values, that individual will be deemed trustworthy, the risk of trusting is assumed and the person will be relied upon. If the other person's represented salient values are dissimilar to one's own salient values, that person will be deemed untrustworthy and the person will be distrusted. One frequently important salient value in human interactions is honesty about one's motives. Perceived efforts to conceal motives produce conclusions of salient value dissimilarity and distrust.

Figure 9.1 presents a general model of relationships between salient value similarity, trust and judgements related to cooperative risk management, such as the acceptability of hazardous technological activities and

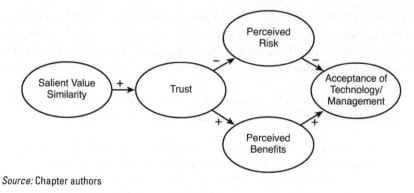

Source: Chapter authors

Figure 9.1 *Model of salient value similarities, trust and evaluations*

the acceptability of risk management policy. Studies examining risk management in a number of different domains provide support for various aspects of this general model of relationships. These include studies on genetic modified organisms (Siegrist, 1999); research on old growth forests (Cvetkovich et al, 1995); protection of threatened and endangered species (Cvetkovich and Winter, 2003); preservation of water quality (Cvetkovich and Winter, 1998); prevention and control of wildfires (Winter and Cvetkovich, 2002); users' fees to support costs of forest management (Winter et al, 1999); government and business responses to the Y2K computer bug (Ottaway et al, 2001); the perception of cancer clusters (Gutscher et al, 2001; Siegrist et al, 2001); and an assortment of other technologies and hazardous activities (Siegrist and Cvetkovich, 2000; Siegrist et al, 2000). Additionally, during the 2000 US presidential election campaign, both implicit and explicit pre-election candidate preferences correlated with trust, and trust correlated with shared value similarity between voters/respondents and candidates and to actual vote (Bain, 2001; Halfacre et al, 2001).

We draw two main conclusions from this research, reflected in Figure 9.1:

1 Trust is important to judgements about characteristics, such as risks and benefits, that relate to conclusions about the acceptability of both potentially hazardous activities and technologies and to management policies.
2 Judgements of trust are based on representations of salient value similarities.[4]

Research has begun to explore some of the complexities of how salient value similarities operate in cooperative risk management situations. These explorations suggest some additions to, and elaborations upon, the general model of trust and evaluations of management policies. We discuss two lines of research:

1   studies on the effects of perceived consistency between values and actions and the legitimacy of inconsistencies; and
2   studies on the effects of new information on judgements of trust.

In both cases, representations of what the risk manager did or did not do are examined relative to representations of how the manager's mind works and resulting trust of risk managers.

Value/action consistency and the legitimacy of inconsistencies

As part of our ongoing collaborative research, we conducted discussions with forest users whose objectives focused on managing threatened and endangered species (Cvetkovich and Winter, 2003). We examined the relationship between trust and management actions taken with California citizens who had a strong interest in forest management. Four patterns of relationships between trust, salient value similarity and perceptions of management actions were identified based on focus group discussions.

Two of the four identified patterns of representations of trust directly parallel expectations based on the SVS model. One of these patterns depicted the Forest Service as not sharing respondent's salient value that human use should be the dominant goal of forest management. Individuals with this pattern of representation distrusted the Forest Service (pattern 1 in Table 9.2). The other expected pattern represented the Forest Service as sharing the respondent's salient value of the primacy of species protection as the dominant goal of forest management (pattern 4 in Table 9.2). Individuals with this pattern of representation trusted the Forest Service.

The two unexpected patterns also represented the Forest Service as sharing the respondent's salient value of species protection as the dominant goal. These two patterns were also similar in that both represented the Forest Service as not always acting based on this shared value. However, people with these two patterns differed in another representation. Individuals with one of these patterns (pattern 3 in Table 9.2) further represented the Forest Service as not being responsible for failures to act on shared values. This representation held that the Forest Service failed to act in a value-consistent way because of budget and other resource constraints, political pressures or other extenuating circumstances. Individuals with this pattern of representations tended to trust the Forest Service. Individuals with the fourth pattern (pattern 2 in Table 9.2) further represented the Forest Service as being responsible for failing to act on shared values. This representation held that the Forest Service was responsible for allowing itself to be affected by influences such as budget and other resource constraints or political pressure. Individuals with this pattern of representations tended to distrust the Forest Service, but not as extremely as those who represented the Forest Service as not sharing their salient values.

Results of telephone interviews with larger representative samples of people living in Arizona, Colorado and New Mexico confirm these patterns of relationships between trust, perceived consistency of values and actions, and judged legitimacy of inconsistencies for managing threatened and endangered species and wildfires (Cvetkovich and Winter, 2004). As shown in Table 9.3, evaluating the effectiveness and approval of management practices for both issues was predicted by salient value similarity, perceived inconsistency between values and actions, followed by the judged legitimacy of inconsistencies. These relationships were consistent across states and genders. Together, the results of the in-depth small sample study and the larger representative sample surveys indicate an important possible addition to the basic SVS model of trust.

Table 9.2 *Four identified patterns of trust*

Pattern of trust	Forest Service shares management values, goals and views	Perceived consistency of Forest Service actions and own values	Legitimacy of Forest Service inconsistency
1 Distrust	No	Inconsistent	(Not relevant)
2 (Dis)trust somewhat	Yes	Inconsistent	Not justified
3 Trust	Yes	Inconsistent	Justified
4 Trust	Yes	Consistent	(Not relevant)

Table 9.3 *Regression analyses of evaluations (effectiveness and approval) of management practices for total samples, states and genders*

		Beta weights		Adj. R	F	p
	Salient value similarity	Forest Service value/action inconsistency	Legitimacy of Forest Service inconsistency			
		Threatened and endangered species survey				
Total sample	.428	−.274	.158	.54	682.20	< .0001
Arizona	.380	−.350	.137	.51	230.93	< .0001
Colorado	.395	−.315	.182	.59	169.11	< .0001
New Mexico	.537	−.294	.181	.63	203.57	< .0001
Males	.428	−.283	.160	.56	233.64	< .0001
Females	.427	−.293	.181	.58	240.12	< .0001
		Wildland and wilderness fires survey				
Total sample	.442	−.286	.169	.58	495.70	< .0001
Arizona	.470	−.198	.202	.54	144.90	< .0001
Colorado	.448	−.248	.144	.51	127.07	< .0001
New Mexico	.372	−.375	.118	.55	147.54	< .0001
Males	.411	−.264	.176	.53	213.25	< .0001
Females	.440	−.287	.134	.52	188.57	< .0001

New information and trust

Risk managers have a strong practical interest in increasing trust. There is more research in demonstrating the difficulty of doing this than there is in examining the processes involved in how trust develops and changes. Not surprisingly, the available research does not yield the sought-after simple answers to the question of how to increase trust. One of the conclusions uncritically repeated in published discussions is Deutsch's (1958) statement that trust is slow to develop. The asymmetry principle also addresses the presumed difficulty of gaining (and retaining) trust (Slovic, 1993). The asymmetrical principle states that it is much easier to lose trust than it is to gain it because negative information triggers a stronger emotional reaction, is more diagnostic and is more credible than positive information. The SVS model, grounded on assumptions about mind-reading processes, indicates that speed of assessing trustworthiness should depend upon the difficulty of reaching conclusions about the other person's salient value similarities. When this is very difficult, as it would be in the minimal social interaction situations studied by Deutsch, trust will be slow to develop. SVS suggests that changes in trust should depend upon existing level of trust, a possibility not suggested by the asymmetry principle.

Bob Woodworth's (2004) book *Plan of Attack* provides an example of how existing levels of trust focus on interpretations of new information. *Plan of Attack* has been described as 'akin to raw intelligence. It is a rough record ... equivalent of the kind of satellite photograph that cannot distinguish objects less than 1 foot in diameter. For this reason, and because neither analysis nor context ... is part of the Woodward package, the information he provides is open to multiple, and highly inconsistent, points of view' (Hertzberg, 2004). Existing trust provides the frame for representing actions depicted in the book. The nature of the resulting representations of actions leads to persistence of initial trust or distrust. The 'Re-elect George W. Bush' website includes *Plan of Attack* on its suggested reading list apparently because the book conveys the image of a strong, decisive leader. John Kerry, the Democratic contender for president, also recommends *Plan of Attack*, apparently because the book conveys an image of impulsive, ineffective leadership.

Two studies support the trust-as-an-interpretive-lens expectation of the SVS model (Cvetkovich et al, 2002). Reactions to positive and negative information about managers involved in two different risk domains – the operation of nuclear power plants, and food production and distribution – indicated what appeared to be a double asymmetry effect. Overall, bad news had a stronger effect on decreasing trust than good news had on increasing trust, as expected by the original asymmetry principle. However, as expected by the SVS model, good news about trusted managers increased trust more than good news about distrusted managers. Bad news about distrusted managers decreased trust more than bad news about trusted managers. Subsequent research replicates both effects and also indicates that what appears in Cvetkovich et al (2002) to be double asymmetry effects may be one effect (White et al, 2003). Bad news may have a stronger effect than good because of people's existing expectations. Bad news has a stronger effect than good news when people expect bad news. If people expect bad news, either about a particular distrusted person or, more generally, about a perceived risky activity, hearing this bad news will have a more profound effect than hearing good news. Bad news about distrusted individuals and good news about trusted individuals is expected as more likely than the reverse. Trust may change the diagnostic value of news by influencing the expectation.

THE 'WHEN' OF TRUST

Most of us who are interested in cooperative risk management have come to see trust as an important determinant of risk perceptions (Slovic, 1999). It has been a rare voice that questions the propositions that:

- It is difficult for people, in general, to directly know about and manage large-scale risks.
- Therefore, trust of those with management responsibilities strongly affects perceptions of risk, safety, benefits and other characteristics of potentially hazardous activities and evaluations of risk management (compare Sjöberg, 1999).

Evidence from SVS studies, reviewed in the previous section, adds to other research that supports this notion.

But trust is not always important. As already noted, and as demonstrated by Yamagishi's research, trust is one of two routes towards reliance. Trust may not be as relevant to perceptions of risk when strong relational assurances of reliable behaviour exist. Institutionalized accountability – a relational assurance – might increase reliance and cooperation (see Frewer et al, 1998).

Trust, and distrust, should also *de*crease in relevance as the need to rely on risk managers *de*creases. The already-noted studies on the relative importance of credibility seem to support this conclusion. As expected, strong correlations between social trust and judged risks and benefits have been found for hazards about which people did not possess much knowledge (Siegrist and Cvetkovich, 2000). No significant correlations between social trust and judged risks and benefits occurred for hazards about which people were knowledgeable.

Nevertheless, the results of our telephone surveys on the forest management policies of representative residents of Arizona, Colorado and New Mexico, as well as California, provide mixed evidence about the importance of self-assessed knowledge (Cvetkovich and Winter, 2004). Evaluations of the effectiveness and acceptability of forest practices to protect species indicate that trust of the Forest Service is more important for those with lower self-assessed knowledge than for those with higher assessed knowledge. But the difference does not seem to be as strong as that reported in the earlier study. Trust was a significant predictor at both high and low levels of self-assessed knowledge of evaluations of the wildfire management practices of mechanical interventions, bans/closures and signs/restrictions on use (Winter and Cvetkovich, 2002). For other wildfire management practices, the amount of variance in evaluations explained by trust differed according to levels of self-assessed knowledge, although not always in the expected direction. Levels of self-assessed knowledge may be an important consideration in determining the role of trust in public attitudes and perceptions. However, these results indicate that its actual weight may sometimes be insignificant, and that its specific role is somewhat unclear. Even a very knowledgeable individual has little direct ability to mitigate many important hazards.

Discussion

What is trust?

Our discussion of the 'what' of trust makes two suggestions: the differentiation of simultaneous modes of processing that contribute to the products of information processing, and differentiation between the two paths of reliance – trust and assurance. In both cases, we believe that there is multiple-level evidence to indicate that these distinctions are valid and useful. Here we deal with a few of the issues and implications of these distinctions.

Modes of information processing

As we collectively build on the distinction of modes of information processing in developing our understandings of trust and cooperative risk management, our conceptualizations should reflect the mutually dependent nature of the two modes of processing. We should avoid thinking of these as two separate dimensions or states.

Significant research measurement and design questions are raised by dual-mode processing. Frequently used paper-and-pencil self-report measures alone may not yield valid measures or produce comprehensive understandings of the mutual functioning of both modes. There are increasing examples of research that uses various qualitative data collection methods (see Walker et al, 1999; Earle, 2004; Chapter 5 in this book; our focus group study mentioned earlier). A combination of types of research across studies, if not in the same study, is required not only to better understand dual-mode information processing, but other aspects of trust, as well.

What practical implications for cooperative risk management in democratic societies do we take from our recognition of dual-mode processing? Paul Slovic and colleagues (Slovic, 1999; Slovic et al, 2004) have begun exploring some of the possibilities. In the past, we often assumed that people should be operating in the explicit mode and, therefore, would be responsive directly to scientific and technical information. As we become increasingly aware of the primary importance of implicit processes, what importance should emotions and heuristic-based judgements have in cooperative risk management? Rejection of the importance of emotion has been referred to as 'Descartes' error' (Damasio, 1994). We should take care not to commit the opposite of 'Descartes' error' (the 'New Age' error?) and allow emotion to become the trump card in our discourses on cooperative risk management. Recognizing the role of both emotion (specifically through personal values and importance of things to self) and fact-based alternatives seems to lead to better collaborative decisions when the public is involved (Arvai et al, 2001).

An interesting and important avenue of future exploration is the examination of the possibility that trust and distrust differ in the balance of implicit and explicit information processing. The agreement underlying trust may induce less active explicit information processing than does the disagreement underlying distrust. Earle (2004) found that the number of comments made while talking aloud about a statement on global climate change, an indication of conscious thought, was negatively related to the level of trust felt for the person making the statement. Individuals described by Pidgeon, Woortinga and Walls (see Chapter 5 in this book) as having 'critical trust' seem to be in an active explicit mode of information processing.

Outside the normal range of psychological functioning, extreme distrust (paranoia) seems to involve both considerable amounts of explicit processing as well as 'hot' emotional implicit processing. Paranoid representations are characterized by an active suspicion that other people not only have ill intentions, but that they are attempting to conceal these intentions (Kramer, 1998, 1999a, 1999b; Kramer and Messick, 1998; Kramer and Wei, 1999). Active suspicion engenders a form of explicit processing that Kramer (1998) refers to as a 'ruminative mode of social interaction'. This information processing is characterized by hypervigilance and 'dysphoric self-consciousness'. Paranoid attributions about other people are self-conscious in that they primarily focus on implications for the person making the attribution. Self-referent content is continuously read into others' thoughts and behaviour: 'Other people are thinking about me and they have hostile intentions.' The attributions are dysphoric in that they produce uncomfortable, often extreme, negative emotional reactions.

By definition, paranoid beliefs are irrational in the sense that the underlying suspicion is not based on evidence. Most of us at some time have irrational thoughts about others' intentions towards us. An important distinction between pathological paranoia and normal paranoia is flexibility. While attributions of distrust, and trust, tend to persist (see the earlier section on 'The "how" of trust: Social psychological processes involved in trusting') most of us do not continue to persist in our irrational suspicions (Kramer, 1998). Investigations of flexibility comparing pathological and normal suspiciousness could yield important practical and conceptual results.

Assurance and trust
Differentiating between the two routes of reliance should promote the development of better conceptualizations and increase the likelihood that applied efforts will be effective. Confusing the routes of reliance leads to using the same terms for processes with different meanings on the functional level.

We think that a strong argument can be made for the theoretical and practical advantages of better understanding the joint operation of relational assurances and trust. While there are active areas of research and conceptualization on relational assurances – for example, on procedural justice – those of us interested in cooperative risk management have not given as much attention to assurance as perhaps we should. The focus of reliance research has been mostly on trust (at least in name). In many studies, trust has been used interchangeably with confidence, overlooking a possible difference between the two. This is one of the consequences of not distinguishing between the two routes of reliance. We discuss some possible ways in which assurance and trust influence each other below in the section on 'The "when" of trust'.

Recognizing the distinction between routes to reliance should lead to an increased focus on relational assurances. We suggest that it is important to redress this imbalance. Further support for greater attention to relational assurances is provided by Haslam and Fiske, who have concluded that mind reading, such as that involved in trust, may not play the dominant role in everyday social cognition. Mind reading 'is only one part of the social cognitive apparatus that underlies social expertise, and is not obviously pre-eminent among the different parts. Shared understandings of relationships ... are distinct and demonstrably important in everyday social cognition' (Haslam and Fiske, 2004, p3).

The 'how' of trust

The 'what' of trust section presents our framework of conceptual terms. Within this framework, our examination of the particulars of trust, the 'how' and the 'when,' uses the salient values similarity model. We have reviewed the results of research using the SVS model as an illustration of one direction that can be taken to reduce the conceptual jumble that the field of trust and cooperative risk management finds itself in.

The SVS-based studies reviewed are illustrative of an identified trend in trust research. Earle (2004) points out that 'the overall trend in studies of trust in risk management over the past decade has been from the general and abstract to the specific and concrete'. Earlier studies assumed that trust is based on universal normative characteristics, such as 'objectivity' and 'fairness'. More recent studies, including the SVS-based studies discussed, recognizing the importance of social identity, have focused on the influences of context-specific characteristics.

The two lines of SVS research discussed focus on representations of the actions of risk managers. One of these lines of research indicates that the similarity of risk managers' values is the basis of trust, and demonstrates the importance of perceived value/action consistency and the judged legitimacy of inconsistency. Perceived value/action inconsistency and judged illegitimacy of inconsistency affect SVS-based attributions about the mind of risk managers by increasing suspicions of motives and morals, thereby decreasing trust. These findings suggest that risk managers should attempt to act consistently with shared values and to make efforts, when possible, to explain apparent inconsistencies.

The other line of research suggests that existing levels of trust affect the representation of new information about a risk manager. The representation, in turn, affects the amount of change in trust. Trust leads to an expectation about risk managers' actions. Actions that are consistent with expectations (trust based or otherwise) have a bigger effect than expectancy-inconsistent actions.

There are important limits on these conclusions. Measured effects in the reviewed studies are immediate, not long term, are self-report questionnaire based, and are often on topics not personally relevant to respondents. Distinctions between trust and confidence are not always made in these enquiries, and may be blended for our respondents. These studies on change focus on process, not static states. An important unanswered question about change is 'Under what circumstances do accumulated inconstancies become sufficient to overcome the inertia of expectancy/trust persistence?' Again, there is a suggested need for multiple innovative research methods and designs that move away from one-time research snapshots of trust.

The 'when' of trust

Trust is an important influence on perceptions of risk and evaluations of risk management policies, but not always. Efforts to identify when trust is important are needed. In this section, we consider two influences: self-assessed knowledge and existing assurances.

Self-assessed knowledge
We have reviewed evidence indicating that risk perceptions and risk management evaluations are more strongly influenced by trust for individuals who assess their personal knowledge as low than for individuals who assess their knowledge as high. Since reliance on trust does not always

decrease with increases in self-assessed knowledge, issue-specific characteristics may have an influence. One possibility that should be explored is the individual's assessment of the functionality of their own level of knowledge. A person may assess his or her own knowledge as high, but not as high as that of the risk managers. Or a person may assess their own knowledge as high, but have concluded that it makes little difference to effective risk management. The nature of the issue, the operation of the political system, or some other factor may 'disconnect' own knowledge from effective actions.

We have measured self-assessed knowledge rather than using objective knowledge tests for two reasons. The first reason is the difficulty of creating an objective measure of knowledge of risk management issues that is manageably short and comprehensive, measures functional knowledge, not esoterica, and is acceptable to 'experts'. On this last point, we have discovered no surer way of inducing critical questions and, at times, heated exchanges at professional meetings than by presenting a risk management knowledge test. The second and more significant reason for measuring self-assessed knowledge is an assumption of the psychological immediacy of self-assessed knowledge. We assume that a person's perception of who she or he is, including assessments of topical knowledge, more immediately relates to psychological functioning than does a measure of knowledge according to an external standard.

Having recognized the bases for our focus on self-assessed knowledge, we hasten to add that our identification of the trust-relevant constituents of social representations includes representations of knowledge as well as attitudes, values and other evaluations. We also recognize that in the context of cooperative risk management, there is often a need for understanding the content of people's knowledge. Simply to know that people think they are very knowledgeable or not, without knowing what the content of their understanding is, is often insufficient. Breakwell (2001) has recently integrated a mental models approach in an analysis of the social identity processes influencing the social representations of health and environmental hazards. She states that 'The particular value of the mental models approach for risk communication is that it requires one to think in terms of a complex interacting system of beliefs which underpins risk appreciation' (Breakwell, 2001, p342). The mental models approach offers suggestions for solving at least some of the measurement problems noted above, including comparisons of the understandings of the public with those of risk managers and others (Atman et al, 1994; Bostrom et al, 1994; Morgan et al, 2001; Niewöhner et al, 2004). We suggest that future research on knowledge and trust includes efforts that combine a focus on social representations and mental models.

Trust and relational assurances
One interpretation of Yamagishi's cross-national research (Yamagishi et al,

1998a; Yamagishi et al, 1999) is that reliance does not always require trust. The Japanese evidence seems to indicate that reliance can be based solely on well-established social relationships or, by extension, other assurances of dependability. In contrast to Americans, Japanese seemingly shun reliance based on (hastily) arrived-at representations of another person's character. This Japanese tendency may be shared by others from collectivistic societies who are much more likely than Americans to attribute the causes of behaviour to social context than to individual characteristics (Miller, 1984; Lee et al, 1996).

Understanding when trust is important requires understanding the range and frequency of occurrence of the possible range of relationships between trust and assurances. At times, as is indicated by the Yamagishi example, a relational assurance is the abiding route to reliance with trust serving, at most, as a fine-tuning function. In other cases, we might rely on others only based on trust without relational assurances, operating like a member of a high-flying trapeze team working without a safety net. As a third possibility, trust and assurance can be convergent routes to reliance. Reliance occurs both because the other individual is trusted and because there are convincing relational assurances present. In this case, there may be trade-offs between trust and assurances where the same level of reliance might be produced by various combinations of the two. The existence of both trust and assurances also raises questions of how the two affect each other. For example, might strong assurances result in a conclusion that a person is less trustworthy than if weaker, or no, assurances are present (compare Frewer et al, 1998)?

The relationship between trust, relational assurances, reliance and behaviour is in need of continuing exploration. The number of unanswered questions emerging in our discussion supports this. The continuing questions of risk managers focused on how to build, maintain and restore warranted trust also evidences this.

NOTES

1 We wish to thank Michael Siegrist, Ann Bostrom and the other participants in the Zurich Trust and Risk Management Conference 2003 for comments on earlier versions of this chapter. The preparation of this chapter was supported by a collaborative project of the US Department of Agriculture, Forest Service, Pacific Southwest Research Station and Western Washington University.
2 Some selected references are Bargh (1997); Chaiken and Trope (1999); Chen et al (1999); Epstein (1994); Hammond (1996); Reyna and Adam (2003); Sloman (1996); Smith and DeCoster (2000); Winston et al (2002).
3 Thomas Hobbes recognized trust as a social emotion during the 17th century (see Hobbes, 1999). We will not provide extensive documentation for the claim that trust is a social emotion. Trust does share characteristics with other social emotions. In addition to preparing us in a holistic way to respond to another, it

often occurs rapidly, and we often cannot specifically explain why we trust someone except in very general terms.

4   It is important to emphasize that our conceptualization is that salient value similarity affects levels of trust and that levels of trust affect evaluations of management policies and the acceptability of technologies. Salient value similarity may, in some (but not all) cases, directly affect evaluations; but this relationship is not part of our model. The finding that salient values also directly influence evaluation does not invalidate the model – compare Poortinga and Pidgeon (2003) and Chapter 5 in this book.

REFERENCES

Allport, G. W. (1954) *The Nature of Prejudice*, Addison-Wesley, Reading, MA

Arvai, J. L., Gregory, R. and McDaniels, T. L. (2001) 'Testing a structured decision approach: Value-focused thinking for deliberate risk communication', *Risk Analysis*, vol 21, no 6, pp1065–1076

Atman, C. J., Bostrom, A. and Morgan, M. G. (1994) 'Designing risk communications and correcting mental models of hazardous processes, Part 1', *Risk Analysis*, vol 14, no 5, pp779–787

Bain, T. W. (2001) *Implicit and Explicit Political and Racial Attitudes, in Relation with Voting Behaviour*, Western Washington University, Bellingham, WA

Bargh, J. A. (1997) 'The automaticity of everyday life', in R. S. Wyer (ed) *The Automaticity of Everyday Life: Advances in Social Cognition*, Lawrence Erlbaum Associates, Mahwah, NJ

Bechara, A., Tranel, D. and Damasio, H. (2000) 'Characterization of the decision-making deficit of patients with ventromedial prefrontal cortex lesions', *Brain*, vol 123, no 11, pp2189–2202

Bostrom, A., Atman, C. J., Fischhoff, B. and Morgan, M. G. (1994) 'Evaluating risk communications: Completing and correcting mental models of hazardous processes, Part 2', *Risk Analysis*, vol 14, no 5, pp789–798

Breakwell, G. M. (1986) *Coping with Threatened Identities*, Methuen, London and New York

Breakwell, G. M. (2001) 'Mental models and social representations of hazards: The significance of identity processes', *Journal of Risk Research*, vol 4, no 4, pp341–351

Cacioppo, J. T., Petty, R. E., Feinstein, J. A. and Jarvis, W. B. G. (1996) 'Dispositional differences in cognitive motivation: The life and times of individuals varying in need for cognition', *Psychological Bulletin*, vol 119, pp197–253

Chaiken, S. (1980) 'Heuristic versus systematic information processing and the use of source versus message cues', *Journal about Personality and Social Psychology*, vol 39, pp752–766

Chaiken, S. and Trope, Y. (1999) *Dual-Process Theories in Social Psychology*, Guilford Press, New York

Chen, S., Duckworth, K. and Chaiken, S. (1999) 'Motivated heuristic and systematic processing', *Psychological Inquiry*, vol 10, no 1, pp44–49

Clayton, S. and Opotow, S. (2003) *Identity and the Natural Environment: The Psychological Significance of Nature*, MIT Press, Cambridge, MA

Cvetkovich, G. T. and Löfstedt, R. E. (eds) (1999) *Social Trust and the Management of Risk*, Earthscan, London

Cvetkovich, G. T. and Winter, P. L. (1998) *Community Reactions to Water Quality Problems in the Colville National Forest: Final Report*, Western Institute for Social Research; Department of Psychology, Bellingham, WA

Cvetkovich, G. T. and Winter, P. L. (2003) 'Trust and social representations of the management of threatened and endangered species', *Environment and Behaviour*, vol 35, no 2, pp286–303

Cvetkovich, G. T. and Winter, P. L. (2004) *Seeing Eye-to-Eye on Natural Resource Management: Trust, Value Similarity, and Action Consistency/Justification*, Paper presented at the 4th Social Aspects and Recreation Research Symposium: Linking People to the Outdoors: Connections for Healthy Lands, People and Communities, Presidio of San Francisco, Golden Gate National Recreation Area, San Francisco, CA, 4–6 February

Cvetkovich, G. T., Winter, P. L. and Earle, T. C. (1995) *Everybody is Talking About It: Public Participation in Forest Management*, Paper presented at the American Psychological Association, New York, 11–14 August

Cvetkovich, G. T., Siegrist, M., Murray, R. and Tragesser, S. (2002) 'New information and social trust: Asymmetry and perseverance of attributions about hazard managers', *Risk Analysis*, vol 22, no 2, pp359–367

Damasio, A. (1994) *Descartes' error: Emotion, Reason and the Human Brain*, Grouset/Putnam, New York

Deutsch, M. (1958) 'Trust and suspicion', *Journal of Conflict Resolution*, vol 2, pp265–279

Earle, T. C. (2004) 'Thinking aloud about trust: A protocol analysis of trust in risk management', *Risk Analysis*, vol 24, no 1, pp169–183

Earle, T. C., Cvetkovich, G. T. and Slovic, P. (1990) 'The effects of involvement, relevance and ability on risk communication effectiveness', in K. Borcherding, O. I. Larichev and D. M. Messick (eds), *Contemporary Issues in Decision Making*, Elsevier Science Publishers, Amsterdam

Earle, T. C., Siegrist, M. and Gutscher, H. (2000) *Trust and Confidence: A Dual-Mode Model of Cooperation*, Western Washington University, Bellingham, WA

Epstein, S. (1994) 'Integration of the cognitive and the psychodynamic unconscious', *American Psychologist*, vol 49, pp709–724

Finucane, M. L., Alhakami, A., Slovic, P. and Johnson, S. J. (2000) 'The affect heuristic in judgements of risks and benefits', *Journal of Behavioural Decision Making*, vol 13, pp1–17

Fischhoff, B. (1999) 'If trust is so good, why isn't there more of it?', in G. T. Cvetkovich and R. E. Löfstedt (eds) *Social Trust in the Management of Risk*, Earthscan, London, ppviii–x)

Frewer, L. J., Howard, C., Hedderley, D. and Shepherd, R. (1998) 'Methodological approaches to assessing risk perceptions associated with food-related hazards', *Risk Analysis*, vol 18, no 1, pp95–102

Fukuyama, F. (1996) *Trust: The Social Virtues and the Creation of Prosperity*, Free Press, New York

Gray, J. R. (2004) 'Integration of emotion and cognitive control', *Current Directions in Psychological Science*, vol 13, no 2, pp46–48

Gutscher, H., Siegrist, M. and Cvetkovich, G. T. (2001) *Social Trust and the Perception of Cancer Clusters*, Paper presented at the Society for Risk Analysis,

Seattle, WA, 3–5 December

Halfacre, A. C., Sattler, D. N., Khan, S. and Cvetkovich, G. T. (2001) *The 2000 Presidential Debates: Effects of Post-Debate Coverage on Voters' Perceptions of the Candidates*, Paper presented at the American Political Science Association, San Francisco

Hammond, K. R. (1996) *Human Judgement and Social Policy: Irreducible Uncertainty, Inevitable Error, Unavoidable Injustice*, Oxford University Press, New York

Hardin, R. (1993) 'The street-level epistemology of trust', *Politics and Society*, vol 21, pp505–529

Haslam, N. and Fiske, A. P. (2004) 'Social expertise: Theory of mind or theory of relationships?', in N. Haslam (ed) *Relational Models Theory: A Contemporary Overview*, Erlbaum, Mahwah, NJ

Hertzberg, H. (2004) 'In the soup: Bob Woodward's *Plan of Attack*', *New Yorker*, 10 May, pp98–102

Hobbes, T. (1999) *Human Nature and De Corpore Politico*, Oxford University Press, Oxford/New York

Kramer, R. M. (1998) 'Paranoid cognition in social systems: Thinking and acting in the shadow of doubt', *Personality and Social Psychology Review*, vol 2, no 4, pp251–275

Kramer, R. M. (1999a) 'Social uncertainty and collective paranoia in knowledge communities: Thinking and acting in the shadow of doubt', in L. L. Thompson and J. M. Levine (eds) *Shared Cognition in Organizations: The Management of Knowledge*, Lawrence Erlbaum Associates, Mahwah, NJ, pp163–191

Kramer, R. M. (1999b) 'Stalking the sinister attribution error: Paranoia inside the lab and out', in R. J. Bies and R. J. Lewicki (eds) *Research in Negotiation in Organizations*, vol 7, Jai Press, Stamford, CT, pp59–91

Kramer, R. M. and Messick, D. M. (1998) 'Getting by with a little help from our enemies: Collective paranoia and its role in intergroup relations', in C. Sedikides and J. Schopler (eds) *Intergroup Cognition and Intergroup Behaviour*, Lawrence Erlbaum Associates, Mahwah, NJ, pp233–255

Kramer, R. M. and Wei, J. (1999) 'Social uncertainty and the problem of trust in social groups: The social self in doubt', in T. R. Tyler and R. M. Kramer (eds) *The Psychology of the Social Self*, Lawrence Erlbaum Associates, Mahwah, NJ, pp145–168

Lee, F., Hallahan, M. and Herzog, T. (1996) 'Explaining real life events: How culture and domain shape attributions', *Personality and Social Psychology Bulletin*, vol 22, pp732–741

Luhmann, N. (1979) *Trust and Power: Two Works by Niklas Luhmann*, John Wiley and Sons, Chichester, UK

Miller, J. G. (1984) 'Culture and the development of everyday social explanation', *Journal of Personality and Social Psychology*, vol 46, pp961–978

Morgan, M. G., Fischhoff, B., Bostrom, A. and Atman, C. J. (2001) *Risk Communication: A Mental Models Approach*, Cambridge University, Cambridge, MA

Moscovici, S. (2001) 'Why a theory of social representations?', in K. Deaux and G. Philogene (eds) *Representations of the Social: Bridging Theoretical Traditions*, Blackwell Publishers, Malden, MA, pp8–35

Nesse, R. M. (1990) 'The evolutionary explanations of emotions', *Human Nature*, vol 1, pp261–289

Niewöhner, J., Cox, P., Gerrard, S. and Pidgeon, N. F. (2004) 'Evaluating the efficacy of the mental models approach for improving occupational chemical risk protection', *Risk Analysis*, vol 24, *no* 2, pp349–362

Ottaway, S., Siegrist, M. and Cvetkovich, G. T. (2001) *The Y2K Bug: Shared Values, Trust and Perceptions of Government and Business*, Paper presented at the Society for Risk Analysis, Seattle, WA, 3–5 December

Petty, R. E. and Cacioppo, J. T. (1986) *Communication and Persuasion: Central and Peripheral Routes to Attitude Change*, Springer-Verlag, New York

Petty, R. E., Cacioppo, J. T. and Goldman, R. (1981) 'Personal involvement as a determinant of argument-based persuasion', *Journal of Personality and Social Psychology*, vol 41, pp847–855

Petty, R. E., Cacioppo, J. T., Strathman, A. J. and Priester, J. R. (1994) 'To think or not to think: Exploring two routes to persuasion', in S. Shavitt and T. C. Brock (eds) *Persuasion: Psychological insights and perspectives*, Allyn and Bacon, Needham Heights, MA, pp113–147

Poortinga, W. and Pidgeon, N. F. (2003) 'Exploring the dimensionality of trust in risk regulation', *Risk Analysis*, vol 23, no 5, pp961–972

Putnam, R. D. (2000) *Bowling Alone: The Collapse and Revival of American Community*, Simon and Schuster, New York

Reyna, V. F. and Adam, M. B. (2003) 'Fuzzy-traced theory, risk communication and product labeling in sexually transmitted diseases', *Risk Analysis*, vol 23, no 2, pp325–342

Satterfield, T. (2002) *Anatomy of a Conflict: Identity, Knowledge and Emotion in Old-Growth Forests*, University of British Columbia Press, Vancouver, BC

Siegrist, M. (1999) 'A causal model explaining the perception and acceptance of gene technology', *Journal of Applied Social Psychology*, vol 29, no 10, pp2093–2106

Siegrist, M. and Cvetkovich, G. T. (2000) 'Perception of hazards: The role of social trust and knowledge', *Risk Analysis*, vol 20, no 5, pp713–720

Siegrist, M., Cvetkovich, G. T. and Roth, C. (2000) 'Salient values similarity, social trust and risk/benefit perception', *Risk Analysis*, vol 20, no 3, pp353–362

Siegrist, M., Cvetkovich, G. T. and Gutscher, H. (2001) 'Shared values, social trust and the perception of geographic cancer clusters', *Risk Analysis*, vol 21, no 6, pp1047–1054

Siegrist, M., Earle, T. C. and Gutscher, H. (2002) 'Test of a trust and confidence model in the applied context of electromagnetic field (EMF) risks', *Risk Analysis*, vol 23, no 4, pp705–716

Sjöberg, L. (1999) 'Perceived competence and motivation in industry and government as factors in risk perception', in G. T. Cvetkovich and R. E. Löfstedt (eds) *Social Trust and the Management of Risk*, Earthscan, London, pp89–99

Sloman, S. A. (1996) 'The empirical case for two systems of reasoning', *Psychological Bulletin*, vol 119, no 1, pp3–22

Slovic, P. (1993) 'Perceived risk, trust and democracy', *Risk Analysis*, vol 13, pp675–682

Slovic, P. (1999) 'Trust, emotion, sex, politics and science: Surveying the risk-assessment battlefield', *Risk Analysis*, vol 19, no 4, pp689–701

Slovic, P., Finucane, M. L., Peters, E. and MacGregor, D. G. (2002) 'The affect heuristic', in T. Gilovich, D. Griffin and D. Kahneman (eds) *Intuitive judgement: Heuristics and Biases*, Cambridge University Press, Cambridge

Slovic, P., Finucane, M. L., Peters, E. and MacGregor, D. G. (2004) 'Risk as analysis and risk as feelings: Some thoughts about affect, reason, risk and rationality', *Risk Analysis*, vol 24, no 2, pp311–322

Smith, E. R. and DeCoster, J. (2000) 'Dual-process models in social and cognitive psychology: Conceptual integration and links to underlying memory systems', *Personality and Social Psychology Review*, vol 4, no 2, pp108–131

Trumbo, C. W. (1999) 'Heuristic-systematic information processing and risk judgement', *Risk Analysis*, vol 19, no 3, pp391–400

Verplanken, B. (1991) 'Persuasive communication of risk information: A test of cue versus message processing effects in a field experiment', *Personality and Social Psychology Bulletin*, vol 17, no 2, pp188–193

Walker, G., Simmons, P., Irwin, A. and Wynne, B. (1999) 'Risk communication, public participation and the Seveso II directive', *Journal of Hazardous Materials*, vol 65, pp179–190

Weidenbacher, D. and Cvetkovich, G. T. (2003) *Assurance and Trust on eBay: Two Routes to Buyers' Reliance on Sellers*, Paper presented at the Judgement and Decision Making Society, Vancouver, British Columbia, 9–10 November

White, M. P., Pahl, S., Buehner, M. and Haye, A. (2003) 'Trust in risky messages: The role of prior attitudes', *Risk Analysis*, vol 23, no 4, pp717–726

Winston, J. S., Strange, B. A., O'Doherty, J. and Dolan, R. J. (2002) 'Automatic and intentional brain responses during evaluation of trustworthiness of faces', *Nature Neuroscience*, vol 5, no 3, pp277–283

Winter, P. L. and Cvetkovich, G. T. (2002) *The Role of Trust, Knowledge, Concern and Sociodemographic Characteristics in the Prediction of Californian's Reactions To Fire Management*, Paper presented at the Association of Fire Ecology: Managing Fire and Fuels in the Remaining Wildlands and Open Spaces of the Southwestern United States, San Diego, CA, 2–5 December

Winter, P. L., Palucki, L. J. and Burkhardt, R. L. (1999) 'Anticipated responses to a fee program: The key is trust', *Journal of Leisure Research*, vol 31, no 3, pp207–226

Woodward, B. (2004) *Plan of Attack*, Simon and Schuster Adult Publishing Group, New York

Yamagishi, T. and Yamagishi, M. (1994) 'Trust and commitment in the United States and Japan', *Motivation and Emotion. Special Issue: Trust and Distrust: Psychological and Social Dimensions*, vol 18, no 2, pp129–166

Yamagishi, T., Cook, K. S. and Watabe, M. (1998a) 'Uncertainty, trust and commitment formation in the United States and Japan', *American Journal of Sociology*, vol 104, no 1, pp165–194

Yamagishi, T., Jin, N. and Miller, A. S. (1998b) 'In-group bias and culture of collectivism', *Asian Journal of Social Psychology*, vol 1, no 3, pp315–328

Yamagishi, T., Kikuchi, M. and Kosugi, M. (1999) 'Trust, gullibility and social intelligence', *Asian Journal of Social Psychology. Special Issue: Theoretical and Methodological Advances in Social Psychology*, vol 2, no 1, pp145–161

10 Getting Out of the Swamp: Toward Understanding Sources of Local Officials' Trust in Wetlands Management[1]

Branden B. Johnson

INTRODUCTION

Risk analysts have increasingly focused on sources of trust in institutional risk managers because trust seems critical to how people perceive hazards (although see Poortinga and Pidgeon, 2005), and risk managers want their messages about risk magnitudes and risk management actions to be credible. Much confusion persists about such sources since few studies have explicitly tested alternative hypotheses and determining how to do so properly is difficult (Johnson, 1999). Yet, understanding bases for trust could be critical in developing effective strategies for cooperative risk management (see, for example, Siegrist et al, 2003, on crisis management).

Early attempts (for example, Earle and Cvetkovich, 1995; Peters et al, 1997; Metlay, 1999) were illuminating, but did not cover all hypothesized sources of trust. For example, while the standard assumption is that performance information (for instance, on the trust target's competence or procedures) drives trust judgements, Earle and Cvetkovich (1995) argued that a sense of shared values is a heuristic cue when people are unwilling or unable to obtain such information. These ideas had not been compared directly (Siegrist et al, 2003). Of three chapters on trust sources in this volume (including Chapter 7), Chapter 5 (see also Poortinga and Pidgeon, 2003) and this one are among the first to conduct that direct comparison.

This chapter illustrates some conceptual and methodological issues with data from a survey of local officials on wetlands management in New Jersey. Its main purpose was not hypothesis-testing about trust sources, but provision of practical information for activist and government use. Thus, the survey instrument could include only a few trust measures and less-than-ideal research designs for the questions discussed here.

However, the resulting data provoke questions worth more systematic investigation about the role of 'familiarity' and various trust-target attributes in trust judgements, and how these might vary across types of judges and trust targets.

BACKGROUND

Determining sources of trust

The most common approach to studying sources of trust has been attribute rating. People are asked to rate a single institution on multiple criteria derived from competing theories of foundations for trust, which might include various measures of competence, care or consensual values (to use summary descriptive terms proposed by Johnson, 1999). These ratings are factor analysed, and then trust ratings of the institutions are regressed upon the criteria. Metlay (1999) used this approach (excluding consensual values) on trust of the US Department of Energy (DOE) among representatives of non-federal and non-contractor organizations, 'many' being local government officials (Metlay, 1999, p115). As stakeholders in DOE site clean-ups and operations, they were more interested and probably more informed than the average citizen. The unadjusted R-square for explaining trust in DOE was large (60 per cent and 46 per cent in 1992 and 1994 surveys, respectively). Poortinga and Pidgeon (2003) used a random sample of the British public to judge trust in a more abstract entity ('the British government') for managing nuclear waste, genetically modified (GM) food, radiation from mobile telephones, genetic testing and climate change. Adjusted R-square ranged from 18 to 53 per cent. Value similarity played a minor role, but was a statistically significant predictor for issues that were highly polarized (GM food) or 'not fully developed' (mobile phone radiation; genetic testing). Both studies produced factors that combined apparently inconsistent measures (for example, competence and care), and that were inconsistent across the two studies, perhaps in part due to use of specific versus abstract trust targets (Poortinga and Pidgeon, 2003) or different judges.

The methodological assumption is that people who believe that the trust target has positive attributes will rate it as more trustworthy, while those who believe it is low on positive attributes or high on negative ones will express less trust. While plausible, cognitive processes underlying the assumption have not been demonstrated. Earle and Cvetkovich (1995) criticized the notion of systematic assessment of attributes of trust targets as a basis for trust judgements. Instead, they argued, people will tend to judge trust heuristically, specifically on the basis of 'salient value similarity' (SVS): if the target is thought to share our values (or at least compatible ones), we will trust it more than if it is thought to have different values. Earle, Cvetkovich and colleagues (see Earle and Cvetkovich,

1997; Siegrist et al, 2003) have used a mix of attribute-rating tasks (analysed with structural equation modelling) and experiments to explore this hypothesis.

Familiarity

A dual-process model of trust based on Earle and Cvetkovich (1995) posits that, for most people, trust judgements will stem from the heuristic of shared values (and honesty, caring and fairness, according to Siegrist et al, 2003, p707). However, if people know about or are familiar with a hazard or trust target, the model posits that competence and other attributes deemed diagnostic under systematic information processing would play a more important role (Siegrist et al, 2003).[2] The role of familiarity is not limited to the dual-process model:

> It may well be that the (trust) level evoked in any particular study will vary as a function of familiarity with the institution being judged. In particular, the more specific the evaluated subject, the more likely it is that someone will have more differentiated views. (Poortinga and Pidgeon, 2003, p969)

No study has yet tested whether trust varies by subjects' knowledge or familiarity with the target.

At least two kinds of 'familiarity' are applicable to the topic discussed here: wetlands management in New Jersey. First, perceived knowledge and objective knowledge about wetlands or wetlands management are pertinent. How much one knows or thinks one knows about the topic could affect one's trust judgements or the criteria for making trust judgements. Second, one could vary the degree to which trust judges work with or otherwise have contact with the entity judged for trustworthiness. Such contact, shaped by both resources and motivation (Verba et al, 1995), can affect beliefs and attitudes about the trust target. At one extreme people might have never, or only, heard of the person or institution; at the other extreme, they might be or have been in the trust target's role themselves. At this end of the spectrum, of course, 'familiarity' may become conflated with one's identity and promote reports of trust to avoid cognitive dissonance. But at any magnitude, neither knowledge nor role association are necessarily accurate about the prudence of trust or the nature of the judged entity. In this study, local officials of various types are the trust judges (that is, they rate trust and trust sources). Some trust targets (entities judged for trustworthiness in wetlands management) are ones in which judges play a role or have in the past, as with local elected officials and planning board members judging trustworthiness of local elected officials and the local planning board. Others are less familiar (local zoning board and environmental commission members judging the two local entities; all subjects judging the state environmental agency).

If 'experience or evidence' are infallible guides to belief that the trust target will behave as one expects, good or bad (Siegrist et al, 2003, p706), why should self-reported knowledge be included with objective knowledge measures? Trust (see Siegrist et al, 2003, p707, on trust as affect), value consensus and other attributes of trust targets are all personal judgements or 'perceptions' in the trust–risk literature. It is not whether the target *does* share one's values that affects trust, only whether one *thinks* it does. A large number of studies show 'experience' and 'evidence' as at least partly perceptions or social constructions (Johnson, 1993; Morgan et al, 2002). Given this pattern, perhaps we should ask why one should expect objective knowledge, rather than one's beliefs about how much one knows, to affect trust. Pending definitive evidence on either knowledge type's relation to trust judgements, it seems prudent to include both here.

This background suggests the following hypotheses about knowledge.

Hypothesis 1. People who are or believe themselves to be knowledgeable about wetlands and wetlands management will exhibit different levels of trust than will people with less perceived or actual knowledge.

Whether trust will be higher or lower among the more knowledgeable is unclear since they may know more about 'skeletons in the closet', as well as about positive attributes of the trust target.

Hypothesis 2. People who are or believe themselves to be knowledgeable about wetlands and wetlands management will rate sources of trust differently, and have different patterns between these ratings and trust judgements, than will those with less perceived or actual knowledge.

Again, the literature offers no guidance on the direction of these effects, so the analysis here will be exploratory rather than confirmatory.

Hypothesis 3. Those who do occupy, or have occupied, the same role as the trust target will report more trust in that target than will those without that experience.

As with knowledge, this role familiarity might expose bad as well as good news, and thus might reduce trust. However, the defensive motivation to avoid denigrating an institution with which one is or has been associated is assumed to elicit more public expressions of trust than distrust, whatever subjects' private thoughts. There are no grounds for hypotheses about the effect such role experience might have on sources of trust, so this will simply be a question to be explored.

Measuring attributes versus measuring importance

The attribute-rating method of Metlay (1999) and Poortinga and Pidgeon (2003) exemplifies one aspect of the 'most popular framework for understanding the relation between attitudes and the evaluative meaning of beliefs': the expectancy-value model (Eagly and Chaiken, 1993, p106). Expectancy is 'the subjective probability that the attitude object has or is characterized by the attribute', expressed in trust studies by rating the degree (without assigning probabilities) to which the trust target is deemed honest or fair or competent, among other attributes. Value 'is the evaluation of the attribute', which might be measured in terms of positive or negative feelings, or its importance, among others. An attitude – such as trust in the target entity – is then 'predicted' by multiplying each attribute rating by its value rating and summing over all attributes. While 'even relatively high correlations between attitudes and summed Expectancy × Value products do not necessarily demonstrate that beliefs determine attitudes rather than vice versa' (Eagly and Chaiken, 1993, p232), the expectancy-value model has been common in psychological research, but absent from trust research. Ideally, then, a study applying the expectancy-value model to trust sources would elicit three types of judgements: the trust rating, the attribute rating (expectancy) and a value rating (for example, ratings of the importance of the attributes to the trust judgement).

However, using importance ratings as value measures alone, as in this study, or in a full expectancy-value model, faces two challenges. First, it requires study subjects to directly link their trust judgements to evaluations of the attributes, and to engage in introspection about how they came to their ratings of trust. We do not know how well people can assess their own thought processes on this topic; the psychological literature suggests that people do well on some, but not all, kinds of introspection.[3] Earle and Cvetkovich (1995) believe trust is mostly heuristic; Sloman (2002) has argued that cognition based on rule-based computation allows awareness of both the result and the process, while associative (heuristic) processing allows awareness only of the result. Thus, if trust judgements are, indeed, heuristic, subjects might offer value (importance) ratings with no patterned relationship to the trust rating, as the true reasons for the latter will be inaccessible to them. Only if trust involves a more systematic processing – which Earle and Cvetkovich (1995) agree can occur when there is sufficient motivation – would introspection on the process by which one decides to trust or distrust be feasible.

Second, any observed pattern between importance and trust ratings might stem from social desirability. Earle (pers comm, 13 May 1996) (see also Johnson, 1999) argued that 'Western' political discourse is likely to make technical competence synonymous with trust because it is socially accepted rhetoric regardless of its true role. In other words, people will justify their level of trust or distrust in a given impersonal entity on the grounds of competence (for example, topic-relevant education) because reasonable people use it as a criterion. Competence might be an even more socially desirable answer if the target is, for example, an environmental agency since science and engineering expertise is critical to its work as it is not for a motor vehicle agency or labour exchange firm. Such justification, if it occurs, obviously would make reliance upon reasons offered for judgements, or judgements of the relative importance of trust sources, a poor strategy for understanding trust sources (although good for identifying social norms for trust criteria). An explicit request for relative-importance ratings would privilege 'legitimate' criteria more than when institutions are indirectly rated on the attributes.[4] Earle (2004, p180) found differing bases of agreement elicited by a think-aloud protocol on climate change (the issue, the policy, inferred values, information quality or organization quality). He interpreted these results as more consistent with the SVS model of variability 'across persons, contexts and time' than with one in which 'people search for signs of objectivity and fairness, and perhaps also for signs of competence and expertise', so that 'a person who is judged trustworthy by one should be judged trustworthy by all' (Earle, 2004, p179). That may be true; but his acceptance of the reasons offered (for example, 'organization quality') seems to undercut the social desirability hypothesis.

Although using importance ratings directly to explore sources of trust has not appeared in the risk–trust literature before, Frewer et al (1996) did ask subjects to generate their own reasons for trust in various entities that could provide information on food risks. These were then used in the standard attribute-rating fashion, but, as Poortinga and Pidgeon (2003, p963) commented, 'on the dimensions they themselves [the subjects] thought were important'. In any event, the validity of the reasons offered by subjects in that study, as with those offered by subjects in Earle's (2004) thought-listing study, were taken seriously by researchers. The question is whether we can take importance ratings seriously as indicators of trust sources.

The aim in the study reported here was to take advantage of these apparent weaknesses in the use of importance ratings in order to generate insights into trust sources. For example, if accurately rating the importance of various criteria for one's trust judgement requires systematic information processing, but the trust judgement itself stems from heuristic information processing, then regressing trust ratings upon importance ratings should yield statistical gibberish. Since heuristic processors should be unable to use introspection to accurately retrieve the bases for their trust judgements, only by pure chance would importance 'predictors' of trust form a coherent pattern.

Hypothesis 4. Importance ratings will exhibit no consistent pattern with trust ratings.
As yet, we have no evidence on whether certain attributes, such as those related to 'competence', are more likely than others to get rhetorical deference due to perceptions of social desirability. The Earle (2004) data suggest otherwise, but have not been analysed explicitly on those grounds. Forcing people to 'justify' their trust judgements with importance ratings seems a prime opportunity for revealing such a tendency. Having local officials rate attributes, with public reputations to maintain or enhance

(even if under the anonymity of a mailed survey), might even enhance the effect over asking ordinary citizens to do so.

Hypothesis 5. Importance ratings of competence attributes will be higher than those for such attributes as shared values or similarity to oneself.

However, we might expect an interaction between knowledge – an often-presumed component of competence (see Johnson, 1999, for a review) – and this rhetorical deference, as in hypothesis 6.

Hypothesis 6. People who are or believe themselves to be knowledgeable about the topic will rate competence attributes as more important to their trust judgements than will those with less perceived or actual knowledge.

Certain kinds of officials may exhibit this effect more strongly than other officials, given expected differences in topical knowledge.

Hypothesis 7. Environmental commission and planning board officials, as the local officials most likely to know about or deal with wetlands issues in their decision-making, will rate competence attributes as more important to their trust judgements than will zoning or elected officials.

But, assuming that competence has only a rhetorical relationship with trust judgements and that people do know the substantive sources of their trust judgements, then the SVS model predicts hypothesis 8.

Hypothesis 8. The relative importance of shared values or similarity to oneself, but not the importance of competence variables, will significantly 'predict' trust ratings.

Only this last hypothesis presumes that people know what factors influence their trust; we may learn something by examining patterns produced by analyses, implying a causal relation of importance ratings and trust. If hypotheses 4 and 8 are rejected here, this would not justify an assumption that introspection does allow accurate access to trust sources. But such results would argue for research designs (such as an expectancy-value model) that could test the assumption.

Elites as judges

The inclusion of elites in this study is a departure from the norm. Almost all trust research (excluding Metlay, 1999), whatever its discipline or topical focus, concerns trust by ordinary people of elites, such as by the 'general public' of institutions or by employees of their superiors. This bias stems from:

- the theory that a democracy requires some citizen trust in leaders for enduring social order;
- research funders' greater interest in public distrust of institutions than in institutions' trust in each other; and
- the greater number and ease of recruiting ordinary people.

But this focus presents certain problems for both theory and practice.

Getting things done among groups with competing goals, or even with goals transcending any one group, can be done with subterfuge, coercion or temporary alliances; but trust can make progress easier. Thus, trust among political, economic, social or cultural elites would seem at least as worthy of study as public trust in elites if such research is to help promote cooperation in risk management or any other societal endeavour. Levi and Stoker (2000, p496) mention that 'trust among political elites' has been largely ignored. The record in the risk field, broadly construed, is little better, though it tends to underline the potential importance of (dis)trust among and by elites (Pierce et al, 1987; Soden and Conary, 1991; Tonn and Peretz, 1997).

In struggles or cooperation among contending groups, 'the general public' often plays only the role of spectator, if that (Renn, 1992). Hardin (2002, p152) suggested that trust in government by citizens is oversold, with military service and taxpaying:

> ... among the relatively rare policy realms in which voluntary compliance by individual citizens is virtually necessary. In many other areas, compliance can simplify the tasks of government but is not so crucial... it simply cannot be true that modern governments are unable to work without such trust... *it is surely more important that government be trustworthy than that it be trusted* [emphasis in original].

Hardin (2002, p163) went on to suggest that:

> ... having only 5 per cent of the citizenry alert and committed to politics is all it takes to make things work. If this is even roughly true, then the survey evidence on declining trust, confidence or faith in government across the general population is of little interest. What we need to know if we are to assess the prospects for government in this era is how the 5 per cent who are alert and active in politics rate government.

The object of the research reported here was to test the relative impact of attribute importance and familiarity (knowledge and role performance) on trust judgements as these might vary across both multiple *elite* trust judges and multiple institutions judged for trustworthiness. The risk field has often sought judgements of trust in multiple targets (for example, firms, agencies, activists and reporters on a particular hazard); but the single target has been the norm in trust-source research (excluding Peters et al, 1997), and researchers have used only one type of judge (Metlay, 1999, employed multiple judges, but did not analyse differences). Research designs with multiple types and numbers of targets and judges will advance understanding of trust sources.

As noted above, ideally one would ask people for both importance and attribute ratings of targets, whether to test if such ratings produce similar results or to generate an expectancy-value model of trust. Unfortunately, the survey instrument used here had room for only one such task; since the trust-source field had not yet tried the relative importance approach, it was adopted here to explore its implications. The answers from this pilot study will not be definitive, but will yield initial insights on the proposed hypotheses and, perhaps, a partial model for future research.

METHOD

A mail survey was conducted in New Jersey in 2003 on local officials' views of wetlands and wetlands management. The benefits of wetlands include water supply and purification, as well as habitat and flood control; their development is regulated in most US states by the national government (the US Environmental Protection Agency or the US Army Corps of Engineers). New Jersey is one of two states with its own, more stringent, freshwater wetlands management law. Although regulated by the state, wetlands are strongly affected by municipal land-use decisions under the 'home rule' tradition. Thus, the views of local officials are important in wetlands protection.

Surveyed 'officials' included all elected legislative (town council), zoning, planning and environmental commission officials in the 26 towns, partly or wholly in the Stony Brook–Millstone watershed in central New Jersey. Planning and zoning board officials have formal responsibilities under state land-use law (for example, producing a master plan and zoning maps for the town; deciding whether a given use fits zoning for the plot). Environmental commissions are authorized by state law, but are optional for municipalities (all but 2 of 26 towns had them) and have only advisory status to elected officials. In some towns, environmental commissions have a substantial impact on land-use planning and management; in others they seem to be ignored on major issues. Elected officials have final authority (for example, they officially adopt the master plan proposed by the planning board) and may exercise informal influence over land use.

The aim was a census of all targeted local officials, given uncertainty on response rate.[5] Some 224 responses from 574 valid addresses gave a response rate of 39 per cent: elected (EL) 30 per cent; planning board (PB) 43 per cent; zoning board (ZB) 35 per cent; environmental commission (EC) 50 per cent. Answers to the few (mostly wetlands regulations knowledge) questions that this survey had in common with an earlier 2001 survey (57 per cent response) from the same population suggested (but could not guarantee) that 2003 respondents were representative. Post-survey checks with a sample of non-respondents indicated that the main reason for non-response was ignorance on the topic of wetlands.

Current officials did not tend to hold more than one office at a time, as indicated by negative correlations among reported offices ($p < 0.05$ unless noted otherwise): –0.41 for zoning and environmental officials;

−0.30 for elected officials versus both zoning and environmental officials; −0.29 for planning and environmental officials; −0.17 for planning and zoning; and −0.08 (not significant) for elected and planning officials. This is affected by rules for office holding (for example, one elected official is required by state law to be appointed to the planning board, which explains the lower negative correlation in that case). With current and former offices combined, correlations were positive (that is, people who hold one office are likely to have held another office earlier), as one would expect for mostly small towns with few likely office candidates: +0.39 for elected and planning; +0.21 for planning and zoning; +0.19 for elected and zoning; and +0.03 to +0.11 (all not significant) for environmental commissioners' overlap with the other three groups (the lower correlations probably reflect both a wider candidate pool for EC members, and the fact that not all towns have ECs).

The targets of trust judgements included the New Jersey Department of Environmental Protection (NJDEP), the state government agency that regulates freshwater wetlands in the state; municipal elected officials; and municipal planning boards. If elite and citizen judges differ in assessing trust, or any judges differ in rating specific versus generic institutions, these wetland data should produce results closer to Metlay's (1999) findings (US site, stakeholder judges and specific agency target) than to Poortinga and Pidgeon's (2003) (UK general public and government, in general).

Table 10.1 shows measures used here, with the identical names seen by participants. Roughly speaking, these can be categorized by a descriptive (*not* conceptual) taxonomy proposed by Johnson (1999) to include 'competence' (a, c, d, i), 'care' (b, g, h, j, k, l, m) and 'consensual values' or salient value similarity (f, n), plus a measure of experience with the target (e). Siegrist et al (2003, p707) also listed honesty (b), caring and fairness (perhaps including l) as SVS attributes. No trust-attribute taxonomy, including this one, should be deemed definitive; the taxonomy simply organized hypothesized factors in trust judgements into apparently coherent categories. Johnson (1999) noted that these criteria largely derived from researchers' concepts and hunches, so it is not yet clear whether they are relevant to those asked for trust ratings (excluding Frewer et al, 1996). Attributes were selected for potential relevance to the topic and the judges to represent as many sub-groups of hypothesized trust sources as possible, and to keep subject workload reasonable. No attempt was made to include all trust criteria suggested in earlier studies.

Officials were first asked to 'rate your trust in each of the following' three institutions 'to make appropriate decisions affecting wetland management in your town, on a scale from 0 (no trust at all) to 9 (total trust)'. The three entities were described as follows: ' Department of Environmental Protection (DEP) wetlands programme staff', 'elected officials in your town' and 'planning board of your town'. Then, officials were asked to rate the relative importance of the attributes to their own trust judgements:

> Now, how important were each of the following factors in your judgement of trust in ___ on a scale from 1 (not at all important) to 7 (extremely important)? Please read the full list of factors before starting to answer; give identical scores only for factors that were equally important in your trust judgement.

These instructions and criteria were repeated for each of the three institutions.

Other than demographic questions, these 45 trust and attribute questions were the final set in an instrument that had already asked 169 questions on knowledge of wetlands and wetland regulations, local and state policies for wetland management, and information needs and sources. Trust-attribute ratings (how much subjects thought the targets were characterized by these attributes) were thus not added to the respondents' workload.

RESULTS

Trust ratings

Mean trust ratings were 6.33 (standard deviation (SD) = 2.35; median = 7) for NJDEP; 5.43 (SD = 2.63; median = 6) for local elected officials; and 6.07 (SD = 2.43; median = 7) for local planning boards. All trust targets thus obtained moderately positive ratings on this 0 to 9 measure, with the state agency a bit more trusted than the planning board, and the elected officials a bit less trusted. This is consistent with 2001 survey findings, in which these groups respectively ranked fifth, seventh (including zoning boards) and tenth in trust ratings by local officials of 14 entities.

Six of 18 differences in trust ratings of the three target groups across the four groups of judges were at least marginally significant (by t-tests of independent groups). Environmental commissioners had less trust in elected officials than did elected (mean = 4.47 versus 5.73; $p < 0.01$); planning board (PB) (mean = 5.95; $p < 0.002$); and zoning board (ZB) (mean = 5.63; $p < 0.05$) officials. Environmental commissioners also had less trust for PB members (mean = 5.16) than did elected officials (mean = 6.12; $p < 0.10$); PB (mean = 6.76; $p < 0.0002$); and ZB (mean = 6.22; $p < 0.05$) officials. Only two of these differences were significant at the Bonferroni-corrected significance level of $p < 0.0085$, which adjusts for the use of six contrasts across the four official types.

As for the role performance hypothesis, former and current elected officials did not differ from never-elected officials in their trust ratings, including those of elected officials; the same was true for former and current zoning officials versus others. Past and present PB members differed from others only in their greater trust of the planning board (mean = 6.41 versus 5.62; $p < 0.05$). Strikingly, anyone who had ever been an environmental commissioner differed markedly from other officials in all three trust judgements. They were more likely to trust the state agency (mean = 6.63 versus 5.98; $p < 0.05$) and less likely to trust either local elected officials (mean = 4.95 versus 6.01; $p < 0.004$) or planning boards (mean = 5.68 versus 6.51; $p < 0.05$).[6]

Criteria ratings

These reveal much apparent consensus (see Table 10.1). For example, honesty, knowledge of local conditions and accountability rank among the top five criteria for all trust targets, while similarity, bargain/compromise and values consensus rank among the bottom five criteria for all. There are also some interesting differences: wetlands-relevant education is highly ranked for NJDEP, almost as high for the planning board, and much lower for elected officials, while analytic ability is deemed very important for NJDEP, but not at all for the local officials.

These initial results offer mixed support for hypothesis 5. If we treat values consensus and similarity as the only criteria exemplifying the salient value similarity theory of Earle and Cvetkovich (1995), as the hypothesis does, then their ranking in the bottom third for all three trust targets supports the notion that other criteria will rank more highly. Yet, the hypothesis explicitly says that competence attributes will be ranked as far more important. This is true for 'wetlands-relevant education', although it barely exceeds the ranking of shared values for elected officials. 'Ability to do technical analysis' ranks second highest for NJDEP, but only above similarity for judgements of local officials. 'Benefits/harms of past decisions', an explicit performance measure, ranks in the middle. 'Knowledge of local conditions' is an ambiguous member (Johnson, 1999) of the competence category (for example, does it only include technical issues, such as ecological quality of local wetlands, or other knowledge as well, such as of local power relations?); but it is the only one of four potential competence attributes ranking consistently high in importance. This ranking may reflect its salience for local officials asked to rate their trust of entities to manage wetlands 'in your town'. If we add items adduced as SVS criteria by Siegrist et al (2003) – honesty, ranked first in all three cases, and 'same participation chance for all' (as a measure of fairness), ranked in the middle – the picture gets more muddled. The hypothesized high importance for competence attributes and low importance for SVS measures works better for NJDEP (if with inconsistencies) than for the local officials. Perhaps subjects distinguished roles among targets (for example, topical knowledge and skills may be deemed more important for NJDEP staff, who directly regulate wetlands use, than for local officials); but this implies cognitive effort at odds with the heuristic hypothesis of trust.

Strikingly, these trust judges assigned different rankings of importance to criteria for trust in NJDEP than for trust in the municipal elected officials and planning board members. While the two local trust targets had rank correlations of 0.83 to 0.94 across the four judging groups (all Bonferroni significant), the rank correlations of elected officials and NJDEP were only 0.23 to 0.29, and of the planning board and NJDEP only 0.45 to 0.49, none significant even at $p < 0.05$. Only the zoning board members showed significant correlations of criteria ranks across the three target groups: the two types of local officials at 0.73 (Bonferroni significant), NJDEP elected at 0.57 and NJDEP planning at 0.63, both of the latter significant only at $p < 0.05$. Pending direct examination of relative knowledge across types of local officials, these results might reflect lower knowledge of the topic among zoning board officials, thus supporting hypothesis 7.

Table 10.1 *Trust criteria ratings of importance*

Criteria	New Jersey Department of Environmental Protection (NJDEP)		Elected officials		Planning board	
	Mean	Rank	Mean	Rank	Mean	Rank
a Wetlands-relevant education	5.62	4	4.95	9	5.35	6
b Honesty	5.96	1	6.24	1	6.28	1
c Ability to do technical analyses	5.95	2	4.17	13	4.64	13
d Knowledge of local conditions	5.77	3	6.01	2	6.23	2
e Your own experience with staff	4.10	13	5.61	5	5.29	7
f Similarity to me	2.88	14	3.70	14	3.72	14
g Speed of their decision-making	4.60	9	4.59	12	4.65	12
h Bargain/compromise in decisions	4.31	12	4.77	10	4.76	10
i Benefits/harms of past decisions	4.75	7	5.19	6	5.40	5
j Pursuit of public interest versus own	5.37	6	5.96	3	5.89	3
k Convenience of taking part in their decision-making	4.35	11	5.05	8	5.10	9
l Same participation chance for all	4.74	8	5.18	7	5.26	8
m Ability to hold them accountable	5.59	5	5.78	4	5.76	4
n Share my values on wetlands	4.52	10	4.71	11	4.73	11

Note: Scale increases from 1 (not at all important) to 7 (extremely important).

Familiarity variables

Three kinds of knowledge were measured:

1. self-reported knowledge on wetlands and related regulations and procedures (mean of five items on a 0 to 100 point scale);
2. objective knowledge of the effect of wetlands on water quality and soils (number of correct answers for five items); and
3. objective knowledge of New Jersey wetland regulations (also five items).

Self-reported knowledge correlated at +0.41 with wetlands knowledge, and +0.12 with regulatory knowledge; the two objective measures correlated at +0.27.

All six contrasts in self-reported knowledge across the four groups of officials were at least marginally significant ($p < 0.10$), with environmental commissioners expressing more belief in their knowledge at the Bonferroni-corrected significance level of $p < 0.0085$ (mean = 3.13) than any other group (ZB = 1.49; EL = 2.02; PB = 2.42). EC members also exceeded other groups in objective knowledge of wetlands' water quality effects, with a mean score of 2.19 compared to 1.84 (PB; $p < 0.10$), 1.64 (ZB; $p < 0.01$) and 1.58 (EL; $p < 0.01$). The other groups did not differ significantly. There were no significant differences in objective knowledge of regulations (mean = 2.06 to 2.29); both PB and ZB members exceeded EC members in this knowledge. With that exception, however, the knowledge aspect of hypothesis 7 (EC and PB members would be most knowledgeable) was confirmed.

Role experience was tapped with questions on offices held. Twenty-one per cent were current and 8 per cent were former elected officials; 35 per cent and 15 per cent were current and former PB members, respectively; 25 per cent and 7 per cent were current and former ZB members; and 34 per cent and 7 per cent were current and former environmental commissioners. Percentages for current offices sum to more than 100 per cent because some overlap in offices is required (see the former section on 'Method') or allowed.

Criteria rankings across office holders

Spearman's rank correlations show that the four groups of officials agreed on relative rankings of criteria for evaluating trust in these three targets, with all correlations significant at $p < 0.05$, except for PB and ZB on criteria for trust in local elected officials ($r = 0.29$). Most correlations were also significant at the Bonferroni-corrected significance level of $p < 0.00183$, which adjusts for the 28 contrasts across all rankings; with one exception, non-significant correlations concerned trust in elected officials. The number of significant differences of mean importance across 14 criteria, 3 targets and 4 types of judges (compared in six contrasts) was 16 of 252 at $p < 0.05$, barely above the number expected by chance (13); 19 differences were significant at $p < 0.10$, fewer than expected. Overall, these findings refute hypothesis 7 about differences in criteria judgements across different types of local officials, although they do not yet account directly for the hypothesized factor of relative knowledge.

Separate analyses tested whether role experience (current or former membership in that office, versus none) affected trust in elected officials or planning boards. The elected official rankings by ever-elected officials and others correlated at $+0.94$; PB rankings by PB members and others correlated at $+0.87$.[7] Ever-elected and never-elected officials differed on mean importance only for 'your own experience' with elected officials (mean = 5.93 versus 5.39; $p < 0.05$). Current and former PB members differed from others on the importance of 3 of 14 criteria for judging trust in planning boards. PB members rated as more important 'your own experience' (mean = 5.59 versus 4.90; $p < 0.01$), 'similarity to me' (mean = 4.01 versus 3.34; $p < 0.05$) and 'share my values on wetlands' (mean = 4.93 versus 4.46; $p < 0.10$).[8]

Effects of knowledge

Differences across trust and importance ratings were examined by t-tests of independent groups for the three knowledge variables. Median splits were between one and two of five wetland knowledge questions correct, between two and three of five regulatory questions correct, and between ratings of 48 and 49 on the 0 to 100 self-reported knowledge scale. Trust in NJDEP was marginally higher among those more knowledgeable about wetlands (mean = 6.52 versus 5.84; $p < 0.10$), and trust in elected officials was marginally greater (both $p < 0.10$) among those with more regulatory knowledge (5.79 versus 5.16) and lower self-reported knowledge (5.79 versus 5.11). These results weakly support hypothesis 1.

In evaluating NJDEP, people with high wetland knowledge rated knowledge of local conditions, similarity and bargain/compromise as less important than did those with low knowledge ($p < 0.10$). Those with high regulatory knowledge rated the ability to do technical analysis and local knowledge lower, and pursuit of public interest and shared values higher, than did low knowledge respondents. Those who thought they were well informed rated the importance of their own experience with the target group, its speed of decision-making and shared values higher than did those who rated their knowledge as low. In evaluating elected officials, experience with the judged entity and benefits/harms of past actions were rated as more important by those low in wetland knowledge, while regulatory knowledge had no effect. Those low in self-reported knowledge rated knowledge of local conditions higher than did the above median group. In evaluating the planning board, benefits/harms rated as a more important criterion for the low wetland knowledge group; again, regulatory knowledge had no effect. Above-median self-reported knowledge led to rating accountability as more important.

Spearman rank correlations showed that the overall relative importance of criteria did not vary across knowledge groups: the lowest such correlation was between high and low median knowledge of wetlands for rankings of NJDEP criteria (r = +0.86). Thus, hypothesis 2 is only marginally supported by these knowledge data, and hypothesis 6 on more knowledgeable people privileging competence attributes must be rejected outright since, in fact, the reverse was true.

Structure in attribute importance ratings

Attribute-rating studies (Frewer et al, 1996; Metlay, 1999; Poortinga and Pidgeon, 2003) identified two factors each, with varying labels and content. White and Eiser (see Chapter 4 in this volume) note that such analyses seem to lack face validity since criteria loading does not always seem consistent, and comment that the Poortinga and Pidgeon (2003) factors seem to reflect informational valence ('generalized trust' and 'scepticism' are their Poortinga–Pidgeon labels). In any event, exploration of potential structure in the current data seems warranted, even if the data do not use the attribute-rating method of the other studies.

Results appear in Table 10.2 of principal axis factoring with varimax normalized rotation, recommended for seeking underlying dimensions (Russell, 2002). Analyses were conducted for NJDEP, and for elected and planning board officials; in the latter two cases, analyses were done both with and without current and former occupants of these offices in order to separate observer from participant structures in trust criteria importance.

This approach extracted only one factor for NJDEP. High-loading items seemed mostly procedural or 'care' oriented: same chance of participation; convenience of participation; ability to hold them accountable; bargain/compromise in decisions; speed of decision-making; and honesty. Only the performance variable of benefits/harms of past decisions violates this pattern. Two factors were extracted for elected officials (unrotated explained variance was 3.88 and 1.05, respectively), with the first factor exhibiting a process focus (again with bargain/compromise loading moderately high). The second factor emphasized salient value similarity, with similarity to the judge and shared values loading highly. If past and present elected officials were excluded from analysis, the first factor recurred; the second factor now combined competence attributes of ability for technical analysis and wetlands-relevant education with similarity, and shared values dropped out. For the planning board target, two factors were extracted (unrotated explained variance of 4.57 and 1.08, respectively). Process items (accountability, public interest and honesty) combined with benefits/harms and knowledge of local conditions in the first factor. Similarity and values loaded on the second factor; so did

Table 10.2 *Principal axis factoring*

Criteria	New Jersey Department of Environmental Protection (NJDEP)	Elected officials					Planning board				
	All	All		Without			All		Without		
	1	1	2	1	2		1	2	1	2	3
a Wetlands-relevant education	−0.19	+0.07	+0.41	+0.09	+0.56		+0.31	+0.15	−0.02	**+0.75**	+0.14
b Honesty	**−0.55**	**+0.58**	+0.10	**+0.61**	+0.09		**+0.68**	+0.11	+0.47	+0.69	+0.05
c Ability to do technical analyses	−0.24	+0.01	+0.45	−0.04	**+0.72**		+0.22	+0.30	+0.11	+0.48	+0.39
d Knowledge of local conditions	−0.32	+0.36	+0.10	+0.38	+0.16		**+0.55**	+0.16	+0.30	+0.49	−0.00
e Your own experience with staff	−0.48	+0.17	+0.38	+0.16	+0.42		+0.23	+0.36	+0.29	+0.06	+0.52
f Similarity to me	−0.46	+0.16	**+0.69**	+0.27	+0.58		−0.06	**+0.90**	+0.13	+0.01	**+0.84**
g Speed of their decision-making	**−0.59**	**+0.58**	+0.19	**+0.57**	+0.24		+0.40	+0.52	**+0.65**	+0.06	+0.29
h Bargain/compromise in decisions	**−0.63**	**+0.59**	+0.14	**+0.56**	+0.20		+0.44	+0.38	**+0.56**	+0.15	+0.32
i Benefits/harms of past decisions	**−0.60**	**+0.59**	+0.12	**+0.58**	+0.18		**+0.56**	+0.27	**+0.65**	+0.10	+0.20
j Pursuit of public interest versus own	−0.41	**+0.52**	+0.08	**+0.54**	+0.08		**+0.67**	+0.15	**+0.71**	+0.25	+0.04
k Convenience of taking part in their decision-making	**−0.74**	**+0.62**	+0.23	**+0.62**	+0.24		+0.35	**+0.67**	+0.51	+0.14	+0.39
l Same participation chance for all	**−0.75**	**+0.78**	+0.15	**+0.81**	+0.17		+0.49	+0.54	**+0.71**	+0.12	+0.24
m Ability to hold them accountable	**−0.67**	**+0.72**	+0.10	**+0.79**	+0.02		**+0.66**	+0.32	**+0.83**	+0.26	+0.10
n Share my values on wetlands	−0.31	+0.23	**+0.61**	+0.33	+0.46		+0.20	**+0.55**	+0.21	+0.31	+0.55
Explained variance	3.89	3.39	1.54	3.65	1.80		2.91	2.73	3.59	1.82	1.89
Proportion of explained variance	28%	24%	11%	26%	13%		21%	20%	26%	13%	13%
Reliability (Cronbach's alpha for boldface items)	0.84	0.84	0.71	–	–		0.77	0.71	–	–	–

Note: All = original data (n = 141 with case-wise deletion); without = omission of current/former elected officials (n = 97) or planning board officials (n = 70), respectively.

convenience of participation, but confounded with the first factor. With past and present PB members excluded, three factors were extracted (unrotated explained variance of 5.10, 1.12 and 1.08, respectively). The first factor was mostly procedural, the second was dominated by wetlands-relevant education (high-loading honesty was confounded with the first factor) and the third factor comprised similarity, experience with staff and shared values (the latter slightly confounded with the second factor).

There are three caveats to be offered on these analyses. First, statistical analysis can allow one to find non-existent patterns, particularly without prior constraints (for example, theory-based hypotheses).[9] Second, extracted factors need not be uncorrelated, as varimax and other orthogonal rotation methods assume, although this is often the simplest way to represent factors. The two factors extracted from the entire sample for elected officials were correlated +0.45, while equivalent factors for the planning board correlated +0.73. As might be expected from these correlations, only 'same participation chance' had a loading of greater than +0.50 (+0.52) on the common variance factor for elected officials. Nine items had such loadings (+0.51 to +0.66) on the same factor for planning officials, suggesting that a general trust factor is likely to affect most of these criteria judgements, in addition to the unique contributions of items on the two extracted factors in Table 10.2. Third, there is no reason why we should expect consistent patterns in explicit ratings of the relative importance of these criteria responses in the same way that patterns were alleged to appear (with some debate) in attribute ratings, as discussed earlier.

Despite those caveats, there is remarkable consistency in factors extracted here, if with modest explanation of variance (28 to 52 per cent), and lack of internal structure in the NJDEP analysis. Process and SVS items dominate. The inconsistencies (the outcome measure of 'benefits/harms' combining with procedural items; components of factors extracted with members of judged groups omitted) seem no greater than those found in attribute-based factorings (Metlay, 1999; Poortinga and Pidgeon, 2003). The small sample sizes when certain officials were omitted (averaging seven and five observations per variable for elected and PB judgements, respectively) may explain face validity issues in those analyses. It is less clear why past performance tended to load with procedural items. This is consistent with the initial dual-process model of salient value similarity for 'confidence', but incompatible with many attribute-based hypotheses (see review in Johnson, 1999) and with the later Siegrist et al (2003) version of the dual-process model, in which fairness is deemed a factor in trust, not performance-driven confidence.[10]

The high-loading items on each factor (boldfaced in Table 10.2) created reliable scales (Cronbach's alpha = 0.71 to 0.84), with omission of any item – including seeming outliers – reducing reliabilities in each case. The reliability of the second planning board scale on similarity would have actually increased to 0.73 if the convenience of taking part in decision-making had been included; but it was omitted due to moderate confounding with the first factor.

Explaining trust judgements

Trust judgements were regressed upon three classes of independent variables: the scales created from boldfaced items in Table 10.2, plus the remaining importance criteria; the three knowledge variables; and four dummy variables indicating whether the respondent had ever held office as an elected, planning, zoning or environmental official (see Table 10.3). For each trust target, a first regression analysis included only the criteria, with the two 'familiarity' types of variables added in a second run to see how they might modify results.

Trust in NJDEP to manage local wetlands was strongly related to the importance of shared values, similarity and wetlands-relevant education, and decreased with the importance of personal experience with agency staff. Adding familiarity variables largely retained these relations (education switched significance with another 'competence' variable: technical analysis). Knowledge of wetlands regulations raised and self-reported knowledge lowered trust, both marginally. Having been an environmental commissioner significantly increased trust.

Trust in local elected officials increased when people put high importance on similarity and knowledge of local conditions, and decreased with the importance of process, wetlands-relevant education and personal experience. Familiarity variables did not alter these relationships; but past or present membership of the environmental commission also reduced trust (having been an elected official reduced trust in elected officials at p < 0.11).

Trust in the planning board rose as similarity became more important and as wetlands-relevant education became less important. Adding familiarity variables added technical analytic ability as a marginally negative predictor of trust. The opportunity of having been a planning board member increased trust, while self-reported knowledge and having been an elected official reduced trust.

The consistent positive impact of 'similarity' (shared values and similarity to self) and the largely negative impact of process-related variables on trust point to a patterned relationship of importance and trust ratings, refuting hypothesis 4. Although slightly less consistent, that the relative importance of the competence variable 'wetlands-relevant education' raised trust in the target likely to have such education and reduced it in targets without such education also raises doubts about the hypothesis. Perhaps desire to appear rational (aided, perhaps, by instructions to rate the importance of the criteria for trust relative to all other criteria) led to consistent patterns without any real or accurate introspection. The drawback of this alternative explanation is that it would require a larger cognitive workload than introspection, which seems inconsistent with Earle and Cvetkovich's (1995) heuristic model of trust, unless these local officials had extraordinary motivation to explain their judgements.

Table 10.3 *Multiple regression analyses of trust judgements on criteria and knowledge variables (betas)*

Independent variables	New Jersey Department of Environmental Protection (NJDEP)		Elected officials		Planning board	
	I	II	I	II	I	II
NJDEP 'process' scale	−0.15	−0.17[1]	−	−	−	−
Elected officials 'process' scale	−	−	−0.27[3]	−0.23[3]	−	−
Elected officials 'similarity' scale	−	−	+0.44[5]	+0.37[5]	−	−
Planning board 'process' scale	−	−	−	−	+0.07	+0.11
Planning board 'similarity' scale	−	−	−	−	+0.28[3]	+0.29[3]
a Wetlands-relevant education	+0.15[1]	+0.11	−0.20[2]	−0.18[2]	−0.28[3]	−0.28[3]
c Ability to do technical analyses	+0.12	+0.19[2]	+0.07	+0.04	+0.14	+0.15[1]
d Knowledge of local conditions	+0.08	+0.00	+0.41[5]	+0.44[5]	−	−
e Your own experience with staff	−0.20[2]	−0.21[2]	−0.16[2]	−0.17[2]	+0.04	−0.01
f Similarity to me	+0.20[1]	+0.30[2]	−	−	−	−
g Speed of their decision-making	−	−	−	−	−0.07	−0.02
h Bargain/compromise in decisions	−	−	−	−	−0.10	−0.08
j Pursuit of public interest versus own	−0.10	−0.07	−	−	−	−
k Convenience of taking part in their decision-making	−	−	−	−	+0.07	+0.11
l Same participation chance for all	−	−	−	−	+0.02	−0.04
n Share my values on wetlands	+0.34[4]	+0.24[2]	−	−	−	−
Self-reported knowledge	−	−0.16[1]	−	−0.07	−	−0.29[3]
Wetlands knowledge	−	+0.12	−	−0.05	−	+0.01
Wetlands regulations knowledge	−	+0.15[1]	−	+0.07	−	+0.04
Elected official, ever	−	−0.09	−	−0.14	−	−0.17[2]
Planning board, ever	−	+0.11	−	+0.08	−	+0.25[3]
Zoning board, ever	−	−0.10	−	+0.07	−	+0.01
Environmental commission, ever	−	+0.17[2]	−	−0.20[2]	−	−0.07

Table 10.3 *continued*

Independent variables	New Jersey Department of Environmental Protection (NJDEP)		Elected officials		Planning board	
	I	II	I	II	I	II
F, p	5.87[5]	3.92[5]	9.76[5]	5.01[5]	2.76[3]	3.66[5]
R-square	27%	34%	28%	33%	15%	31%
Adjusted R-square	23%	25%	26%	26%	9%	22%
n	135	130	154	147	154	149

Notes: For definitions of scales, see text or Table 10.2; for definitions of knowledge variables, see text. Case-wise deletion: for NJDEP, five outliers (>2 SD) were deleted for I and four in II; for elected officials, six and five omissions, respectively; for planning board, six and six.

[1] $p < 0.10$
[2] $p < 0.05$
[3] $p < 0.01$
[4] $p < 0.001$
[5] $p < 0.0001$

The regression results also refute hypothesis 8. While shared values and similarity match and even exceed, in some cases, the 'predictive' power of competence and process variables, the latter provided significant explanation of trust judgements as well. Clearly, the judged importance of similarity variables was able to exhibit a strong correlation with trust ratings even for a group (local officials) likely to select socially endorsed explanations of trust judgements, and with a task of explicit explanation (importance) of such judgements likely to maximize the salience of socially desirable explanations. Equally clearly, any merely rhetorical appeal of competence and process variables could not obscure a statistically significant relationship with trust.

Familiarity variables added only modest explanatory power over criteria-only analyses, except for the planning board regression. Knowledge had relatively few significant effects on trust – mostly the negative impact of self-reported knowledge. All else equal, people who thought they knew a lot about wetlands and wetland management were less trusting of the three trust targets than more humble judges (since self-reported knowledge correlated only moderately with wetlands knowledge and very weakly with knowledge of wetlands regulations). Perhaps negative stereotypes of government were encoded as perceived knowledge. If 'government employees are no more likely [than others] to have confidence in the people running government institutions' (Lewis, 1990, p223), people who are local officials might not tend to be positive about even their own colleagues on the rarefied topic of wetlands management.

It is striking that, in every case, having been a local elected official reduced trust; planning board membership at any time increased trust;

and environmental commission experience reduced trust in local officials (who, as noted earlier, are free to ignore commission advice), but increased trust in NJDEP. These findings provide mixed support for hypothesis 3: that people with experience in a role will exhibit more trust in occupants of that role than will others.

Although it is not a formal familiarity variable here, the relative importance of 'your own experience with' NJDEP and elected officials significantly predicted lower trust in those targets. Apparently, familiarity (or at least the reported relative importance of familiarity as a trust criterion) does, indeed, breed contempt.

Variance explained here is lower than in the Metlay (1999) study, and equals or exceeds that for climate change and mobile phones, but is lower than for radioactive waste, genetically modified food and genetic testing in Poortinga and Pidgeon (2003) (compare adjusted R-square scores). Shared values (and its hypothesized parallel: similarity to self) were more important in trust (but not dominant) than in Poortinga–Pidgeon findings. Their methodological differences should preclude claims that any one study is 'correct', as each might be correct for its situation.

DISCUSSION

Findings emerging from this pilot study included the following:

- Local government officials gave moderate trust ratings for official bodies to manage local wetlands, with slightly more trust for the state agency officially authorized for such management than for local government colleagues (elected and planning board officials).
- Environmental commissioners, the officials with no formal government authority (except as advisers) and the most wetlands-relevant jurisdiction, had less trust in their local colleagues than did other types of officials (particularly planning board members). They also had greater trust in the state agency's wetlands staff.
- The local officials were able to report what they claimed was the relative importance of criteria for their judgements of trust in the three government entities.
- These officials tended to agree on the criteria important for judging trust in a given type of official entity. A given trust criterion varied in reported importance by which government entity was being evaluated, with a strong distinction between the state agency with formal authority for wetlands management and the two local actors.
- 'Similarity' variables of shared values and similarity to self were rated relatively low in importance by these officials, as predicted by Earle's (2004) 'rhetorical' hypothesis. However, other variables, such as those that Siegrist et al (2003) used to predict trust (honesty and equality of participation opportunity as a measure of fairness) in consort with

similarity ranked much higher. The core of the rhetorical hypothesis – that competence variables would rank high – was refuted. The hypothesis was truest of the state agency; but even for it the evidence was mixed, at best. Explicitly asking local officials about relative importance of trust criteria for wetlands management appears to elicit only modest amounts of social desirability bias, at least as predicted by a salient value similarity theorist.

- Principal axis factoring of trust criteria importance data produced only one 'process' factor for the state agency, the entity least familiar to trust judges. For the local trust targets, two factors each were extracted, largely driven by process and similarity variables. These analyses did not produce factors like those from Metlay (1999) and Poortinga and Pidgeon (2003), although the three studies varied in methods and criteria.
- Objective knowledge about the benefits of wetlands for water quality and about wetlands regulations were little correlated with self-reported knowledge or each other.
- Differences in objective and subjective knowledge had only modest effects on criteria ratings or rankings. Those differences that did occur seemed to contradict a social desirability bias explanation, since people higher in knowledge were less likely to rate knowledge high, and more likely to rate shared values high, than were others.
- Differences in experience in the two local offices evaluated for trust had only weak effects on ratings of trust criteria.

These findings have implications for future research on sources of trust judgements, without a need to believe that ratings of attributes' relative importance accurately tap the sources of judges' trust ratings. First, the expectancy-value model could be applied with these importance ratings in the 'value' role. Relative importance is only one way of operationalizing value, so studies that test alternative rating methods against it for consistency would help to determine whether this is the one that is best to use. Since the paper-and-pencil method for assessing value is far cheaper for researchers to implement than experimental manipulation of hypothesized dimensions, use of recall, or information search (Mathew P. White, pers comm, 10 September 2004), testing the consistency of results across these value methods would be helpful.

Second, the rhetorical hypothesis on self-reports of trust sources deserves more testing than provided by this study, which found only modest evidence to support it. Rhetoric in a somewhat different sense – the rationales that public and private entities, institutions and activists offer in public discourse for trust or distrust in themselves or others – is a constant element in risk management, which appears to have real effects (although that question merits attention). As yet, such evidence has not been tapped for trust research, whether to treat it only as persuasive arguments or also as sincere beliefs of their promulgators. No one has asked people to report what they think are the social norms for criteria of trust either. Examination of importance or other value ratings in this larger context of public discourse and social norms would promote understanding of the strengths and limitations of value approaches to trust sources, and the dynamics by which public debates and private decisions on trust interact.

Third, theory and research in this field (Poortinga and Pidgeon, 2003; Siegrist et al, 2003) is beginning to raise the potential effects of 'familiarity' without yet specifying how this concept should be operationalized or exactly how it might affect trust, confidence or trust sources. In this study, three kinds of knowledge measures (two objective, one subjective) and role performance (whether one had served in the evaluated office or not) had modest effects on criteria ratings, rankings and factorings. Whether this is peculiar to this study, or a more general finding, cannot be determined without more attention to familiarity in trust research.

Fourth, this study found that trust criteria varied across targets and somewhat less across judges; it was not designed to find out if they also might vary across topics. But this was the first study (excluding Peters et al, 1997, on targets, and potentially Metlay, 1999, on judges[11]) with a design allowing such distinctions to be explored. It should not be the last, and further research should explain why such differences might occur. The one explanation tested here (that trust judges did or did not have experience in the role being judged) will not apply to many cases. One broader possibility for distinctions across targets is that people have different expectations of different authorities. For example, if state government has the only direct legal authority on wetlands management, judges might expect expertise on wetlands more from state officials than from local government officials, and have their trust in the former raised and in the latter lowered as the relative importance of wetlands-relevant knowledge increases.[12]

The findings from the multiple regression analyses have not yet been mentioned because they imply that officials were privy to and reported the actual relative importance of attributes driving their trust judgements. As noted earlier, no such assumption can be directly refuted or supported by the current data. That relationships between trust and importance ratings were not 'gibberish' (hypothesis 4) suggests that either people did have some access to their own decision processes or exerted considerable effort to make their trust and importance ratings 'agree' – neither option is consistent with a purely heuristic trust process. But that needs to be confirmed.

If we do assume accurate introspection, however, the results of these multiple regression analyses do not deviate sharply from prior findings from attribute-rating studies. The relative importance of similarity, process and knowledge criteria all played a significant role in trust judgements, thus supporting expectations of both SVS and attribute-rating researchers.[13] Although not as strongly, the familiarity variables (particularly self-reported knowledge) also had significant effects on trust. Given the modest trust in target entities reported by these judges, it is not surprising that the importance of certain attributes (for example, process and experience with the entity) undermined trust. Wetlands-relevant education, likely only in NJDEP wetlands programme staff, raised trust in those staff as it became more important, but lowered trust in local officials.

These regression findings also raise important issues for future trust source research, not all predicated on assuming accurate introspection. First, self-reported knowledge had much more effect on trust ratings than did objective knowledge, and is conceptually more akin to the 'familiarity' hypothesized to lower salience of both trust and values consensus (Poortinga and Pidgeon, 2003; Siegrist et al, 2003). Moreover, it had these effects with similarity variables (including shared values) *also* significantly predicting trust. Besides underlining the importance of clarifying familiarity in both theory and measurement, this finding points to the need for more explicit use of the heuristic systematic model (HSM) of information processing (Eagly and Chaiken, 1993) in trust source theory. The dual-process SVS theory assumes that trust must be affective and value driven, because performance information about competence and process would not be cognitively available except with more thought than seems likely for most people in most situations. People familiar with a hazard manager or hazard topic, however, might have the needed knowledge already, so a separate 'confidence' process might use performance data. But HSM assumes that heuristic and systematic processing are often complements, not just substitutes, and any kind of evidence (for example, expertise) could be used in both kinds of processing. If risk researchers are going to use ideas from psychology, we should use (and test) them systematically ourselves. A small emerging literature on survey applications of HSM could be used to good effect here (for example, Trumbo, 1999; Kahlor et al, 2003; Johnson, 2005).

Second, relative importance ratings of criteria that current theories (especially SVS) assume are not compatible jointly 'determined' trust judgements here. It must be reiterated that it is impossible to test hypotheses that trust is one-dimensional (whatever the criteria assumed to comprise it) without including more diverse criteria in research designs (Johnson, 1999).

Third, no two studies in this emerging field have used the same measures, although there is certainly overlap and apparent equivalence to some degree. Without the same measures, we cannot tell whether disparate findings across different studies are methodological artefacts or substantive differences; without assurance that our research measures mean the same thing to us and to our study participants, it is a daring act to reach any conclusion at all on trust. If subject-derived criteria (Frewer et al, 1996; Earle, 2004) supplement valid researcher-derived criteria, and we are sure that they mean the same thing to all concerned, we may resolve the question of whether there is, in fact, consistent structure to the bases for trust judgements.

Fourth, we also need to resolve whether relative-importance ratings are valid and reliable measures of sources of trust judgements. This need not be done in an expectancy-value framework, although that is one approach; just about any method and theory used now could be adapted to compare its standard measures against importance substitutes. Controls for social desirability bias might be useful. Theoretical and methodological clarity would be enhanced whatever the outcome of these tests.

Fifth, Hardin's (2002) claim that few citizens' trust judgements have a substantive effect on government operations raises the question of whether more trust studies should focus on institutionally 'attentive' citizens and other elites, as did this study, whether government is the trust target or not. Certainly, we cannot assume that sources of trust important for the general public, much less for college students and other opportunity samples, will generalize to elites.

This study was conducted with local government officials in a region of the US (and specifically in New Jersey) under intense development pressure affecting wetlands management; results might vary with other trust judges in other states or nations on other topics and under other conditions. Use of different instructions, criteria and targets might have altered the results, as well. Its findings need confirmation across such variations.

CONCLUSIONS

Building cooperation in risk management requires an understanding of the critical factors that might bolster or impede either cooperation in, or the effectiveness of, such management. Past findings that trust seems to be a significant factor in beliefs about risk have stimulated a growing body of research to understand the nature of this effect and the sources of trust. Aside from theoretical interest, implicit in this focus is the notion that one result might be advice to risk managers on how to generate trust in themselves or others by influencing a salient criterion or two. The still-scanty literature in this area and results reported here suggest that a single 'solution' may be unlikely. Criteria important in trust judgements may vary widely across types of risk, trust judges and trust targets. Even without this variance, there might be wide divergence on how to evaluate performance on these criteria. It is not even clear whether a focus on trust judgements of ordinary citizens will be as effective as a focus on more elite groups, or on factors other than trust. If scholars design their studies to take these possibilities into account, they may find themselves at least able to reach consistent conclusions, whatever their practical implications for risk management.

NOTES

1 The wetlands research reported here was funded in part by grant R823-7288 from the US Environmental Protection Agency's (EPA's) Science to Achieve Results (STAR) programme to the Stony Brook–Millstone Watershed Association (SBMWA). This research has not been subjected to any EPA review and therefore does not necessarily reflect the views of the agency, and no official endorsement should be inferred. The views expressed here also may not reflect those of the SBMWA or the New Jersey Department of Environmental Protection.
2 These performance attributes in the model produce 'confidence' ('belief, based on experience or evidence, that certain future events will occur as expected') rather than 'trust' ('willingness to make oneself vulnerable to another based on a judgement of similarity of intentions or values'); see Siegrist et al (2003, p706). It is not clear from these definitions whether confidence is different from trust as an entity in itself, or only due to its different sources. Siegrist et al (2003, p706) claim that 'trust involves risk and vulnerability, but confidence does not'; but it is not clear why experience or evidence would make perceived risk and vulnerability vanish. If trust and confidence lie on a continuum, there is no basis yet for defining a threshold level of experience or evidence beyond which trust becomes confidence. Given this confusion, here no such distinction will be made in regressing trust variables on attributes.
3 People might unconsciously switch to an easier question, rating attributes by how positive they feel rather than by actual introspection on the bases for their trust judgements. This switch would still satisfy the expectancy value model, while leaving researchers thinking that subjects had answered the possibly less valid importance question instead.
4 Earle and Cvetkovich (1995) also criticized procedural attributes, such as fairness of decision-making or the openness of participation, as requiring more cognition than most people are willing to exert in most cases. While this claim suggests that process attributes might be subject to the same social desirability bias as competence attributes, for simplicity this chapter will focus upon the latter in evaluating social desirability.
5 For all but two towns, surveys were mailed to officials' homes from addresses provided by the town clerk. In those two towns, the town clerk required that she send surveys to officials.
6 Environmental commission (EC) members, past and present, had served for a median of four years (range 1 to 23). Regressing EC trust judgements on their years in office had effectively zero explanatory power, implying that these attitudes were not a result of socialization while in office, although that explanation cannot be ruled out entirely.
7 Despite significant differences in trust ratings by ever EC and never EC officials, they did not differ in rankings of criteria importance, with Spearman rank correlations of +0.93 (NJDEP), +0.96 (elected) and +0.92 (PB).

8   In Table 10.1, the importance rank of 'your own experience' with the target is very low for NJDEP, while it is of at least middling rank for local officials. As noted earlier, one should not confuse self-reported importance of criteria with attribute ratings of the targets. But if one plausibly assumes that local officials are more familiar with their own colleagues than with the state agency, and if their importance ratings accurately reflect the criteria they use for judging trust, then familiarity should be a less important criterion for less familiar trust targets, such as NJDEP.

9   Confirmatory factor analyses, which require theory-based constraint on pattern 'invention', suggested structure like that found by Poortinga and Pidgeon (2003). Three 'general trust' and two 'scepticism' attributes from this study fit the data better than either a competence–care–values structure (Johnson, 1999) or a knowledge–process–ethics–outcomes–values structure (face validity of measures). As with confirmatory factor analyses (CFA) using Metlay's (1999) equivalent structure, fit did not reach acceptable levels, although it was perhaps confounded by odd parameterization (Statistica 6.0's SEPATH module was used) or incomplete replication of variables from earlier studies. Ambiguity in classification (for example, is 'pursuit of public interest versus own' here equivalent to 'acting in the public interest' in Poortinga and Pidgeon's 'general trust' factor, or 'too influenced by industry' in its 'scepticism' factor?) also limited the value of CFA.

10 Factoring within each group of officials found that 9 of the resulting 12 analyses produced three-factor solutions, and the nature of the solutions seemed to differ across official types. While the small samples involved (n = 30 to 48) do not warrant detailed attention here, the results suggest that larger studies should not overlook the possibility that different judges will vary in the implied similarity of various trust attributes.

11 Metlay (1999) used diverse judges but did not test for sub-group differences since they had less than 100 members each (D. Metlay, pers comm, 19 November 2003).

12 In a duplication of Table 10.3 analyses including only judges correctly identifying the state as the sole authority, wetlands-relevant education failed to significantly predict NJDEP trust. Technical analytic ability was insignificant unless familiarity variables were included (beta = +0.38; $p < 0.05$). Since the overall equations were not significant at $p < 0.05$, with n = 44, this 'refutation' of the role expectation hypothesis must be treated with caution.

13 Except for the high-ranked honesty criterion in this study, its design does not facilitate a test of or comparison to the intuitive detection theory of trust sources advanced in Chapter 4 of this volume by White and Eiser. While competence items here seem related to its discrimination ability concept, and similarity variables to its response bias concept, intuitive detection theory has yet to grapple explicitly with these apparent parallels. For example, are knowledge and prior decision outcomes sub-classes or cues for discrimination? Does the very specific idea of response bias co-vary with the more general notion of shared values in SVS research? Are the process variables so prominent in attribute-rating literature irrelevant to intuitive detection theorists, as SVS theorists say they are irrelevant in their work?

References

Eagly, A. H. and Chaiken S. (1993) *The Psychology of Attitudes*, Harcourt Brace College Publishers, Fort Worth

Earle, T. C. (2004) 'Thinking aloud about trust: A protocol analysis of trust in risk management', *Risk Analysis*, vol 24, pp169–183

Earle, T. C. and Cvetkovich G. (1995) *Social Trust: Toward a Cosmopolitan Society*, Praeger, Westport, Connecticut

Earle, T. C. and Cvetkovich G. (1997) 'Culture, cosmopolitanism and risk management', *Risk Analysis*, vol 17, pp55–65

Frewer, L. J., Howard, C., Hedderley, D. and Shepherd, R. (1996) 'What determines trust in information about food-related risks? Underlying psychological constructs', *Risk Analysis*, vol 16, pp473–485

Hardin, R. (2002) *Trust and Trustworthiness*, Russell Sage Foundation, New York

Johnson, B. B. (1993) 'Advancing understanding of knowledge's role in lay risk perception', *Risk: Issues in Health and Safety*, vol 4, pp189–212

Johnson, B. B. (1999) 'Exploring dimensionality in the origins of hazard-related trust', *Journal of Risk Research*, vol 2, pp325–354

Johnson, B. B. (2005) 'Testing and expanding a model of cognitive processing of risk information', *Risk Analysis*, vol 25, pp631–650

Kahlor, L., Dunwoody, S., Griffin, R. J., Neuwirth, K. and Giese, J. (2003) 'Studying heuristic-systematic processing of risk communication', *Risk Analysis*, vol 23, pp355–368

Levi, M. and Stoker, L. (2000) 'Political trust and trustworthiness', *Annual Review of Political Science*, vol 3, pp475–507

Lewis, G. B. (1990) 'In search of the Machiavellian milquetoasts: Comparing attitudes of bureaucrats and ordinary people', *Public Administration Review*, vol 50, pp220–227

Metlay, D. (1999) 'Institutional trust and confidence: A journey into a conceptual quagmire', in G. Cvetkovich and R. E. Löfstedt (eds) *Social Trust and the Management of Risk*, Earthscan, London, pp100–116

Morgan, M. G., Fischhoff, B., Bostrom, A. and Atman, C. J. (2002) *Risk Communication: A Mental Models Approach*, Cambridge University Press, New York

Peters, R. G., Covello V. T. and McCallum, D. B. (1997) 'The determinants of trust and credibility in environmental risk communication: An empirical study', *Risk Analysis*, vol 17, pp43–54

Pierce, J. C., Lovrich Jr., N. P., Tsurutani, T. and Abe T. (1987) 'Environmental policy elites' trust of information sources: Japan and the United States', *American Behavioral Scientist*, vol 30, pp578–596

Poortinga, W. and Pidgeon, N. F. (2003) 'Exploring the dimensionality of trust in risk regulation', *Risk Analysis*, vol 23, pp961–972

Poortinga, W. and Pidgeon, N. F. (2005) 'Trust in risk regulation: Cause or consequence of the acceptability of GM food?', *Risk Analysis*, vol 25, pp199–209

Renn, O. (1992) 'The social arena concept of risk debates', in S. Krimsky and D. Golding (eds) *Social Theories of Risk*, Praeger, Westport, Connecticut, pp179–196

Russell, D. W. (2002) 'In search of underlying dimensions: The use (and abuse) of factor analysis in *Personality and Social Psychology Bulletin*', *Personality and Social Psychology Bulletin*, vol 28, pp1629–1646

Siegrist, M., Earle, T. C. and Gutscher, H. (2003) 'Test of a trust and confidence model in the applied context of electromagnetic field (EMF) risks', *Risk Analysis*, vol 23, pp705–716

Sloman, S. A. (2002) 'Two systems of reasoning', in T. Gilovich, D. Griffin and D. Kahneman (eds) *Heuristics and Biases: The Psychology of Intuitive Judgment*, Cambridge University Press, New York, pp379–396

Soden, D. L. and Conary, J. S. (1991) 'Trust in sources of technical information among environmental professionals', *The Environmental Professional*, vol 13, pp363–369

Tonn, B. E. and Peretz, J. H. (1997) 'Field notes on using risk in environmental decision making: Lack of credibility all around', *Risk Policy Report*, vol 4, pp33–36

Trumbo, C. W. (1999) 'Heuristic-systematic information processing and risk judgment', *Risk Analysis*, vol 19, pp391–400

Verba, S., Schlozman, K. L. and Brady, H. E. (1995) *Voice and Equality: Civic Voluntarism in American Politics*, Harvard University Press, Cambridge

11 Antecedents of System Trust: Cues and Process Feedback

Peter de Vries, Cees Midden and Anneloes Meijnders

INTRODUCTION

Trust is generally acknowledged to play an important role in our interactions with other people and, as we will argue, also with systems. Presumably, however, the antecedents of the trust vested by one person in another vary, at least to some extent, with the duration and the intensity of the relation. Someone buying his or her first car, for example, will have little to go on when judging the trustworthiness of the salesperson, whereas someone who buys a new car at that particular sales point once every couple of years may have a more sizeable information pool to draw from.

Gauging the persons we interact with is vital for interactions to run to a satisfactory completion. After all, the salesperson may pursue intentions that are exclusively self-serving, such as trying to make a profit selling the ignorant buyer a ramshackle old car or negotiating an unreasonably high price. On the other hand, he or she may be sincerely friendly, and be sympathetic to the buyer's wishes not to overspend on a car. Clearly, forming correct expectations about the behaviour and intentions of an interaction partner provides a considerable benefit. Contrarily, not knowing what goes on in the mind of the salesperson is likely to cause a great deal of uncertainty on the part of the intended buyer.

In first-time interactions, however, the available information may be very little. In the case of buying one's first car, the only information available may be the salesperson's appearance, or recommendations (or dissuasion attempts) of friends that have negotiated with this particular person in the past. Negative experiences, such as feeling cheated, or positive ones, such as the conviction that a good deal was made with this particular salesperson, will undoubtedly reverberate in the advice given to a friend who is thinking about buying his or her first car. However, since buying a car for most people constitutes an infrequent and costly affair, and, hence, is probably accompanied by feelings of considerable uncertainty, such advice alone may provide insufficient insight into the salesperson's trustworthiness. If the need for more information about the intentions of the seller cannot be satisfied at the appropriate moment, as is often the case, the only thing that may reduce such feelings of uncertainty and, thus, prevent the buyer from discontinuing the negotiation is a sufficient level of trust.

Indeed, trust effectively limits the vast number of possible future interaction outcomes to only a relatively small number of expectations. This may allow for a more careful investigation of the remaining options, thus reducing both uncertainty and risk of the actor (Luhmann, 1979). As Anthony Giddens (1990) argued, there would be no uncertainty and, hence, no need for trust if all the trustee's activities and inner processes were perfectly clear to the truster. In addition, Luhmann argued that trust constitutes a continuous feedback loop that indicates whether or not trust is justified. More specifically, there is an object at which trust is directed – the referee or trustee – and this object provides feedback on the basis of which trust might be built up or broken down. So, in terms of interpersonal trust, the trusted person's behaviour may be closely watched by the truster to see if the trust placed in the trustee was justified. If the trustee performs according to the truster's expectations, trust may be maintained or increased; not living up to expectations will result in a breakdown of trust, possibly to the extent that trust is replaced by distrust.

Naturally, one may select dealers from the *Yellow Pages*, and enquire about prices or special offers; but if effort is undesirable or does not allow for easy discrimination, advice provided by others may actually help make up one's mind. Recommendations from a trustworthy source may set initial trust to a level high enough for the truster to start the process of buying a car, whereas negative information will likely cause the truster to abstain from going to a particular sales point. In the former case, the truster enters Luhmann's (1979) feedback loop; trust levels, initially based upon the recommendations of others, will be constantly updated by what the salesperson says and does. Aspects in the dealer's behaviour that are indicative of being sympathetic to the buyer's wishes, or, on the other hand, of solely having self-serving intentions, will continuously be fed back to increase or decrease trust levels, thus incorporating direct experiences in the process of forming a judgement.

System trust

When interacting with systems, such as decision aids, people may have similar experiences as when interacting with human counterparts. Users, too, may lack information concerning a system's behaviour and the outcomes that it provides. As with interpersonal trust, meaningful interaction requires sufficient levels of trust to enable reductions of uncertainty regarding the functioning of this particular system and its capabilities. Hence, the concept of system trust is crucial in understanding how people interact with systems,[1] an idea that has firmly taken root in research in this field (for instance, see Halpin et al, 1973; Sheridan and Hennessy, 1984; Muir, 1988; Zuboff, 1988; Lee and Moray, 1992).

Important issues concern the development of system trust. For instance, some difference of opinion exists on whether the knowledge about the development of trust acquired in the field of interpersonal research can simply be applied to trust in machines, or whether a theoretical distinction between system trust and interpersonal trust is justified. Furthermore, the development of trust is not a blind process and, as such, needs some kind of informational input. Besides the experience people gain from their own interactions with a system, we will argue that information obtained from other people will also provide such input.

Trust in humans versus systems
Generally, models of system trust used by many researchers do not explicitly distinguish between humans and automation, probably because these were adopted from existing models concerning trust in interpersonal relationships. However, some researchers have put this assumed equality of interpersonal trust and system trust to the test.

Lerch and Prietula (1989), for instance, investigated how attributions of qualities to agents influenced trust on the part of the operator. They found that source pedigree – that is, the source being a human novice, a human expert or a computer – played an important part in the formation of trust. Interestingly, their results suggest that although participants' levels of confidence (trust) in an expert system did not differ from their confidence in a human novice offering the same advice, the information used to form these judgements somehow differed. In addition, although confidence in the human expert's advice was higher than in the human novice, the information used seemed to be the same. If the source was human, participants did not seem to use their judgements of agreement with each individual piece of advice to update confidence levels when an unpredictable negative event occurred. Contrarily, agreement judgements were incorporated if the source of advice was an expert system.

Waern and Ramberg (1996) conducted two studies in which they compared trust in advice given by humans or by automation, but found contradictory results. In a study concerning a matrices test, they found that human advice was trusted more than computer advice, whereas in a study concerning car repair problems, they found opposite results. Waern and Ramberg argued that these findings may well be explained by differences in the particular task and participants' background knowledge. In contrast to the first study, the task in the second study was more difficult and required domain-specific, rather than general, knowledge about cars, which may have caused participants to place more trust in computer-generated advice than in advice from humans.

Lewandowsky et al (2000) acknowledged human operators and automation to be interchangeable at the task level, but hypothesized that control allocation and trust in human–human interactions and human–automation interactions may differ at the social level. Similar to human operators interacting with a human collaborator, a person's trust in automation would be positively linked to performance. However, Lewandowsky et al (2000) argued that, contrary to control delegation between humans, switching to automatic control implies that the person delegating control still bears ultimate responsibility for the quality of the process's outcomes. Because this responsibility is recognized by operators, the occurrence of errors may affect self-confidence. Contrarily, in situations in which humans interact with each other, this responsibility is shared.[2] This diffused responsibility may cause self-confidence to be more resilient to the occurrence of errors. Indeed, in their study they found that, in contrast to the automation condition, self-confidence remained largely unaffected by errors during manual operation in the human–human condition. Lewandowsky et al (2000, p121), furthermore, found no evidence indicating that people are more reluctant to delegate control to a human collaborator than to automation, and concluded that 'the moment-to-moment dynamics of trust between people who share tasks within a complex environment resemble those observed between human operators and automation'.

Chapter 1 of this book proposed a dual-process model of cooperation, separating the concepts of trust and confidence. Whereas trust is based on social relations, group membership and shared values, confidence is proposed to be a belief concerning the occurrence of expected future events, based on experience or evidence. Specifically, trust is taken to be a relation between an agent and another (presumed) agent, whereas confidence concerns agent–object relations. The distinction between trust and confidence should therefore be made on whether agency is conferred on the other (trust), or whether the other is objectified (confidence). Conference of agency implies, among other things, inferring values and reasons, rather than causes, behind behaviour. Another key difference, according to Chapter 1, is the centrality of emotions to trust and the attempt to avoid them in confidence.

Central to some of the ideas mentioned above is the contention that the difference between trust in humans and non-humans lies in the attribution of concepts as traits, reasons, intentions and values to the entity to be trusted. Lerch and Prietula (1989), who found that the same advice was trusted more when it was given by a human expert rather than a computer or a human novice, argued that this phenomenon was caused by users' attributing a trait as dependability to human experts, but not to human novices and expert systems. In a similar vein, Lewandowsky et al (2000) argued that trust between humans and automation is asymmetrical because people may not be willing to attribute values, motivation and personal goals to machines. In Chapter 1, although the possibility that people may take certain objects to be agents is not excluded, the distinction between confidence and trust appears to favour a similar distinction between trust in other humans and trust in non-human entities based on the inference of agency. Given sufficient interaction, people attribute agency (value similarity, intentions and reasons) to other people, but probably not to systems; interaction with systems will probably not encourage judgements to go beyond the mere prediction of behaviour from objective evidence, such as perceived causes.

However, empirical evidence has yet to show that the attribution of concepts as values, intentions or goals allows for a valid distinction between system trust and interpersonal trust. One could argue that the development of system trust to the point where such attributions are made is a mere matter of interaction duration and complexity. Rempel and colleagues (1985) argued that the level of dependability is reached only after the truster has observed a sufficient amount of predictable behaviour; as such, interpersonal trust is assumed to develop from observation of objective information (the other's behaviour) to a stage in which attributions are made (dependability). In light of this notion, the contention that trust in systems seems to be based on different information than trust in humans, as implied in Chapter 1, may not necessarily stem from conceptual differences, but rather from differences in developmental stages. In other words, system trust may, indeed, be based on different information, but perhaps only because it has not yet had the opportunity to evolve into the same developmental stage as interpersonal trust. Not only may our interactions with systems be less frequent than those with other people, trust-relevant information may also be more available in social interaction than in human–system interactions (for a discussion, see Ronald and Sipper, 2001). Both differences in the frequency of interactions as well as the amount of trust-relevant information available per interaction may cause system trust to progress more slowly through Rempel et al's (1985) developmental stages than interpersonal trust. Given sufficient time and interaction, system trust, too, may become based on trait inference. The development of system trust, therefore, should be allowed sufficient time to develop before possible conceptual differences with trust in humans can be established.

In addition, perceptions of causes, as a proposed antecedent of confidence rather than trust (see Chapter 1) can probably only be made when the system's inner working are relatively straightforward; systems that are considerably more complex may make it hard, if not virtually impossible, for a user to establish cause-and-effect relations. Causes being obscured, users may turn to less objective information, such as emotions, and may be less reluctant to attribute traits as dependability and competence, and agency to the system, thus transcending Rempel et al's (1985) stage of predictability. In fact, this is in keeping with Dennett's (1987) idea of the intentional stance. Dennett argued that behaviour of not only other people, but also complex artefacts such as chess computers, could be predicted and interpreted much more easily when these entities are treated as rational agents. Hence, this stance is frequently adopted with regard to systems.

The observation that system trust comes down to predicting future outcomes after observation of behaviour does not exclude the possibility that, given time and complexity, it evolves to a stage analogous to trust in a human actor. The fact that most people are aware that systems cannot actually hold traits, values or intentions in the same way that humans do is by no means detrimental to this conclusion. In fact, research by Nass and Moon (2000) clearly indicates individuals mindlessly apply social rules and expectations to computers, although every single one of them articulated awareness that a computer is not a person and does not warrant attributions or treatment, as such. For instance, after exposure to a social category cue (a Korean or Caucasian video face), persons with the same ethnical background perceived the agent to be more attractive, persuasive, intelligent and trustworthy, compared to participants with a different ethnicity, just as they would if they had been dealing with a real person. As Nass and Moon (2000, p86) state: 'Once categorized as an ethnically marked social actor, human or non-human was no longer an issue'.

Besides, value similarity as a basis for trust may not be restricted to interpersonal relationships. Similar phenomena can be found outside this context. It may, for instance, be comparable to selectivity, a principle thought to underlie trust in media, as some media researchers argue (Kohring and Kastenholz, 2000). Thus, one may trust the content of a particular paper because its perceived social or political stance, which becomes apparent from the selection of and reporting on news items, matches one's own. In a similar vein, the output of relatively complex systems may depend upon a hierarchy of decision rules; a route planner, for instance, may favour a particular strategy for determining routes during rush hours, whereas another strategy is selected in the quiet evening hours. A change in the prioritization of decision rules that causes output patterns to change could be interpreted by the user in the same way that he or she interprets changes in the behaviour of another human to be indicative of intentions.

With most systems, the trust a user has in it could reasonably be labelled competence – that is, an expectation of 'competent role performance' (Barber, 1983). In other words, most of these systems will not invoke the user to attribute such concepts as agency, intention or values to it. That does not mean that system trust should, therefore, be considered a special case of confidence; as was argued earlier, systems that are highly complex may actually very well cause users to attribute such concepts to the system. Additionally, the lower frequencies with which users interact with systems compared to fellow humans may also influence whether trust is based on past behaviour or on attributions of any kind. Indeed, system trust probably does not depend solely upon whatever information becomes available from past behaviours; it may also find root in simple cues, similar to group membership in value similarity.

In sum, system trust does not necessarily rely on more and objectified information, or solely on past behaviour, but may also be based on simple cues and, possibly, on inferred agency, intentions or values, analogous to trust in a social context. Thus, the only difference between trust in human–system interactions and that in interpersonal interactions may well be the actual balance between perceptions of competence, on the one hand, and perceived agency, values or intentions, on the other. Perhaps trust in human actors may put more emphasis on the latter, while also incorporating the former, whereas with trust in systems this may be the other way around. Therefore, willingness to rely on the advice or help of a system is not necessarily a matter of mere confidence; rather, interactions with systems may involve both confidence and trust. No formal distinction between trust in humans and non-humans, or between trust and confidence, will, therefore, be made in this chapter.

Trust based on indirect information

In April 2003, visitors of online bookstore Amazon.com were greeted with the message that the retailer 'continues to show remarkably high levels of customer satisfaction. With a score of 88 (up 5 per cent) (American Customer Satisfaction Index), it is generating satisfaction at a level unheard of in the service industry.'

Information such as this is commonly used in the world of e-commerce. Online bookstores, for example, often provide the possibility to check other readers' satisfaction, not only with the product itself, but also with the services provided by the affiliated bookstore to which customers are directed following a query. In addition, previous customers' satisfaction levels with both book and affiliated bookstore are often summarized by displaying scores on five-point scales. Thus, online bookstores try to reduce any uncertainty associated with buying products from the internet.

Research in consumer behaviour has shown that recommendations or satisfaction reported by others are highly influential in choosing between product alternatives. A survey by Formisano et al (1982) (see also Aggarwal, 1998), for example, suggested that most of the targeted consumers (as much as 75 per cent), who all had purchased a life insurance policy relatively recently, reported not to have engaged in any pre-purchase information acquisition about other insurers. Assuming that these consumers did not already have such information available in long-term memory, Formisano et al (1982) concluded that they had found strong support for their notion that consumers' choices were predominantly based on recommendations made by others. The main source of such recommendations, in this case, was the salesperson, although some respondents mentioned other sources as well, such as parents, friends and relatives.

Although recommendations are frequently used to influence consumer behaviour, the effects on trust have largely escaped attention in experimental research on system trust. Yet, some researchers have theorized about different kinds of indirect information in relation to trust. Barber (1983), for instance, noted fiduciary expectations as a possible basis for trust; a user who may not yet have had the opportunity to form trust towards a system based on prior interactions may hold fiduciary expectations towards a system or its designers. Similar to a patient trusting the skills of a previously unknown physician on the basis of the latter's presumed thorough education, a user may expect the system to function as it should, merely because it was designed by people who are trained and employed to build correctly functioning systems. This constituent of trust offers the possibility of trusting a system based on the moral obligations of its designers, instead of on assessments of reliability (see also Lee and Moray's, 1992, concept of purpose).

Numan and colleagues (Arion et al, 1994; Numan, 1998) are among the very few to explicitly acknowledge the role of other people's experiences as a basis for system trust. According to them, merely observing others in interaction with the system may lead the observer to draw conclusions about the system's trustworthiness; observing successful interactions may increase the observer's trust, whereas observing the occurrence of output errors, for example, is likely to decrease trust. This idea is similar to Bandura's (1977) notion of vicarious learning, or learning by observation, which emphasizes the role of models in the transmission of specific behaviours and emotional responses (see also Green and Osborne, 1985). According to this principle, people may, for instance, exhibit aggression because it is a behavioural pattern that is learned from observing the behaviours of others. Numan and colleagues specifically applied observation of others' behaviours to the formation of trust. Moreover, Numan and colleagues also noted the possible influence on trust of the reported experiences of others – for example, in the form of recommendations (Arion et al, 1994; Numan, 1998).

In the light of the attention for such cues in consumer research and in persuasion and attitude literature, it is surprising to notice the scant attention that recommendations have received in experimental studies on system trust. Virtually nothing is known about the short-term and long-term effects of such recommendations on trust, and the interaction with other types of information, such as one's own experiences with a system.

Trust based on direct information
Direct experience is gained by actually interacting with the system and may yield information about the system's behaviour over time. Repeatedly yielding satisfactory output, the system may be perceived as predictable,

consistent and stable, thus enabling users to anticipate future system behaviour (see Rempel et al, 1985; Zuboff, 1988; and Lee and Moray, 1992, in the context of interpersonal trust).

As opposed to indirect information, direct information or the information obtained by one's own interactions with the system has attracted considerable interest in experimental studies on system trust. These studies have mostly focused on a single type of direct experience – namely, output feedback. Typically, the focal system produces systematically varied numbers of output errors, such as underheating or overheating of juice in a pasteurization plant (Muir, 1989; Lee and Moray, 1994), or the incorrect classification of characters as either letters or digits (Riley, 1996), which are subsequently shown to influence trust and reliance on automation.

Such unequivocal errors, however, may not be the only trust-relevant information obtainable from direct experience. Woods et al (1987), for instance, found that when technicians do not trust a decision aid, they either reject its solution to a problem or try to manipulate the output towards their own preconceived solutions. In their study, they found evidence that occasionally technicians, working with a system designed to diagnose faults in an electromagnetic device and suggest repairs, simply judged themselves whether the system's pending advice was likely to solve the problem, rather than implementing the suggested change and subsequently checking whether it provides the desired results. In other words, these technicians apparently did not wait until unequivocal right–wrong feedback became available to them to form a trust judgement, but rather followed their own judgements on the plausibility of the system's 'line of reasoning' as it was fed back to them. Apparently, people sometimes judge the quality of system advice on feedback regarding the process that led to that advice.

Lee and Moray (1992) hypothesized that, besides automation reliability, 'process' should also be considered as a trust component of direct experiences. Process is used to denote an understanding of the system's underlying qualities or characteristics. Whereas in humans this might encompass stable dispositions or character traits, in a more technological domain this could be interpreted as rules or algorithms that determine how the system behaves. As such, 'process' bears a close resemblance to Zuboff's (1988) 'understanding' and Rempel et al's (1985) 'dependability' as an aspect of interpersonal trust. Others have come up with mental models to denote understanding of a system. Mental models refer to representations that capture the workings of a device (Sebrechts et al, 1987). As such, they represent knowledge of how a system works, what components it consists of, how these are related, what the internal processes are, and how they affect components (Carroll and Olson, 1988). Mental models allow users to explain why a particular action produces specific results; however, they may be incomplete or internally inconsistent (Allen, 1997).

Such understanding of how a system arrives at a solution to a problem presumably increases user trust. Presumably, one aspect of such process feedback that instils trust is consistency; users may conclude that there is a reason for the system's process feedback to show a particular recurrent pattern. For example, a user may request system advice on a number of different routes and subsequently find that the system persists in favouring routes that use a ring road over those that take a short cut through the city centre (or vice versa). Despite an initial belief that suitable routes should more or less follow a straight line, repeatedly encountering this preference may cause the user to conclude that the system's advice is not so inaccurate after all. To the user, there must be a reason why this seemingly inaccurate route subsection is favoured, and he may start conjecturing what that reason might be. Eventually, the user may infer that although the short cut through the centre seems faster, the system may discard it because it is prone to dense traffic. Although such explanations do not necessarily match the system's actual procedures, they may facilitate the formation of beliefs about what is happening 'inside the computer'. This more profound insight in the system's inner workings (comparable to, for instance, Zuboff's, 1988, 'understanding', Lee and Moray's, 1992, 'process' and Rempel et al's, 1985, dependability') may reduce the user's uncertainty even further and, thus, lead to a greater willingness to rely on the system's advice. Indeed, research by Dzindolet et al (2003) has shown that participants working with a 'contrast detector' to find camouflaged soldiers in terrain slides trusted the system more, and were more likely to rely on its advice when they knew why the decision aid might err, compared to those who were ignorant of such causes.

Although Dzindolet et al's (2003) studies provide additional empirical support for the idea that a sense of understanding is beneficial for trust, their participants did not obtain this information from their own direct experiences with the device, as both Lee and Moray's (1992) concept of 'process' and mental model theory entails, but rather received that information from the experimenter. As such, the assumption that users gain understanding by observing system behaviour remains untested.

Combined effects of indirect and direct information
It is not improbable that, in daily life, users have multiple concurrent types of information available to help them form a trust judgement about a particular system; besides their own experiences, based on process and outcome feedback, the opinions of others may also be used. Someone whose own experiences are somewhat limited may also take into account what friends or colleagues have to say about it before deciding to use the system another time. Potentially important in this regard may be the amount of information obtainable per type; indirect experiences may yield far less information than direct experience. Indeed, the recommendations that someone receives concerning a system are sometimes made by others

simply expressing a mere final judgement in terms of suggested use or disuse, unaccompanied by reasons or considerations for arriving at this judgement. Hence, such recommendations may be less influential than direct experience (Arion et al, 1994; Regan and Fazio, 1977) and are presumably easily overridden by information with a higher information density.

However, direct experiences may not always contain more information than indirect experiences. Consider, for example, drinking wine; one may have read an expert's view on why a particular wine is excellent, or one may try to taste why it has such a good reputation. In the case of reading an expert's comments, one is likely to obtain more information than after tasting it; the first mode may yield information about tannin content, how oak casks have contributed to its taste, the excellent quality of its grapes, the carefully controlled yeasting process, etc., whereas the latter may yield a judgement based on far fewer dimensions – for example, sweetness and 'body'. Likewise, the process feedback of a system may not necessarily provide a user with information to enable him or her to form beliefs about its functioning. When a system's feedback is rather consistent, this will allow users to generate a line of reasoning to explain the regularities. We can therefore consider this type of feedback as highly informational and being able to override the influence of the recommendations. Contrarily, inconsistent feedback may contain far less information that will be instrumental in the formation of such beliefs. As such, it will probably result in decreased levels of trust, while it may not be able to cancel the effect of recommendations. Probably, the presence of recommendations will also be needed to interpret the process feedback. One may also argue, however, that inconsistent feedback does provide information – namely, information that is used to form distrust, instead of trust. If this were, indeed, the case, inconsistent feedback would, in addition to decreased trust levels, cancel the impact of recommendations on trust as well.

Two experiments were conducted using a route planner as the focal system. The effects of recommendations (consensus information) and mere process feedback on system trust will be established in study 1. Study 2 will deal with consistency as an aspect of process feedback, its influence on system trust and its interaction with consensus information.

STUDY 1: CUE EFFECTIVENESS AND MERE PROCESS FEEDBACK

Participants were requested to interact with four supposedly different route planners, two with route display in automatic mode, and two without. Prior to the interaction phase they were given high- and low-consensus cues. Trust measurements took place both before and after the actual route planning phase in order to capture both the effect of the cue on trust, as well as a possible effect of the mere presence of process feedback. All participants had to complete the route planning trials in automatic mode.

We expected that, in the absence of direct information, participants would rely on consensus information to form trust. Specifically, participants who receive high-consensus information (majority endorsement) will have more trust in the system. Contrarily, participants who receive low-consensus information (minority endorsement) are expected to have less trust. Furthermore, we expected that the influence of this cue on after interaction trust measures would depend upon the availability of other concurrent information – that is, whether or not process feedback was made available.

Design

Twenty-four undergraduate students participated in this study. The experiment had a 2 (consensus: minority endorsement versus majority endorsement) × 2 (process feedback: yes versus no) within-participants design.

Procedure

On arrival at the laboratory, participants were seated in separate cubicles, where the experiment was run at computers. First, all participants were instructed to carefully read the instruction before starting the route planning programme. From the instruction they learned that they would participate in research concerning the way in which people deal with complex systems. Specifically, they would have to interact with route planners capable of determining an optimal route by estimating the effects of a vast number of factors, ranging from simple ones, such as obstructions and one-way roads, to more complex ones, such as (rush-hour) traffic patterns. Furthermore, they were told that the computer had a database at its disposal, containing route information based on the reported long-time city traffic experiences of ambulance personnel and policemen from that city. These experiences supposedly constituted a reliable set of optimal routes, against which both manually and automatically planned routes could be compared and subsequently scored. Thus, the route planning capability of both human and machine could be validated; feedback regarding the quality of the routes was given after completion of the entire experiment.

During the experiment, a map of London was shown on the screen (see Figure 11.1). Using this map, participants were requested to perform a professional route dispatcher's task by sending the quickest possible routes to imaginary cars, the current location and destination of which were indicated on the screen. The route planning phase consisted of five trials with each of the three route planners. By clicking the 'automatic' button, the route generating process was started; by clicking 'accept,' it was supposedly sent.

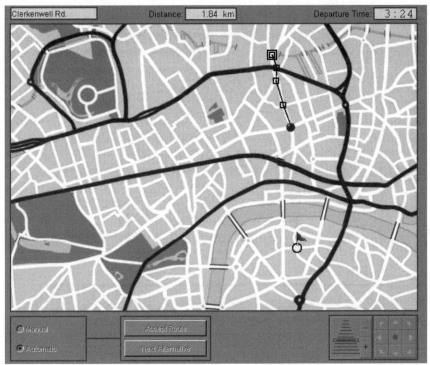

Source: Chapter authors

Figure 11.1 *Route planner interface*

Before actual interacting with a route planner, high-consensus participants learned that a majority (either 'more than 83 per cent' or 'approximately 88 per cent') of the students who had participated in a recent pre-test were extremely satisfied. In the low-consensus condition, participants were told that this percentage was either 'less than 17 per cent' or 'approximately 12 per cent'. Process feedback was manipulated during interaction with each of the route planners. In the process feedback condition, the route planner could be observed to find a route by connecting crossings from start to finish, whereas in the no process feedback, this information was entirely absent.

We assumed that participants would be more committed to the task if a certain risk were to be associated with their choices. Thus, we designed the experiment so that they were allotted ten credits per route planning trial, which, either entirely or partially, could be put at stake. Directly after a route was indicated on the map, a dialogue box would appear on the screen, asking participants to enter any number of the allotted ten credits as stakes. When an automatically generated route, after comparison with the database with reported routes, was judged slower, participants lost the credits that they had staked; a quicker route resulted in a doubling of the staked credits. The number of credits that participants accumulated was

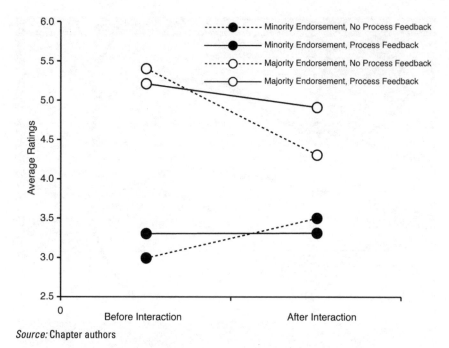

Source: Chapter authors

Figure 11.2 *Effects of consensus and process feedback in before and after interaction measurements*

updated after each trial and was visible on the screen throughout the experiment. They were told that the actual amount of money that they would receive in return for participating would depend upon this accumulated sum of revenues. Besides committing participants to their task, the procedure of staking credits presumably also provided information regarding the level of system trust. After all, high stakes indicate equally high levels of trust that the system would not betray participants' expectations regarding its performance.

Both before and after interaction with the route planner, participants were required to rate the extent to which they trusted the system – a seven-point scale, ranging from 'very little' (1) to 'very much' (7). Thus, we obtained self-reports of system trust, in addition to the measure of trust derived from the staking of credits.

Finally, participants were debriefed, thanked and paid. However, as the programme only gave bogus feedback, all participants were rewarded equally with 3 Euros (approximately US$3.50).

Results

Before and after interaction trust measures

An analysis of variance (ANOVA) was run in which both before and after interaction trust measures were inserted as a separate variable: time. Results are shown in Figure 11.2.

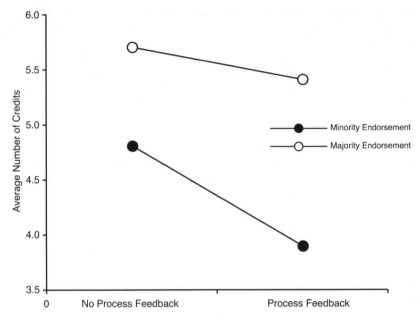

Source: Chapter authors

Figure 11.3 *Effects of consensus and process feedback on staked credits*

The overall main effect of consensus turned out to be highly significant: $F (1, 23) = 38.4; p < 0.01$. Both process feedback and the interaction between consensus and process feedback did not yield significant effects: $Fs < 1$. The relative timing of the trust measures did produce a marginally significant main effect, indicating that before interaction measurements of trust were slightly higher than after interaction measurements: $F (1, 23) = 3.6; p < 0.08$. However, the differences between both averages were very small: mean = 4.2 versus mean = 4.0. In addition, a significant interaction between consensus and time was found: $F (1, 23) = 11.1; p < 0.01$; the effect of the consensus manipulation (majority versus minority endorsement) was largest in the before interaction measurements. In addition, a significant three-way interaction between consensus, process feedback and time was found: $F (1, 23) = 6.0; p < 0.03$ (see Figure 11.2). Compared to the before interaction measurements, both in the process feedback and no process feedback conditions the after interaction trust measurements tended to shift towards the centre of the scale, especially when a majority endorsement cue had been given. This effect was most pronounced in the no process feedback conditions.

In addition to a marginally significant main effect of time and a highly significant effect of consensus – $F (1, 23) = 3.8; p < 0.07; F (1, 23) = 25.7; p < 0.01$, respectively – additional analyses revealed a highly significant interaction between consensus and time when process feedback was absent: $F (1, 23) = 18.8$; $p < 0.01$. Contrarily, when process feedback was given, only a significant effect of consensus was found: $F (1, 23) = 35.7$; $p < 0.01$; no effects of time or an interaction emerged: all $Fs < 1$. Apparently, the previously mentioned three-way interaction is accounted for by a significant interaction between consensus and time in the no process feedback condition.

Staked credits
An ANOVA revealed a significant within-participants effect of consensus: $F (1, 23) = 7.2$; $p < 0.02$ (sphericity assumed) – indicating that participants had entered less credits in trials preceded by a minority cue than in trials preceded a majority cue. However, no significant effects of process feedback or of an interaction were found: $F (1, 23) = 2.3$; $p < 0.15$; and $F (1, 23) < 1$. Results are shown in Figure 11.3.

Conclusions

In this experiment we have been able to show the effect of recommendations on system trust. Negative recommendations in terms of minority endorsement information caused people to be less willing to put something at stake regarding the outcome of the route planning process than when majority endorsement information was given. Similar effects were found on the trust ratings that were taken before and after interacting with the system: when participants were led to believe that only a minority of previous participants were satisfied with the system, they rated trust lower than when majority endorsement information was given.

We expected that this cue effect would at least be attenuated by the availability of process feedback – that is, display of the calculated route. However, this expectation was not supported by our data: no main effect of process feedback or a significant interaction was found. Instead, we found results that suggest that when no process feedback was given, the after interaction trust measurements tended to shift towards the centre of the scale, compared to before interaction measurements.

STUDY 2: CUE EFFECTIVENESS AND PROCESS FEEDBACK CONSISTENCY

This experiment was conducted to study the effects of consensus information in combination with consistent versus random route generation. We expected that when a random process determines the displayed route, the effect of a consensus manipulation would show no signs of fading over time. Specifically, we expected no differences between the before or after interaction trust measurements in terms of cue effectiveness, as participants will constantly recall the consensus information from memory in order to be able to interpret the indefinite visual information, as was found in the previous study. However, as random process feedback may be interpreted as system inadequacy, we could reasonably expect to find an additional decrease in trust ratings. Conversely, when process feedback is consistent in that it displays routes with a distinct preference – say, for arterial roads instead of routes through the centre – this may allow participants to generate a line of reasoning to explain the regularities. As such, consistent process feedback contains more information than random process feedback and causes trust to increase, while overruling the influence of consensus information.

Design

Thirty-two undergraduate students participated in this study. The experiment had a 2 (consensus: minority endorsement versus majority endorsement) × 2 (process feedback: random versus consistent) within-participants design.

Procedure

Except for manipulations of process feedback and the addition of manipulation checks, the procedure followed in this experiment was similar to that of study 1. Again, trust was measured by before and after interaction measurements, as well as by logging the number of credits staked at the beginning of each trial. Furthermore, process feedback was manipulated by displaying routes that either favoured arterial roads (consistent process feedback) or displayed routes that were randomly selected from a subset of alternatives (random process feedback); manipulations of consensus did not differ from study 1. The manipulation checks targeted the extent to which participants could predict the generated routes, the generated routes matched the way in which they themselves would have planned them, they thought the generated routes displayed a certain pattern, and they thought that the generated routes were based on fixed rules.

Results

Manipulation checks

Our manipulation of process feedback proved quite successful. Analyses of the manipulation checks indicated that consistent process feedback allowed participants to get an idea of how the system functioned. First, participants clearly noted the consistency in the process feedback; in the consistent condition, they rated the extent to which they thought the process feedback showed a certain pattern significantly higher than in the random condition: mean = 6.5 versus mean = 4.5; $F(1, 31) = 22.7$; $p < 0.01$. Second, consistency enhanced participants' ability to predict route generation, as indicated by higher predictability ratings in the consistent condition compared to the random condition: mean = 6.9 versus mean = 5.4; $F(1, 31) = 44.3$; $p < 0.01$. Finally, in the consistent condition, they

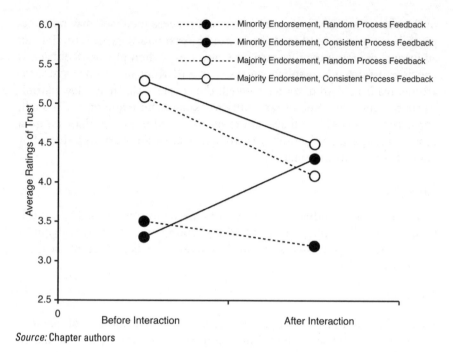

Figure 11.4 *Effects of consensus and process feedback in before and after interaction measurements*

believed more strongly that fixed rules were the basis for the generated routes than in the random condition: mean = 6.9 versus mean = 5.5; $F(1, 31) = 15.3$; $p < 0.01$.

In addition, participants rated a better match between the generated routes and the way in which they would have planned it themselves, when consistent process feedback had been given: mean = 5.6 versus mean = 4.6; $F(1, 31) = 8.3$; $p < 0.01$, which may indicate that randomness in process feedback is interpreted negatively.

Before and after interaction trust measures
A repeated measures ANOVA was performed on the before interaction trust measure (for both consistent and random process feedback). A highly significant effect of the consensus manipulation on the before interaction trust measure was found, indicating that trust was rated higher after a majority cue than after a minority cue: $F(1, 31) = 73.8$; $p < 0.01$.

The after interaction trust measure showed a similar effect of consensus: $F(1, 31) = 5.1$; $p = 0.03$; a majority endorsement message led to higher trust ratings than did minority endorsement. Furthermore, a significant main effect of process feedback was found: $F(1, 31) = 7.8$; $p < 0.01$, indicating that trust measures were higher after consistent process feedback than after random process feedback. Finally, no interaction between consensus and process feedback was found: $F(1, 31) = 2.3$, not significant, suggesting that the effects of consensus and process feedback were additive.

In sum, before interaction trust measures proved to be affected by consensus information. After interaction trust measures were affected additively by consensus and process feedback (see Figure 11.4).

Another repeated measures ANOVA revealed a significant three-way interaction between consensus, process feedback and time: $F (1, 31) = 4.4; p = 0.04$. Subsequently, we tested our specific hypotheses by performing two separate two-way repeated measures ANOVAs with consensus and time (before, after), one for random, and one for consistent process feedback.

With regard to random process feedback, the hypothesis stated that both before interaction, as well as after interaction, trust would be influenced by consensus information and that, possibly, trust ratings would decrease over time as random process feedback was interpreted negatively. Indeed, both before and after interaction measures were affected by the consensus manipulation; ratings were significantly higher following a majority endorsement message than they were after a minority endorsement: $F (1, 31) = 25.7; p < 0.01$. No interaction was found: $F (1, 23) = 2.9; p = 0.1$. Additionally, a significant main effect of time was found – after interaction measures were lower than before interaction measures: $F (1, 31) = 9.1; p < 0.01$.

Concerning the effects of consistent process feedback over time, two different hypotheses were generated. We expected consistent process feedback either to have an additive effect on the after interaction measurements or to interact with consensus. Again, a two-way repeated measures ANOVA was performed, revealing a main effect of consensus: $F (1, 31) = 27.9; p < 0.01$, indicating that trust ratings were higher when a majority cue had been given than they were after a minority cue. No main effect of time was found: $F (1, 31) < 1$, not significant. Moreover, a highly significant interaction was found: $F (1, 31) = 25.5; p < 0.01$, indicating that the consensus manipulations only had an effect on the before interaction trust measurement, but not on the after interaction measurement.

Staked credits
The number of stakes entered showed a significant effect of consensus: $F (1, 31) = 5.3; p < 0.03$, as well as a marginally significant interaction between consensus and process feedback: $F (1, 31) = 3.1; p = 0.09$. Process feedback did not produce a significant effect: $F = < 1$, not significant (see Figure 11.5).

Summary

In sum, consensus information proved to have an effect on both before and after measures of trust. As in the previous experiment, majority endorsement provided participants with information that was subse-

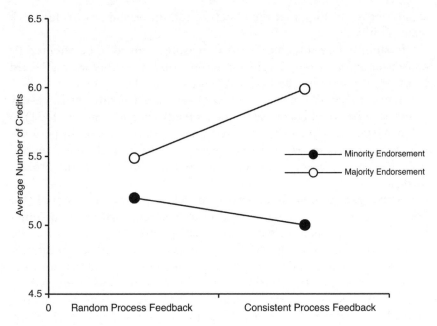

Figure 11.5 *Effects of consensus and process feedback on staked credits*

quently used upon which to build trust. Furthermore, process feedback also provided participants with information that was used to judge the system in terms of trustworthiness, depending upon the type of that information (random or consistent).

With regard to random process feedback, our data show that the differences in trust between majority and minority endorsement treatments did not change over time, an indication that participants, indeed, needed the consensus information stored in memory to interpret the randomized information presented on the screen. Additionally, after interaction measures of trust were lower than before interaction measures in the random process feedback conditions; apparently, randomized process feedback was used to decrease trust.

In the consistent process feedback treatments, a different pattern emerged. Although consensus information influenced participants to rate their initial levels of trust as high (majority information) or low (minority information), this effect was absent in the after interaction measure, which was in line with one of the two hypotheses.

The credits staked provide partial support for these findings. Our results show that these stakes, too, were influenced by the consensus information. In addition, a trust-enhancing effect of consistent process feedback seems present in the majority endorsement condition; however, this could not be ascertained at the required level of significance.

Discussion

The two reported studies present a number of interesting phenomena pertaining to situations in which users of systems have different kinds of information at their disposal. Study 1 shows that cue effectiveness can be moderated by process information. When process feedback was omitted, the influence of the cue, given beforehand, diminished over time. Our participants clearly rated their trust as less extreme afterwards than they did beforehand: very low initial trust ratings increased somewhat, whereas high trust decreased with interaction. Presumably, because the elapsed time between the cue manipulations and the first trust measurement was much smaller than that between the cue manipulation and the second trust measurement, the latter could have caused participants' recollection of the cue content to become less salient, resulting in less extreme trust ratings. On the other hand, if process feedback was present, trust ratings remained fairly stable over time. As the corresponding routes probably did not provide clues as to how fast they actually were, they provided rather ambiguous information. Possibly, this necessitated participants to use the only other information available (the consensus manipulation) in order to decide whether a particular route should be classified as either slow or fast. In other words, the cue content was probably used to fill in the gaps that the visual information provided. As a result, the cue content would have to be recalled from memory with each new displayed route, thus decreasing the possibility that the memory of the cue content would fade.

Providing participants with process feedback with a random appearance, we observed a pattern of trust ratings similar to the first experiment. Presumably, random process feedback, too, provided indefinite visual information that required cue content for interpretation. Hence, the effect of consensus information was present both before and after the interaction stage. Nevertheless, randomized routes also resulted in a general decrease of trust ratings. In all probability, randomness was taken as a sign of a system's incompetence. Consistent process feedback, on the other hand, apparently provided information upon which trust judgements were subsequently based. The regularities that could be observed in the displayed routes, such as a preference for routes along arterial roads, led to higher levels of trust. In fact, its effects were strong enough to completely annihilate the effects of consensus information.

The finding that indirect information is overruled by direct information seems to be consistent with Numan et al's (Arion et al, 1994; Numan, 1998) notions. In their view, indirect experiences simply yield less information than direct experiences. However, we would like to argue that this is not entirely accurate. Whether or not direct experiences provide more information than indirect experiences depends, in part, upon the nature of the former. If process feedback is consistent, a user may form expectations and beliefs about the system's functioning, which may, indeed, overrule the less informational indirect information. However, our data seem to indicate that in the case of random feedback, indirect information is actually necessary for interpretation. Although this contention may be speculative, our data revealed that this particular combination of direct and indirect information resulted in persistence, rather than extinction, of the effect of consensus information.

Although the results of study 2 show its trust-enhancing capacities, consistency alone may not provide sufficient grounds for trust to arise. Indeed, one may think of a system yielding output that consists of consistent, yet unlikely or unacceptable, advice. If the user wants advice on how to travel from, say, the Royal Albert Hall to Piccadilly Circus, and from Piccadilly Circus to Blackfriar's Bridge, a route planner that consistently incorporates Hyde Park in its suggestions is not very likely to instil trust in the user. In the research presented here, the consistent process feedback likely consisted of fairly agreeable routing advice; questions remain as to what will happen to a user's trust in a system when process feedback is far less agreeable or likely. Specifically, the question is what the influence of consistency will be when the routes displayed are highly unlikely as correct solutions. Perhaps only consistently agreeable process feedback will cause trust to rise. Consistent process feedback that is unlikely to be correct, on the other hand, may actually cause a decrease of trust.

The concept of process feedback may be operative in different ways as well as different contexts. First, processing duration, such as the search times of an internet search engine, may also be indicative of underlying processes. If a search engine or a medical database consistently takes a long time to answer a query, this may be interpreted as the content of the database being enormous and, therefore, the answer being trustworthy. Varying search times, on the other hand, may not. Likewise, the automatic conversion of word-processor documents into PDF files, or the automatic generation of a reference list in a research paper, is usually accompanied by phenomena that convey information about the ongoing process, such as duration, either visualized in progress bars or not, indications of the subtask currently performed and flickers on the screen.

Process feedback, however, is not exclusive to the interaction between humans and systems; it may also play a role in the interactions of individuals with other people or organizations. In the context of risk management, this implies, for instance, that risk regulatory institutions should be especially aware of the information concerning their functioning that is available to the general public. The trust of the public in the various kinds of governmental or scientific bodies that are involved in risk regulation may have roots in far more subtle and implicit information than the information that is conveyed through official channels. The eventual conclusions of a committee appointed to investigate such issues as noise levels generated by aircraft approaching a nearby airport, the possible negative impact on health of a nuclear power plant, or the risk of electromagnetic fields caused by overhead power lines are probably the result of many months of investigation and deliberation. During this period of retreat, however, members of the general public who are interested in the subject matter do not have a lot of trust-relevant information at their disposal; yet, a judgement in terms of trust is preferable in order to correctly assess the forthcoming research results. To satisfy their desire for information, these members of the public may turn to what they hear from other people or, as is suggested by the results of these experiments, from more subtle and implicit information about the functioning of the committee. The second experiment shows that such process information may contain trust-enhancing cues and that these may interact with other available information, such as someone else's positive or negative opinions about the issue at hand.

Indeed, the research presented here suggests that consistency in process feedback may increase trust. However, as was argued earlier, consistency may not always be interpreted positively. Therefore, members of investigatory committees who seek to convince the public that they can be trusted to draw the right conclusions from their research should consider explicitly informing people about exactly how they operate, instead of forcing them to piece it together on what little process information they can find. This may increase the likelihood that the public will accept conclusions that do not meet their expectations – for instance, when a relation between the presence of electromagnetic fields and occurrences of cancer, assumed by a worried public, is disproved. Dzindolet et al's (2003) findings that the occurrence of system failures may not harm trust levels if users learn why such failures occur, compared to when users are left in the dark about system functioning, provide support for this.

According to Yamagishi and Yamagishi (1994), trust involves the judgemental implications of insufficient information to be overestimated in order to reduce uncertainty. This notion typifies our focal phenomenon quite well. After all, the participants in these experiments were supplied with as little information as possible. First, the map of London that was displayed on the screen during the experiments did not contain information about, for instance, traffic lights, tourist highlights or one-way streets, which would have aided them in their tasks. Furthermore, the staking of credits took place before the route of the specific trial was shown on the screen. Finally, outcome feedback was given after all the route planning trials were completed. In sum, all experiments reported here concerned situations of uncertainty or, as Numan (1998) called it, partial evidence. As such, the mental states that were manipulated and measured here would be more in line with Lewis and Weigert's (1985) notion of trust; according to them, trust involves a leap of faith, contrary to confidence.

One could argue that the system utilized here was probably not complex enough to warrant the inference of human-like phenomena, such as intentions, benevolence or value similarity, as a basis for trust. Nevertheless, trust as observed in these experiments bears a close resemblance to some of Rempel et al's (1985) notions. These results suggest that inferences were made about the system, indicating that system trust went beyond the level of mere predictability of system behaviour. As such, it would be analogous to Rempel et al's (1985) stage of dependability, which also goes beyond mere predictability, involving the inference of general traits, such as dependability or reliability. In addition, as argued in the 'Introduction' to this chapter, inference of such concepts as shared values and intentions from system behaviour is by no means an unrealistic possibility (compare Dennett 1987; Nass and Moon, 2000).

The studies reported here show that system trust does not necessarily rely on more and objectified information, or solely on past behaviour, but may also be based on simple cues and, possibly, on inferred agency, intentions or values, analogous to trust in a social context. Perhaps the only difference between trust in human–system interactions and that in interpersonal interactions may well be the actual balance between perceptions of competence, on the one hand, and perceived agency, values or intentions, on the other. Maybe trust in human actors may put more emphasis on the latter, while also incorporating the former, whereas with trust in systems this may be the other way around.

Although the importance of information from others seems to be acknowledged in the field of marketing, as the examples in the section on 'Trust based on indirect information' illustrate, the research presented is, at least as far as we know, the first to experimentally test its effects on system trust. In these experiments, consensus information was used as indirect information; but naturally there are many other elements worth studying with regard to system trust, either alone or in concurrence with other types of information. Besides recommendations, one could, for example, also concentrate on the effects of applying humanlike aspects, such as speech and emotions, to system interfaces.

Likewise, the concept of process feedback, and its possible beneficial effects on understanding, has received only scant attention since it found its way into theories on system trust a few decades ago (see Zuboff, 1988; Lee and Moray, 1992). Although Dzindolet et al (2003) quite recently tested the effects of understanding of a system's processes on trust, the emergence of understanding from actually observing process feedback remained obscured. Therefore, these experiments are also the first to try and uncover how process feedback plays a role in the development of system trust via understanding.

Needless to say, much work needs to be done to fully uncover all trust-relevant aspects of direct information. Since interactions with systems normally entail both outcome as well as process feedback, we would like to stress the importance of studying not only the former, which yields feedback in clear right–wrong verdicts, but also the latter, which may contain far subtler clues. In order to fully understand users' perceptions of, and interactions with, systems, what matters is not just what these systems do, but how they do it.

NOTES

1 In this context, the term 'system' is used to denominate automated computer-based systems (applications), notably decisions aids, and will henceforth be considered synonymous with 'automation'. 'System trust' implies trust in one-on-one interactions between users and such systems.

2 However, we would like to argue that this is only the case if the task to be delegated can also be performed manually; a task that one cannot perform oneself, such as performing surgery on oneself, requires the delegation of both control and responsibility to someone else – in this example, a surgeon.

REFERENCES

Aggarwal, P. (1998) *Deciding Not To Decide: Antecedents of Consumer Decision Delegation*, PhD thesis, University of Minnesota, Minneapolis, MN

Allen, R. B. (1997) 'Mental models and human models', in M. Helander, T. K. Landauer and P. Prabhu (eds) *Handbook of Human-Computer Interaction*, Elsevier Science, Amsterdam, The Netherlands, pp49–63

Arion, M., Numan, J. H., Pitariu, H. and Jorna, R. (1994) 'Placing trust in human–computer interaction', in *Proceedings of the 7th European Conference on Cognitive Ergonomics (ECCE 7)*, Gesellschaft fur Mathematik und Datenverarbeitung, Bonn, Germany, pp353–365

Bandura, A. (1977) *Social Learning Theory*, Prentice-Hall, Englewood Cliffs, NJ

Barber, B. (1983) *The Logic and Limits of Trust*, Rutgers University Press, New Brunswick, NJ

Carroll, J. M. and Olson, J. R. (1988) 'Mental models in human-computer interaction', in M. Helander (ed) *Handbook of Human-Computer Interaction*, Elsevier Science, Washington, DC, pp49–65

Dennett, D. C. (1987) *The Intentional Stance*, MIT Press, Cambridge, MA

Dzindolet, M. T., Peterson, S. A., Pomranky, R. A., Pierce, L. G. and Beck, H. P. (2003) 'The role of trust in automation reliance', *International Journal of Human-Computer Studies*, vol 58, pp697–718

Formisano, R. A., Olshavsky, R. W. and Tapp, S. (1982) 'Choice strategy in a difficult task environment', *Journal of Consumer Research*, vol 8, pp474–479

Giddens, A. (1990) *The Consequences of Modernity*, Stanford University Press, Stanford, CA

Green, G. and Osborne, J. G. (1985) 'Does vicarious instigation provide support for observational learning theories? A critical review', *Psychological Bulletin*, vol 97, no 1, pp3–17

Halpin, S., Johnson, E. and Thornberry, J. (1973) 'Cognitive reliability in manned systems', *IEEE Transactions on Reliability*, vol R-22, no 3, pp165–169

Kohring, M. and Kastenholz, H. (2000) *Vertrauen in Medien: Forschungsstand, theoretische Einordnung und Schlussfolgerungen für Vertrauen in Technologie*, Friedrich-Schiller-Universität Jena, Germany

Lee, J. and Moray, N. (1992) 'Trust, control strategies and allocation of function in human-machine systems', *Ergonomics*, vol 35, pp1243–1270

Lee, J. D. and Moray, N. (1994) 'Trust, self-confidence, and operators' adaptation to automation', *International Journal of Human-Computer Studies*, vol 40, pp153–184

Lerch, F. and Prietula, M. (1989) 'How do we trust machine advice?', in G. Salvendy and M. Smith (eds) *Designing and Using Human-Computer Interfaces and Knowledge-Based Sytems*, Elsevier Science, Amsterdam, pp410–419

Lewandowsky, S., Mundy, M. and Tan, G. P. A. (2000) 'The dynamics of trust: Comparing humans to automation', *Journal of Experimental Psychology: Applied*, vol 6, pp104–123

Lewis, D. and Weigert, A. (1985) 'Trust as a social reality', *Social Forces*, vol 63, pp967–985

Luhmann, N. (1979) *Trust and Power: Two Works by Niklas Luhmann*, John Wiley and Sons, Chichester

Muir, B. M. (1988) 'Trust between humans and machines, and the design of decision aids', in E. Hollnagel, G. Mancini and D. D. Woods (eds) *Cognitive Engineering in Complex Dynamic Worlds*, Academic Press, London, pp71–84

Muir, B. M. (1989) *Operators' Trust in and Use of Automatic Controllers in a Supervisory Process Control Task*, PhD thesis, University of Toronto, Toronto, Canada

Nass, C. and Moon, Y. (2000) 'Machines and mindlessness: Social responses to computers', *Journal of Social Issues*, vol 56, pp81–103

Numan, J. H. (1998) *Knowledge-Based Systems as Companions: Trust, Human Computer Interaction and Complex Systems*, PhD thesis, Rijksuniversiteit Groningen, The Netherlands

Regan, D. T. and Fazio, R. H. (1977) 'On the consistency between attitudes and behavior: Look to the method of attitude formation', *Journal of Experimental Social Psychology*, vol 13, pp28–45

Rempel, J. K., Holmes, J. G. and Zanna, M. P. (1985) 'Trust in close relationships', *Journal of Personality and Social Psychology*, vol 49, no 1, pp95–112

Riley, V. (1996) 'Operator reliance on automation: Theory and data', in R. Parasuraman and M. Mouloua (eds) *Automation and Human Performance: Theory and Applications*, Lawrence Erlbaum Associates, Mahwah, NJ, pp19–36

Ronald, E. M. A. and Sipper, M. (2001) 'Intelligence is not enough: On the socialization of talking machines', *Minds and Machines*, vol 11, pp567–576

Sebrechts, M. M., Marsh, R. L. and Furstenburg, C. T. (1987) 'Integrative modeling: Changes in mental models during learning', in W. Zachary and S. Robertson (eds) *Cognition, Computing and Cooperation*, Ablex, Norwood, NJ, pp338–398

Sheridan, T. B. and Hennessy, R. T. (1984) *Research and Modeling of Supervisory Control Behavior*, National Academy Press, Washington, DC

Waern, Y. and Ramberg, R. (1996) 'People's perception of human and computer advice', *Computers in Human Behavior*, vol 12, pp17–27

Woods, D. D., Roth, E. M. and Bennet, K. (1987) 'Exploration in joint human–machine cognitive systems', in W. Zachary and S. Robertson (eds) *Cognition, Computing and Cooperation*, Ablex, Norwood, NJ, pp123–158

Yamagishi, T. and Yamagishi, M. (1994) 'Trust and commitment in the United States and Japan', *Motivation and Emotion*, vol 18, pp130–166

Zuboff, S. (1988) *In the Age of the Smart Machine: The Future of Work and Power*, Basic Books, New York

12 Trust and Confidence in Crisis Communication: Three Case Studies

Michael Siegrist, Heinz Gutscher and Carmen Keller

The concepts of trust and confidence may be key to a better understanding of why crisis communication sometimes works and sometimes fails. A short theoretical introduction to this chapter outlines the difference between trust and confidence. Three crisis case studies illustrate how trust and confidence can influence people's willingness to cooperate. Furthermore, the factors that may have been crucial for the success or failure of the measures taken during the crisis cases are identified. Finally, based on the theoretical considerations and the three case studies, we derive some general recommendations.

KNOWLEDGE AND CRISIS COMMUNICATION

In risk communication or crisis communication the focus is often on facts and knowledge. However, this strategy may not work very well. Empirical studies suggest that there is little connection between knowledge and acceptance of a hazard. Large gaps in knowledge force us to accept the assessments of experts and opinion leaders. In contrast, if we are well informed on a particular topic, we can make use of that available knowledge when making decisions, and trust becomes superfluous. A study by Siegrist and Cvetkovich (2000) showed that trust plays an important role when knowledge is limited. For technologies and activities that participants in the study had little knowledge of, the researchers observed correlations between trust and risks. Participants did not even attempt to make their own assessments of the risks from such hazards. However, when participants had relatively comprehensive knowledge of a topic, they did not have to rely on trust and were able to evaluate the risks themselves. As a consequence, no correlations were observed between trust and perceived risk.

In a crisis, most people do not have the knowledge that they need for making an informed decision. Here, people may need trust in order to reduce the complexity that they face (Luhmann, 1989).

THE IMPORTANCE OF TRUST IN A CRISIS

Different types of trust should be distinguished. For crisis communication, the proposed difference between trust and confidence seems to be especially crucial (Earle et al, 2002). Confidence is based on familiarity, experience and past performance. Social trust, in contrast, refers to the willingness to rely upon others. We trust a person whose values are compatible with our own main goals. Other researchers have used the term 'shared values' to describe this notion (Earle and Cvetkovich, 1995). Confidence, on the other hand, is based on past experience and evidence. We do not have to wonder each morning whether there is electricity or whether it could be dangerous to switch on the coffee machine. For technologies that we are accustomed to and that are not the focus of public discourse, this confidence plays an important role.

Based on an extended review of the literature on trust, a dual-mode model of cooperation based on social trust and confidence has been proposed (Earle et al, 2002). The causal model depicted in Figure 12.1 is derived from that model. According to the model, judgements of value similarity determine judgements of social trust. This assumption is supported by several studies (such as Earle and Cvetkovich, 1999; Siegrist et al, 2003). The other path to cooperation is via confidence. Perceived past performance determines confidence. A number of social psychological studies suggest that value information tends to predominate over performance information, meaning that morality information is the more important in forming an impression and that it provides the context for, and affects the interpretation of, performance information. Given positive value information, negative performance is judged much less harshly than it would be if the value information were negative. Furthermore, social trust can have a direct effect on confidence. The proposed dual-mode model was tested in the applied context of electromagnetic fields (EMFs) (Siegrist et al, 2003). Results of the study suggested that the dual-mode model of social trust and confidence was useful as a common framework in the field of trust and risk management.

For the functioning of most institutions, the absence of social trust is not critical; the absence of mistrust suffices. Social trust is not a necessary condition for people's cooperation. As long as uncertainty is low (indicating constancy), confidence is sufficient for cooperation. What is frequently overlooked, however, is that the situation becomes very different when disruptions turn into crisis situations and lead to the shattering of confidence.

Occasionally, certain performance information exceeds the limits of familiarity and past experience, and demands our attention. Losing confidence is experienced as negative affect. In psychological terms, we experience stress. In a desire to get relief from that stress, we turn to members of our group, to people for whom similar values are important.

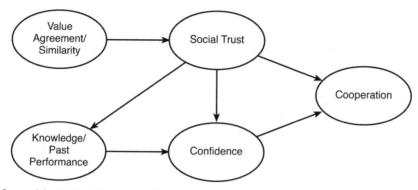

Source: Adapted from Siegrist et al (2003, p708)

Figure 12.1 *The core parts of the dual-mode model of trust and confidence*

Restoration of confidence is not possible if a fundament of social trust has been established earlier on.

In the three case studies that follow, we will use the distinction between trust and confidence to analyse Shell Oil Corporation's problems with the disposal of the Brent Spar oil platform, as well as two food-related cases: the failure of Coca-Cola Company in handling the European contamination crisis and Sara Lee Corporation's successful handling of the listeriosis crisis in the US.

SHELL'S PLANNED DEEP-SEA DISPOSAL OF BRENT SPAR

In the spring of 1995, there were extensive reports in the European media on Shell Oil's plan for abandonment of the Brent Spar floating oil storage and tanker off-loading buoy through deep-sea disposal in the North Atlantic. Greenpeace fuelled the media coverage, sent activists to occupy Brent Spar and provided the media with video coverage. Greenpeace accused Shell of committing an environmental crime. As a result, consumers across Europe boycotted Shell products. In the end, the oil giant was forced to give up deep-sea disposal of the Brent Spar installation, even though the disposal plan had been approved by government officials in the UK and in Norway.

A number of studies have examined Shell Oil's decision to dump Brent Spar at sea and identified various factors that may have led Shell to neglect public attitudes towards deep-sea disposal of the installation. Shell Oil's scenario planning did not take into account the shift in values that had occurred in the public, which perceived environmental problems as more urgent than in the past (Elkington and Trisoglio, 1996). In addition, during the decision-making process, Shell focused on the dollar cost of disposal and did not take other costs into sufficient account: the cost of losing its reputation, the long-term ecological dangers of deep-sea dumping, and stricter regulation as a response to its own actions (Nutt, 2004).

Others criticized Shell's risk communication strategy. Löfstedt and Renn (1997) suggested that involving representatives from environmental organizations in the decision-making process could increase the acceptance of a selected option. Grolin (1998) also emphasized that corporations would need to develop new and trustworthy forms of dialogue with the public. Neale (1997) noted that collaboration with outsiders should be initiated before this is imposed by forces outside of a company's own control.

Some background information on Brent Spar

Brent Spar, a floating oil storage facility and tanker off-loading buoy, was installed in the North Sea in 1976 and taken out of commission in 1991 in preparation for abandonment. Shell examined various offshore and onshore abandonment options for Brent Spar. Dismantling and onshore disposal were one alternative to deep-water dumping. However, according to a risk analysis done by Aberdeen University Research and Industrial Services, the onshore disposal option was four times as expensive, and hazard to the workforce and risk of fatal accidents was six times higher than deep-water disposal (Abbott, 1996). Even the environmental risks could have been higher for the onshore disposal option because of the danger that the buoy might break up.

Several studies of the abandonment methods suggested that deep-water disposal of Brent Spar in the Atlantic was the best practicable environmental option (BPEO). The UK government approved deep-water disposal in February 1995. No objections were raised against the disposal as 'an expert group set up by the British government said that the global environmental impact of deep-sea disposal for Brent Spar would be "very small, roughly equivalent to the impacts associated with the wreckage of a fairly large ship"' (Masood, 1998, p407).

Greenpeace activists occupied the Brent Spar in April 1995 and launched a campaign via the mass media (broadcast worldwide). Greenpeace argued that the planned deep-sea dumping was harmful to the environment and that the only advantage of that option was its cost-effectiveness.

During the summer of 1995, public protest against the planned deep-water disposal of Brent Spar arose in several European countries. Consumers boycotted Shell products. Notably in Germany, the Greenpeace campaign led to consumer boycotts and firebomb attacks on, and vandalism of, Shell gas stations. The media and the public – strongly supported by environmental organizations – spoke out against deep-sea disposal of the Brent Spar. The public outcry took Shell by complete surprise. Several factors then convinced Shell not to dispose of the Brent Spar in the Atlantic: opposition from European governments, consumer boycotts, and the risks for retailers and staff from bombings and shootings (Side, 1997).

Once Shell decided not to dump Brent Spar, it emerged that Greenpeace had mistakenly issued false information about the buoy's contents (Abbott, 1996). Shell received an apology from Greenpeace for grossly overestimating the amount of toxic chemicals and oil in the Brent Spar tanks (Side, 1997). The mistake notwithstanding, the environmental organization insisted that deep-water disposal was not an option.

According to Gage and Gordon (1995), Shell's impact assessments of disposing of the Brent Spar showed important errors in the appraisal of the environments at the proposed disposal sites. Nevertheless, they conceded: 'a one-off deep-sea disposal of Brent Spar might not be the environmental catastrophe suggested by Greenpeace' (Gage and Gordon, 1995, p772). Interestingly enough, research completed after Shell's capitulation supported Shell's original proposal and not Greenpeace's assessments of the option (Pulsipher and Daniel, 2000).

Media coverage

Jensen (2003) analysed mass media coverage of the Brent Spar conflict by media in Denmark. The media discourse was dominated by the arguments used by Greenpeace and prominent politicians and ministers opposed to deep-sea disposal of Brent Spar. Danish newspapers helped to create in the public the perception of an ecological disaster. The conflict between Greenpeace and the giant oil company was portrayed as a battle between David and Goliath. The public and the mass media, both of which adopted Greenpeace's position, emphasized the same set of values in fighting against an alleged environmental offender. Danish newspapers interpreted the outcome of the conflict, Shell bowing to pressure and not dumping Brent Spar in the Atlantic, as a victory for Greenpeace and the environment. The Danish media was probably not an exception in Europe in depicting Shell as the culprit and Greenpeace as a white knight.

Value implications

Had Greenpeace activists not occupied the Brent Spar, the public would not have been concerned about its disposal in the North Atlantic. The deep-water option favoured by Shell was legal, and the UK government supported the decision. However, Greenpeace's actions against Shell undermined public trust in the oil giant. Both actors, Greenpeace and Shell, presented completely different information to the public. Shell emphasized that the chosen option was the BPEO. Greenpeace, on the other hand, argued that disposing of the Brent Spar would be very harmful to the ocean. People in Europe were not able to double-check the facts provided by Greenpeace and by Shell. And in the absence of knowledge, judgements are guided by social trust (Siegrist and Cvetkovich, 2000).

According to the dual-mode model of trust and confidence, value similarity determines social trust, and the situation determines what values will be relevant. For the UK government and Shell, the costs associated with disposing of the Brent Spar were an important factor. The impression that Shell was more concerned about the costs of disposal than about protecting the environment was nourished by risk assessments presented by Shell itself. In the evaluations of the disposal options, technical issues dominated (Nutt, 2004). The costs associated with technical difficulties, risks to the workforce and environmental impacts were estimated. Even the UK energy minister emphasized the cost implications of the decision (Neale, 1997). For the public, on the other hand, the salient value was protection of the environment – not cost or technical issues. Whereas the public may trust Shell with regard to the economical implications of a chosen option, most people in Europe have more value similarity with Greenpeace than with Shell regarding environmental protection. Therefore, it is not surprising that people placed more trust in Greenpeace than in Shell.

Trust dominates confidence. In other words, if you trust in an organization, some performance aspects might be of less importance. Although Shell had the facts right and Greenpeace had the facts wrong, this was of little importance to the public or the mass media. Shell's strategy was confidence based: it was a presentation of facts along with the argument that Greenpeace had used the wrong facts. The phenomenon that in a crisis situation, trust – and not necessarily confidence – is important showed up clearly when neither Greenpeace nor the public altered their positions after Greenpeace admitted that they had issued erroneous information.

What went wrong?

Since the early 1970s, Shell has instructed its workers not to throw waste into the sea from platforms (Side, 1997). Shell was evidently quite aware that a segment of the public is concerned about how a company handles waste and waste disposal. It is therefore surprising that Shell did not fully take into account the effect on the public that dumping the Brent Spar at sea would have.

Public opinion was not considered in Shell's impact assessments of the different disposal options. Therefore, values that are important to the public were missing in the Shell scenarios. Possible actions of environmental organizations did not enter into the equation for calculating the cost-benefit effects of the various abandonment methods.

The public did not believe that it was a coincidence that the cheapest option was also the one judged to be associated with the lowest risk. By selecting the most cost-effective option, Shell came to be perceived as an organization that was concerned about profit and not about the environment.

Furthermore, Shell did not take into account that people might be against disposing of the Brent Spar in the ocean in any case, no matter how harmful or not this might be to the environment. John Shepherd of the Southampton Oceanography Centre acknowledged: 'If people have an emotional response to pristine areas like Antarctica or the deep sea, and want them to remain unpolluted, it is not up to scientists to say this is irrational' (cited in Abbott, 1996, p14). Values of this kind have been called absolute values (Baron and Spranca, 1997). When people possess absolute values, cost-benefit analyses are impossible because a given value will always be more important than any other value. How society should deal with such absolute values is still an open question.

ANATOMY OF A RECALL: A REPUTATION AND MARKETING NIGHTMARE FOR COCA-COLA IN BELGIUM

In the summer of 1999, a number of children in Belgium fell ill, reportedly after drinking Coca-Cola soft drinks. As a feverish search for the cause of the mysterious epidemic got under way, Coca-Cola asserted in its communications over some time that there was nothing wrong with its products and that it could not have caused the symptoms. Nonetheless, Coca-Cola issued various ever-larger recalls of individual production batches. When more people fell ill in other parts of the country, the Belgian government, under public pressure, placed a total ban on the sale of all Coca-Cola soft drinks. Further countries also imposed selective restrictions and sales bans. Within days, the events had led to the worst public relations disaster and largest recalls in Coca-Cola's over 100-year company history. By the end, Coca-Cola had recalled and destroyed 25 million gallons of soft drink products, at a total cost of US$103 million.

The sequence of events

On 8 June 1999, pupils at a secondary school in Bornem, Belgium, complained that Coca-Cola soft drinks purchased at the school cafeteria smelled and tasted bad. Some children drank the soft drinks anyway, some returned their bottles half full and some received replacements. Because so many pupils complained about the Cokes and Fantas, the school stopped their sale (Gallay and Demarest, 1999). Within a short time, a number of children in different classes had reported themselves ill. To be on the safe side, by the end of the school day, 33 pupils experiencing abdominal discomfort, headache, nausea, malaise respiratory symptoms, trembling and dizziness had been admitted to the local hospital for observation on the advice of the Medical School Inspection (MSI) (Nemery et al, 2002). Because the complaints had begun at lunchtime, Coca-Cola soft drink consumption was quickly identified as a plausible cause of the symptoms. Enquiries of the pupils at school regarding drinking Coke or feeling unwell led to a second wave of hospital admissions, with some children entering the hospital on the evening of 9 June and others on the morning of 10 June. In all, 37 children in Bornem, with an average age of 13, entered the hospital for treatment.

Clinical tests revealed nothing conspicuous, and blood and urine samples yielded no indications of a toxicological syndrome. Most of the pupils were released from the hospital on the same day. The national health authorities were informed, and Coca-Cola Antwerp was contacted to ask if it was aware of the students' complaints. The director of Coca-Cola Antwerp came to the school and collected two crates out of the 15 remaining at the school for examination (Gallay and Demarest, 1999). The events in Bornem were reported in the evening news on television and figured prominently on the radio and in the newspapers.

Coca-Cola decided on 9 June to recall bottles of Coke stemming from the same production batch as the soft drinks at the school. In a press release, the Coca-Cola Company admitted that it had pinpointed 'quality problems' in its bottling plant at Antwerp; but at the same time it asserted that there was no health risk. It did mention that its product, however, could cause headache, dizziness and abdominal pain (Nemery et al, 2002).

In the days following, a total of 75 further cases of symptoms in schoolchildren were reported in various parts of the country. Again, children were transported to the hospital by ambulance, accompanied by media attention. As before, no serious illness was found in the new cases. This time, the products consumed were not bottled soft drinks, but drinks in cans traced to a Coca-Cola bottling plant in Dunkirk, France. On the evening of 10 June, Coca-Cola recalled 2.5 million bottles and cans with specific production codes (NZZ, 1999a; Nemery et al, 2002). Cases of illness continued to be reported. On Saturday, 12 June, the Belgian health authorities banned the sale of all products from Coca-Cola bottling plants to take effect on Monday, 14 June. On 15 June, Luxembourg followed suit with a total ban, and France and The Netherlands issued a partial sales ban on Coca-Cola products (NZZ, 1999c, 1999d). Because of grey market imports, various other countries issued warnings about Coca-Cola products from Belgium (BBC-Online-Network, 1999b; NZZ, 1999b, 1999d).

Schoolchildren were not the only ones reporting symptoms. Following extensive media coverage of the first cases of 8 and 9 June, wide segments of the general population also became ill. By 20 June, the Belgian Poison Control Centre had registered more than 1000 calls, of which 50 per cent reported complaints about Coca-Cola products.

On 15 June, the director-general of Coca-Cola Enterprises Belgium, Philippe Lenfant, announced that they had identified two causes for the outbreak of illness. In the bottles from one Belgian plant, 'bad carbon dioxide' was to blame, and a 'fungicide' applied to transport pallets had contaminated the outside surface of some cans from another plant (BBC-Online-Network, 1999a; NZZ, 1999e).

On 16 June, Coca-Cola chief executive officer (CEO) Douglas Ivester offered a belated public apology to the people of Belgium via newspapers, television and radio. Ivester offered to reimburse all medical treatment costs; a doctor's certification would be accepted as sufficient evidence of Coke-related illness (NZZ, 1999d). In addition, Ivester promised a free Coca-Cola beverage to every resident of Belgium, about 10 million people, as soon as production in Belgium had been resumed.

On 21 June, the company submitted further analysis results to the authorities in an effort to speed up the lifting of the ban on sales. The ban was finally lifted on 24 June under stringent conditions; but the ban on imports from the Dunkirk bottling plant remained in effect. In the following months, Coca-Cola mounted a relaunch campaign, hiring 6000 Belgians to distribute 4.4 million redeemable coupons door to door. Coca-Cola gave away 8 million free soft drinks. The campaign had become necessary because starting as early as 14 June, Coca-Cola's major competitors, Pepsi and Virgin Cola, had experienced a sixfold increase in demand for their products.

The context

It is very important to consider the situation in Belgium at the time when the Coca-Cola contamination crisis occurred. The public was in the middle of dealing with a dangerous food crisis and health scare: the 'dioxin crisis'. In February of the same year, polychlorinated biphenyls (PCBs, dioxin precursor), dioxins and dibenzofurans had been found in animal feed (Nemery et al, 2002). The Belgian public only learned of the danger, however, on 25 May, when first reports leaked through to the media. A major health scare ensued, and a massive blanket recall was ordered first for egg and chicken products and then for most meats and dairy products. The food crisis precipitated a massive political crisis. Belgium's minister of health and minister of agriculture were denounced for having kept information about dioxin contamination secret for months. They were forced to resign shortly before the general election of 13 June 1999.

The dioxin crisis was the main topic of discussion for weeks; the issues of the lack of official information and of food safety were all-pervasive. In particular, the news from researchers that even the smallest amounts of chemicals can cause significant damage to health was foremost in the public mind. Trust in politicians and foodstuff producers was at an absolute low. These circumstances must have had a great influence on the public's perception of risk. Coca-Cola's problems must have fed into this great food and information sensitivity.

Was it the real thing? The mass sociogenic illness (MSI) hypothesis

Toxicological analyses of the soft drinks at the Bornem school revealed a weak contamination of carbon dioxide (CO_2) by carbonyl sulphide (COS).

In the other cases, it was found that the outer surface of soft drink cans had been contaminated by a fungicide. In both instances, the concentrations of the compounds were so low that, while they could have caused the 'off' taste or odour, they could not have caused toxicity and the symptoms reported (Gallay and Demarest, 1999; Nemery et al, 2002).

An *ad hoc* working group in the ministry of health put forward a different hypothesis in a letter to the *Lancet* as early as 3 July: mass sociogenic illness (Nemery et al, 1999), which is described as 'a constellation of symptoms of an organic illness, but without identifiable cause, which occurs among two or more persons who share beliefs related to those symptoms' (Philen et al, 1989).

The Unit of Epidemiology of the Scientific Institute for Public Health also conducted an investigation of the outbreaks at the schools and published the findings in *Lancet* (Van Loock et al, 1999). The researchers ascertained that the consumption of Coca-Cola beverages was clearly and importantly associated with the outbreaks of illness. For the cases in Bornem, intoxication was not excluded as a possible cause (Gallay et al, 2002). However, all of the following classical risk factors of MSI were found to be present (Gallay and Demarest, 1999): girls were more likely to be ill than boys; the outbreak occurred in a school setting; those affected in the schools were teenagers; symptoms were variable and non-specific; there was no consistency between observed symptoms and symptoms that can be expected in response to toxicity; the 'off' odour could be a trigger factor responsible for anxiety symptoms; the context of stress caused by the food safety scare following the dioxin crisis, the upcoming elections and end-of-year examinations were cumulative risk factors; the media (radio, television and newspapers) played a role by widely diffusing the Bornem incident before the second and following outbreak occurred in Brugge; clusters (non-consistent patterns) in classrooms were identified in at least two schools (Lochristi and Harelbeke); and clusters in a classroom would satisfy the classic requirement for MSI to be amplified when occurring 'in line of sight'. Children with low mental health scores (an MSI indicator) showed more than twice the probability of developing symptoms than children with high mental health scores (Van Loock et al, 1999). Altogether, then, it is reasonable to conclude that the clearly discernible 'off' odour and taste of the soft drinks – in the context of anxiety and upheaval about food safety in Belgium and the specific school situation – worked as classical triggers for acute mass sociogenic illness.

Coca-Cola's response

Although Coca-Cola reacted quickly by partially recalling specific production batches on the day following the first cases (9 June) and by issuing a more comprehensive recall on 10 June, it was apparent to the public that Coca-Cola was taking pains to minimize any potential profit losses. Confusing communications concerning production codes did not help to retain public trust in the remaining stock of non-affected Coca-Cola products. The school involved was given corrected production codes for the affected products twice (Deogun et al, 1999). Coca-Cola massively underestimated the public uncertainty and sensitivity with regard to food safety issues in the wake of the dioxin contamination scandal in the previous months, during which time the public had been thoroughly informed about the hazards of even the smallest concentrations of toxic substances in foodstuffs. Coca-Cola's misjudgement of the context is clearly revealed in its press release on 9 June, which acknowledged 'quality problems' in a bottling plant in Antwerp, yet at the same time denied any severe health or safety issues. The release also stated, however, that Coca-Cola products could possibly cause headache, dizziness and abdominal pain (BBC-Online-Network, 1999b; Nemery et al, 2002)!

Even when the Belgian authorities banned the sale of all Coca-Cola products, the company did nothing but attempt to play down the repercussions. Coca-Cola spokeswoman Shelag Kerr declined to confirm or correct the estimated figure of 15 million bottles and cans that would be recalled, stating merely that the recall represented only a 'very limited proportion of our production' in Belgium. Kerr reported that tests by the company had found no toxins in the beverages. She said there was an 'off' taste or odour to the drinks (CNN-Interactive, 1999), but no risk to health (Nemery et al, 2002).

New but sparse information from the company on 15 June, identifying 'bad carbon dioxide' and 'fungicide' without specifying the substances or concentrations, was seen by the authorities and the public as insufficient and lacking in credibility (BBC-Online-Network, 1999c; NZZ, 1999f). On the contrary, it now appeared to most people that Coca-Cola knew more than it was willing to communicate. CEO Ivester made his first public apology on 16 June, which the Belgian public dismissed as belated and lacking in credibility. Coca-Cola's poor timing and the type and manner of its communications had bungled its handling of the crisis, and public trust was lost. Coca-Cola continued to suffer the consequences of its delayed response. While the company insisted that sales rebounded, it admitted that consumer confidence in the product was damaged and that it has had far higher marketing costs in Europe as a result of the incident (Reid, undated).

Conclusion

Since Coca-Cola in the initial phase of the crisis had itself already conducted practically all chemical analyses of the substances found, it apparently knew that the concentrations did not yield a plausible explanation of the illness symptoms. This probably put it at a loss, and its subsequent handling of the crisis was indecisive for too long. The numerous recalls citing complicated codes for individual production batches gave the public the impression that Coca-Cola was only playing for time and that its interest in minimizing economic losses took priority over public health. In addition, Coca-Cola's premature statement prior to the chemical analyses that its products posed no health risks must have only increased public mistrust, considering public awareness of the cover-ups that had taken place during the dioxin crisis. Thus, Coca-Cola's persistent focus on providing confidence-related information did not serve to offset or cushion the confidence lost. Coca-Cola even made reference to suppliers outside Coca-Cola's sphere of influence, which was clearly a reference to confidence-related past performance. In contrast, the correct strategy would have been for Coca-Cola to base its communication efforts on trust, even if it was probable that mass sociogenic illness was at hand. Certainly, the belated appearance of Coca-Cola CEO Ivester was a big mistake in this regard (Cobb, 1999). Coca-Cola handled the crisis according to textbook solutions. The question is: was it the right book? Katharine Delahy Paine of Delahay Medialink, a renowned media research company, said the problem was that the solution was North American, while the problem was not: 'An American solution to a European problem isn't going to work; just because it looks and feels like a familiar crisis doesn't mean that the rules we use in America will work elsewhere' (Cobb, 1999). Coca-Cola, working out of headquarters in Atlanta, was not able to comprehend the events in Belgium and the context in which they were being played out.

In contrast, trust-related communication would have required Coca-Cola to take a different approach:

> When children say they have some illness because of your product, the company had to send the message that this was important to them. They had to say to the Belgian people: 'We don't know whether there is anything wrong here; but we are sending our top guy to make sure it's handled right.' That is such a powerful image. (Cobb, 1999)

Coca-Cola's failure to understand the social and political environment of the Belgian market when the contamination crisis hit led the company to respond in a wholly inappropriate way. Admittedly, inadequate handling of mass sociogenic illness is certainly documented (negatively) far more frequently than appropriate handling (Wessely, 2000). It is, indeed, very difficult to deal successfully with the dilemma that an outbreak of mass sociogenic illness always presents. Too much attention directed towards symptoms and conspicuous and long drawn-out investigations can, in fact, intensify the situation and the frequency of symptoms. Early and firm statements indicating that the symptoms have a socio-psychological cause tend to isolate and blame the people already affected and their families (Wessely, 2000).

We think that it would have been possible to communicate in a neutral manner the fact that fears can be expressed in physical symptoms –

without accusing people of mass hysteria and without denouncing subjectively experienced symptoms as figments of people's imagination. In such cases, the issue is not who is right, and the point is not to prove 'mass hysteria' (Wessely, 2000). Instead, what is needed is a way to slow down and stop the spread of somatization of fear and people's associated real suffering. The goal of communication has to be to find a balance between taking the fears, along with the symptoms, seriously and promising to conduct thorough investigation, on the one hand, and explaining in a neutral manner, on the other, that fear can produce physical symptoms, that symptoms are real and a burden, but they do not have to mean that something terrible has happened. It is important not to suggest that somatization of fear is 'imaginary illness' or 'imaginary symptoms' and to emphasize that this phenomenon is a normal and (usually appropriate) *modus operandi* of healthy people.

In addition, in this case, it would have been appropriate for Coca-Cola Company to issue a very early apology – namely, for giving rise to fears whose justifiability would be thoroughly investigated. After all, the trigger of the outbreak – the 'off' taste and odour of the soft drinks – was real. Coca-Cola should have accepted responsibility very early on since its own quality control measures had obviously failed.

CASE HISTORY OF SARA LEE CORPORATION

Sara Lee Corporation, with headquarters in Chicago, manufactures and markets brand-name products in the food and beverage industry. In December 1998, Sara Lee faced a deadly situation when its Bil Mar Foods plant in Zeeland, Michigan, producing hot dogs and deli meats, was linked to an outbreak of *Listeria monocytegenes*. The outbreak of listeriosis claimed 15 lives and more than 100 people fell ill. Because of its fast, effective and thorough crisis management, Sara Lee survived the deadly crisis with few scars (Hollingsworth, 1999). How did Sara Lee handle this crisis?

On 18 December 1998, the daily press in Chicago reported that federal health officials were investigating an outbreak of illness from listeria. Four people had already died, and more than 35 were ill. The prime suspects were hot dogs and cold cuts (*Chicago Tribune*, 1998a).

Four days later, on 22 December 1998, Bil Mar Foods, a Sara Lee food plant, voluntarily recalled more than 300 varieties of hot dogs and other packaged meat products because of possible contamination with the potentially deadly bacteria. Sara Lee informed the public about the possibly affected brands and said it would take a pretax charge of US$50 million to US$70 million to cover all of the costs of the recall (*Chicago Tribune*, 1998b).

One day after the recall, the Centers for Disease Control and Prevention (CDC) announced that initial test results indicated a link between Bil Mar Foods and the bacterial illness. Sara Lee announced that it was cooperating with the CDC and the US Department of Agriculture (USDA) (Bigness, 1998). On 31 December 1998, the cases of listeriosis were officially traced to the Bil Mar plant (*Chicago Tribune*, 1998c). One week later, eight fatal cases were reported (*Chicago Tribune*, 1999).

Sara Lee spokesperson Theresa Herlevson announced again on 8 January 1999 that Sara Lee was 'continuing to work very closely with CDC'. She appealed to customers to remove all contaminated products from their refrigerators. Bil Mar Foods halted production. Sara Lee hired quality-assurance experts from universities and private industry to develop new solutions for reducing post-production risk of contamination (Bigness, 1999a).

On 21 January, the number of fatalities had climbed to 14 persons. In a public relations offensive, Sara Lee placed full-page advertisements in 60 daily newspapers repeating the information on the recall of the meat products, which was estimated to involve 15 million pounds of meat products. 'We deeply regret this situation, and we continue to make the safety and wholesomeness of our products our number one priority' was the message of Sara Lee CEO John H. Bryan and chief operating officer (COO) C. Steven McMillan in the advertisements (Bigness, 1999b).

At the beginning of February, Sara Lee had developed plans for new procedures and technologies at Bil Mar Foods (Bigness, 1999c), which were implemented on 4 March with the following company statement: 'We believe we are taking unprecedented safety measures.' The USDA confirmed that meat inspectors and safety experts were on site at Bil Mar Foods to control the new safety procedures (Singhania, 1999).

'Our first priority is always the quality and wholesomeness of our products,' reiterated COO C. Steven McMillan on 27 April, and he announced various initiatives for assuring the highest quality (Center for Food and Nutrition Policy, 1999). Sara Lee Corporation named Dr Anne Marie McNamara vice president for food safety in a newly created position. McNamara, a leading researcher in food safety and technology, was director of the microbiology division of the USDA's Office of Public Health and Science in the Food Safety and Inspection Service. Additionally, Sara Lee Corporation established a US$1 million safety research fund at the Center for Food and Nutrition Policy at Georgetown University in Washington, DC, to identify new methods for improved food safety. In a multi-year US$100 million plus programme, Sara Lee went on to implement new state-of-the-art technology, and new processing, cleaning and testing procedures throughout its US packaged meats operations.

The reasons for Sara Lee's crisis communication success

Prior to the outbreak of listeriosis, customers had had high confidence in the quality and performance of Sara Lee. They were confident that they could buy Sara Lee meat products at the supermarket and eat them without suffering negative consequences. It had never occurred to them that Sara Lee's meat manufacturing system could be unstable. Knowing or assuming that the authorities regularly control meat-manufacturing systems served to support their certainty. It is this confidence that provides the necessary basis for consumers' daily meal planning and organization.

With the news of the listeriosis outbreak, Sara Lee's meat production system was instantly rendered unstable. Confidence in Sara Lee was destroyed. The risk involved in eating Sara Lee packaged meats became a question of trust. Trust was required in order to create once more a new and stable state of confidence in consumers. The main reason for Sara Lee's success in handling the crisis was that from the very start, the measures that Sara Lee undertook and the values-oriented information that it communicated were directed towards the building and retaining of trust.

Even before official test results had established a link between the listeriosis outbreak and Bil Mar Foods, Sara Lee voluntarily recalled packaged meat products of the value of US$50 million to US$70 million. The recall of the contaminated meats and the production halt immediately minimized the risk of infection. The voluntary natures of the measures signalled to the public that Sara Lee takes the safety and health of its customers seriously. Sara Lee's prompt acceptance of financial responsibility demonstrated that the company was applying these values generously. As the first indications were discovered that the listeria outbreak was connected with Sara Lee products, the company cooperated without delay with the control authorities. This prompt cooperation underlined Sara Lee's honourable intentions to protect consumers' safety and health.

Safety and health were precisely those customer values that had been shaken by the listeriosis outbreak. The swift measures taken by Sara Lee provided exactly the required values-related information. The similarity between customers' salient values and the values that Sara Lee focused on engendered trust. Customers were able to summon up trust that Sara Lee was undertaking measures in order to protect safety and health and not in order to maximize company profit.

All of Sara Lee's further measures contained the same values-related information directly or indirectly. Direct and clear reference to safety and health was made in the prompt information about the contaminated products, in Sara Lee's appeal to consumers to dispose of those products, in their large public relations offensive, and in the announcements of safety initiatives. Indirectly, the company's measures for medium-term quality and performance improvement also contained values-related information. Broad-based cooperation with researchers, private industry and the authorities for development of cutting-edge technologies and the introduction of new safety procedures at Bil Mar Foods signalled consistently that Sara Lee took safety and health seriously and thus engendered trust in the company. In turn, trust led consumers to view Sara Lee's subsequent measures and performance in a positive light. This opened up the way for the re-establishment of confidence.

In addition to values-related information, Sara Lee's medium-term measures also provided performance-related information. Competency, experience and control were involved in developing and implementing new high-quality and performance standards. And with its very costly long-term safety initiatives, Sara Lee signalled as early as four months after the crisis that it would, indeed, apply those values of safety and health, as well as high-quality and performance standards sustainably. With this, Sara Lee Corporation laid the foundation for the emergence of new, stable consumer confidence in its packaged meat products.

Conclusion

Every risk or crisis communication faces the dilemma that withholding information can lead to a loss of credibility and trust, and the provision of comprehensive risk information can trigger unnecessary fears and anxiety. Moreover, communication of certain risks creates the impression that the risk can be controlled. For these reasons, it is difficult to formulate general advice on how to handle a crisis. Recommendations can be too specific to be useful across situations, or advice may be so abstract and general as to be useless. Nevertheless, there is a lesson to be learned from the analysis of the three case studies.

In all of the three cases, crisis communication took place in an environment in which confidence was destroyed. Shaken confidence, as the Shell and Coca-Cola cases show, cannot be restored through more or better or more intensive communication at the confidence level. According to the dual-mode model of trust and confidence, trust is the safety net counterbalance for the normal, or default, mode of human behaviour, which is confidence. Without trust, confidence disappears, and it is, as a rule, very time consuming and costly to start all over again from this low point.

Credible and convincing communication in crisis situations should be able to be supported by trust: existing or newly formed. Trust is based on communicator and target persons sharing similar values, assumed or explicit. Here we do not mean inter-subjective, objective similarity, but rather our subjective perception of similarity to our own values that are relevant in a given situation. Communicators should not only be knowledgeable about 'the facts of the case', they should also, and primarily, be knowledgeable about the concerns and values of their target audience. In practice, to achieve value similarity at all, international companies, in particular, cannot do without knowledge of local value preferences. This is a problem that Coca-Cola neglected, or underestimated, with punitive consequences.

All three of the cases were conflicts between companies and consumers, or an environmental organization. One problem is that health and environmental issues were important from the perspective of consumers, whereas companies were automatically assumed to seek profit as their highest and only priority.

With Shell, the profitability of a means of disposal was emphasized. In the Coca-Cola crisis, very limited recalls followed by more and more extensive recalls and a much too late apology from the CEO created the impression that profit was the main concern. In such cases, the disparity of values is great and, as a consequence, trust is forfeited. For this reason, it was easier for Greenpeace to win the race to trust, and this remained the case even when it became clear that Greenpeace had provided erroneous figures on the contents of the Brent Spar tanks.

The Sara Lee case is very different. Clear declaration of responsibility, the expression of regret and sorrow, early and expensive recalls to minimize risk, broad public warning campaigns, a readiness to work with the authorities, and the long-term implementation of new and drastic safety measures all made it possible for consumers to perceive Sara Lee's values as similar to their own. For this reason, trust in Sara Lee was, to a large extent, upheld, and thus a basis existed for new confidence in Sara Lee's production processes. One year after the listeriosis crisis, Sara Lee had emerged basically unscathed from a very serious and dramatic situation. In contrast, in a similar case, Hudson Foods went completely out of business, its finances and reputation in tatters, after poor management of an *E. coli* contamination crisis in 1999 (Hollingsworth, 1999).

In sum, the differentiation of trust and confidence according to the dual-mode model is useful when applied to crisis communication. Moreover, we are convinced that the distinction between trust and confidence will support the correct setting of priorities in crises and, possibly also, in crisis planning.

REFERENCES

Abbott, A. (1996) 'Brent Spar: When science is not to blame', *Nature*, vol 380, pp13–14

Baron, J. and Spranca, M. (1997) 'Protected values', *Organizational Behavior and Human Decision Processes*, vol 70, pp1–16

BBC-Online-Network (1999a) *Business: The Company File/Belgium Bans Coca-Cola*, 14 June, www.news.bbc.co.uk/1/hi/world/europe/369089.stm, accessed 15 July 2003

BBC-Online-Network (1999b) *Business: The Company File/European Warning over Coca-Cola*, 16 June, www.news.bbc.co.uk/1/hi/business/the_company_file/369684.stm, accessed 15 July 2003

BBC-Online-Network (1999c) *World/Europe/Belgium Considers Lifting Coke Ban*, 16 June, www.news.bbc.co.uk/2/hi/europe/370681.stm, accessed 15 July 2003

Bigness, J. (1998) 'Sara Lee recalling packaged meats hot dogs, other items may be contaminated', *Chicago Tribune*, 23 December, p1

Bigness, J. (1999a) '4 more die from tainted meat toll at 8; bacteria found in Sara Lee deli meat', *Chicago Tribune*, 8 January, p1

Bigness, J. (1999b) 'Meat toll 14; Scope of recall disclosed', *Chicago Tribune*, 21 January, p1

Bigness, J. (1999c) 'Sara Lee plant to adopt new food-safety technique', *Chicago Tribune*, 9 February, p1

Center for Food and Nutrition Policy (1999) *Sara Lee Listeria Grant Announcement*, 27 April, www.ceresnet.org/ViewEntry.cfm?ID=289andSection=research, accessed 17 February 2006

Chicago Tribune (1998a) 'Food poisonings leave 4 dead; hot dogs, cold cuts suspected', *Chicago Tribune*, 18 December, p8

Chicago Tribune (1998b) 'Sara Lee unit recalling meats', *Chicago Tribune*, 22 December, p1

Chicago Tribune (1998c) 'Outbreak traced to Sara Lee unit', *Chicago Tribune*, 31 December, p1

Chicago Tribune (1999) 'Bacterial outbreak death toll at 8', *Chicago Tribune*, 7 January, p1

CNN-Interactive (1999) *Belgium Widens Coke Recall as More Children Fall Ill*, 14 June, www.cnn.com/Food/news/9906/14/cocacola.belgium/, accessed 15 July 2003

Cobb, C. (1999) *The Aftermath of Coke's Belgian Waffle*, September, www.prsa.org/_Publications/magazines/Tactics/9909views1.html, accessed 8 August 2003

Deogun, N., Hagerty, J. R., Stecklow, S. and Johannes, L. (1999) 'Cola stains – Anatomy of a recall', *The Wall Street Journal*, 29 June, ppC1, C10

Earle, T. C. and Cvetkovich, G. T. (1995) *Social Trust: Toward a Cosmopolitan Society*, Praeger, Westport, CT

Earle, T. C. and Cvetkovich, G. (1999) 'Social trust and culture in risk management', in G. Cvetkovich and R. E. Löfstedt (eds) *Social Trust and the Management of Risk*, Earthscan, London, pp9–21

Earle, T. C., Siegrist, M. and Gutscher, H. (2002) *Trust and Confidence: A Dual-Mode Model of Cooperation*, Unpublished manuscript

Elkington J. and Trisoglio, A. (1996) 'Developing realistic scenarios for the environment: Lessons from Brent Spar', *Long Range Planning*, vol 29, pp762–769

Gage, J. D. and Gordon, J. D. M. (1995) 'Sound bites, science and the Brent Spar: Environmental considerations relevant to the deep-sea disposal option', *Marine Pollution Bulletin*, vol 30, pp772–779

Gallay, A. and Demarest, S. (1999) *Case Control Study among Schoolchildren on the Incident Related to Complaints following the Consumption of Coca-Cola Company Products*, November, www.health.fgov.be/CSH_HGR/Francais/Avis/Coca-Cola%20Report%20Final.htm, accessed 4 August 2003

Gallay, A., Van Loock, F., Demarest, S., Van Der Heyden, J., Jans, B. and Van Oyen, H. (2002) 'Belgian Coca-Cola-related outbreak: Intoxication, mass sociogenic illness or both?', *American Journal of Epidemiology*, vol 155, no 2, pp140–147

Grolin, J. (1998) 'Corporate legitimacy in risk society: The case of Brent Spar', *Business Strategy and the Environment*, vol 7, pp213–222

Hollingsworth, P. (1999) 'Crisis management: Planning for the unthinkable', *Food Technology*, vol 53, no 10, p28

Jensen, H. R. (2003) 'Staging political consumption: A discourse analysis of the Brent Spar conflict as recast by the Danish mass media', *Journal of Retailing and Consumer Services*, vol 10, pp71–80

Löfstedt, R. E. and Renn, O. (1997) 'The Brent Spar controversy: An example of risk communication gone wrong', *Risk Analysis*, vol 17, pp131–136

Luhmann, N. (1989) *Vertrauen: Ein Mechanismus der Reduktion sozialer Komplexität*, Enke, Stuttgart

Masood, E. (1998) 'Ministers block disposal of oil rigs at sea', *Nature*, vol 394, p407

Neale A. (1997) 'Organisational learning in contested environments: Lessons from Brent Spar', *Business Strategy and the Environment*, vol 6, pp93–103

Nemery, B., Fischler, B., Boogaerts, M. and Lison, D. (1999) 'Dioxins, Coca-Cola and mass sociogenic illness in Belgium', *Lancet*, vol 354, p77

Nemery, B., Fischler, B., Boogaerts, M., Lison, D. and Willems, J. (2002) 'The Coca-Cola incident in Belgium, June 1999', *Food and Chemical Toxicology*, vol 40, pp1657–1667

Nutt, P. C. (2004) 'Averting decision debacles', *Technological Forecasting and Social Change*, vol 71, pp239–265

NZZ (*Neue Zürcher Zeitung*) (1999a) 'Coca-Cola Belgien ruft 2,5 Millionen Flaschen zurück/Erkrankung von Schulkindern', *Neue Zürcher Zeitung*, 11 June, p63

NZZ (1999b) 'Coca-Cola Dosen/in der Schweiz beschlagnahmt', *Neue Zürcher Zeitung*, 19 June, p64

NZZ (1999c) 'Coca-Cola in vier Ländern zurückgezogen/Verbot des Verkaufs auch in Luxenburg', *Neue Zürcher Zeitung*, 16 June, p64

NZZ (1999d) 'Ersatz von Coca-Cola Dosen in der Romandie/Fünf Forderungen des Bundesamtes für Gesundheitswesen', *Neue Zürcher Zeitung*, 18 June, p64

NZZ (1999e) 'Falsche Kohlensäure als Ursache?', *Neue Zürcher Zeitung*, 16 June, p64

NZZ (1999f) 'Wenig Detailliertes von Coca-Cola', *Neue Zürcher Zeitung*, 17 June, p21

Philen, R. M., Kilbourne, E. M., McKinley, T. W. and Parrish, R. G. (1989) 'Mass sociogenic illness by proxy: Parentally reported epidemic in an elementary school', *Lancet*, vol ii, pp1372–1376

Pulsipher, A. G. and Daniel IV, W. B. (2000) 'Onshore disposition of offshore oil and gas platforms: Western politics and international standards', *Ocean and Coastal Management*, vol 43, pp973–995

Reid, J. (undated) 'Now you've really blown it! How to keep a crisis from going from bad to worse', www.janinereid.com/csw.htm, accessed 5 August 2003

Side, J. (1997) 'The future of North Sea oil industry abandonment in the light of the Brent Spar decision', *Marine Policy*, vol 21, pp45–52

Siegrist, M. and Cvetkovich, G. (2000) 'Perception of hazards: The role of social trust and knowledge', *Risk Analysis*, vol 20, pp713–719

Siegrist, M., Earle, T. C. and Gutscher, H. (2003) 'Test of a trust and confidence model in the applied context of electromagnetic field (EMF) risks', *Risk Analysis*, vol 23, pp705–716

Singhania, L. (1999) 'Partial restart at Sara Lee plant Michigan unit taking "unprecedented" safety steps', *Chicago Tribune*, 4 March, p3

Van Loock, F., Gallay, A., Demarest, S., Van Der Heyden, J. and Van Oyen, H. (1999) 'Outbreak of Coca-Cola-related illness in Belgium: A true association', *Lancet*, vol 354, pp680–681

Wessely, S. (2000) 'Responding to mass psychogenic illness', *New England Journal of Medicine*, vol 342, no 2, pp129–130

Index

absolute values 273
abstract targets 21, 24
acceptability of risk 136–137, 151–152
acceptance of technology 145–146, 147
accountability 97
advice automation 243–244
affective components 97, 113, 130, 136
African-Americans 83
agency 20, 22, 244–245, 247, 264
agricultural level 130–134, 159
altruism 73, 89–90, 124–126
altruistic trust *see* moralistic trust
amplification of risk 95, 106, 108, 109–110, 160
  see also mass sociogenic illness
analysis of risk 143–155
antecedents of trust 241–264
appropriate response bias 103–104, 111
  see also response bias
arbitrators of risk 106, 109–110
associative processing 32–33
assurance 190–192, 201, 203–204
  see also confidence
asymmetry principle 2, 21, 34, 168, 197–198, 202
attenuators of risk 106, 109–110
attitudes 85–88, 164, 166, 215
attributes
  importance measurement contrast 214–217
  judgements 20, 23, 25
  ratings 212, 215, 218–219, 220–223, 226–228, 233
  systems trust 244, 245, 246
automated advice 243–244

beef consumption 159, 160, 169
Belgium 273–279
belonging to a group 51–64
benevolence 163
beverage industry 273–282
bias 32, 95, 97
  see also response bias
Bil Mar Foods 279–282
biotechnology 130–134
bioterrorism 173, 176, 177, 181, 184
Birkenstock 89–90
black sheep 61
boundary organizations 148–149
bovine spongiform encephalopathy (BSE) 118, 159, 160, 169
Brent Spar oil platform 269–273, 283
BSE *see* bovine spongiform encephalopathy

carbon dioxide 274, 275, 277
care 97, 119
car purchase 241–242
categorization 55–56, 58, 59, 63, 64
charity *see* altruism
chat rooms 84
chicken consumption 165
clarification 6
climate change 129
Coca-Cola (Coke) 273–279, 282, 283

cognitive styles 112–113
coherence 190
cohesion 57
commitment 253–254
commons dilemma 54
  communication
    credibility of 189–190
    crisis 267–283
    food safety 144, 168
    governance of science 118
    openness and transparency 101–102
    process information 263
    public dialogue 136
    smallpox vaccination 184–185
    technology acceptance 145–146, 147
    trust relationship 1–2, 17
    variability 152–154
community level 31–32, 33
competence
  attribute ratings 215
  dimensionality 96, 97, 119
  food safety 163
  health and safety 125
  IDT 95, 101
  system trust 246, 264
  technological survey 128, 130
  wetlands management 217, 222, 229
competition 218
compliance 80, 81, 218
conceptual levels
critical trust 117–138
  IDT 96, 97–98
  social reliance 187, 188–192
  TCC model 3–6, 7, 11–13, 31, 35
concrete targets 21, 23–24
confidence
crisis communication 267–283
  strategic trust 74–75
  system trust 244, 246, 247
  TCC model 1–49
  see also assurance
conflicts 52, 53
conjunction fallacy 181–182
consensual values 97, 119
consensus information
  interaction measurements 254, 255–256, 258, 259
  process feedback 252, 253, 260, 261, 264

consistency 119, 195–196, 202, 256–260, 261, 262, 263
consumers 149, 150, 159–169, 247–248, 270
contending groups 218
corruption 82
cost-effectiveness 270, 272, 283
credibility 151, 189–190, 277
crisis 159–169, 267–283
criteria ratings 222–223, 224–225, 230–231
critical sentiments 126–128
critical trust 117–138, 134–135, 137, 138, 200
cue effectiveness 251–260, 261
cultural level 64, 278

danger 5, 85, 98–99
decision bias 95
decision-making
  consumers 160, 161
  experts 95, 98–102
  food biotechnology 133, 134, 135
  public consultation 150
decision theory *see* game theory
deep-sea waste disposal 269–273
defensive avoidance 113
deficit model 146
deli meats 279
dependency 5, 34, 51, 264
  see also reliance
dependent variables 24–25, 36–39
depersonalization 56
desirability 7, 215, 233
  see also morality information; performance information
deviance 61
differentiation 193
diffused responsibility 244
dimensionality 96–98, 112, 119–121, 127, 135
dioxin contamination 164–167, 275, 277
direct information 248–251, 252, 261–262
discrimination ability 98–100, 101, 102, 103–113
  see also knowledge
disregard of non-members 137

dissidents 61
diversionary reframing 137

eBay 191
education 222, 229
elected legislative officials 219–232
elites 217–219
emancipation theory of trust 14
emerging technologies 14, 151–152
emotions 16, 188–189, 200
encapsulated experiences 85, 86–88
enforcement 123
environmental commission officials 217, 219–232
environmental level 105, 106, 107, 108, 272, 283
    see also Greenpeace
ethnicity 62, 246
evaluative–instrumental attitudes 164
evaluative self-definition 59
expectancy-value model 214–215, 236
expectations
    asymmetry principles 197–198, 202
    systems trust 241, 242, 246, 248
    wetlands management 234
    see also strategic trust
experience
    confidence 268
    encapsulated 85, 86–88
    familiarity 214
    food safety 161
    organizational invisibility 138
    other people's 248
    role 224, 225, 232, 233
    systems trust 251, 261–262
    varieties of trust 76–77, 78
    worldviews 80
    see also information
experts
    decision-making 95, 98–102, 174
    human versus systems 243
    public contrast 1, 145–146, 153
    reliance on 149
    smallpox 179–180, 184
explicit modes of information processing 188–190, 200
expressed trust 180

fairness 35, 119
faith 119
false alarms 98, 99, 112
false positives and negatives 98–99
familiarity
    role of 213–214
    tentative knowledge 122–123
    wetlands management 224, 229, 231, 232, 234, 235
feedback 241–264
first-time interactions 241
flexibility 200–201
focus groups 121–122
folded response bias 108–109
    see also response bias
food industry 133, 134, 135, 144, 159–169, 275, 279–282
forest management 195–196, 198–199
fragile trust 11, 13
freedom of the other 4, 5, 6, 33–34
free-riding 54
functionality 203
fungicide contamination 274, 276, 277

game theory 53–54
general trust (GT)
    dimensionality 97
    legal systems 81, 82
    moralistic trust comparison 76–77
    optimism 73
    out-groups 83
    particularized trust contrast 75
    TCC model 4, 6, 9, 13–14, 30, 34–35
    worldviews 79–80
    see also moralistic trust
genetically modified (GM) foods
    acceptability 151–152
    consumers 150, 159
    food scares 166
    institutional survey 129, 130–134, 135
GM see genetically modified foods
governance 118, 132, 134
government level
    generalized trust 81
    mobile phone industry 105, 106, 107, 108
    process feedback 262–263

technological surveys 128–134
wetlands management 218, 236
Greenpeace 269, 270, 271, 283
grounded theory approaches 122
group level
  focus 121–122
  GT 13, 14
  particularized trust 79
  prediction and control 31–32
  shared membership 16
  social identity 51–64
  SVS 192–194
  validity 33
GT *see* general trust

harmonization of assessments 145
health and safety 102, 121–128,
    135–136, 169, 283
  see also safety
heuristic systematic model (HSM) 235
hierarchical relationships 163–164
homosexuals 84
honesty 97
hot dogs 279
HSM *see* heuristic systematic model
human–system interactions 243–247
hypervigilance 113

identity 15, 16, 51–64, 192–193
idiosyncrasy credits 60
IDTs *see* intuitive detection theorists
immigrants 73, 83, 84
implicit modes of information processing 188–190, 200
importance 35, 214–217, 222, 223,
    226–228, 233–234
inconsistencies 195–196, 202
increasing trust 63–64
indirect information 247–248,
    250–251, 261–262, 264
individual trust 162
inferences 138, 245, 263–264
information
  communication 136
  crisis communication 281–282
  dimensionality 112
  first-time interactions 241
  food safety 161, 164, 168
  GM foods 131–132, 134, 135

  mobile phones and health 102
  new 197–198
  processing 32–33, 188–190, 199–201
  smallpox 178, 179–183, 184–185
  strategic trust 77
  systems trust 245, 247–251, 252,
    261–263, 264
  TCC model 17, 30
  see also experience
in-groups
  GT 13, 14
  jury service 83
  particularized trust 79
  reliance 192, 193
  social identity 52, 56, 57–62, 64
innovation 60, 88, 150–151
inspections 123
institutional level
  confidence 12
  critical trust 117–138
  distrust 146–147
  food safety 144
  process feedback 262–263
  smallpox 179
  social trust 268
  see also organization level
instrumental attitudes 164
integrated approaches 31, 121
integrity *see* shared values
intellectual desirability *see*
    performance
intentional stance 245–246
interactions
  human–system 243–247
  interpersonal 241–242
  trust measures 254–256, 258–259,
    261
interdependence 193
inter-group trust 52, 62–64
internal standard of coherence 190
internet 84–85, 191, 247
interpersonal level 4, 30, 51, 242, 243,
    244
  see also inter-group trust
inter-subjectivity 57, 59, 64
intra-group trust 52, 59–62
introspection 215, 234, 235
intuitive detection theorists (IDTs)
    95–114

investigation 123
invisibility 138
Italy 83

Japan 191, 204
judgements
  elites 217–219
  IDT 95–114
  moral importance 35
  public-expert gaps 1
  TCC model 20–21, 22, 23, 25, 27–28, 36–39
jury service 83

knowledge
  bias 97
  crisis communication 267, 282
  familiarity 213–214
  functionality 203
  health and safety 122
  mobile phones and health 102
  public versus expert 146, 153
  self-assessed 198–199, 202–203, 217
  self-reported 214, 224, 225–226, 231, 233, 235
  smallpox 175
  TCC model 25–26, 26, 27, 28, 29
  tentative 122–123
  wetlands management 222, 225–226, 230–231
  see also discrimination ability; strategic trust

labelling schemes 169
lay perceptions 145–146, 153
leadership 54, 60, 61
learning by observation 248
legal levels 73–74, 80–83
legitimacy 150–151
listeriosis 279–282
local officials study 211–236
Luhmann's feedback loop 242

mad cow crisis 118, 159, 160, 169
Maghribi people 79
majority endorsement *see* consensus information
manipulation checks 257–258
marginal members 59, 61, 64

mass sociogenic illness (MSI) 275–276, 278–279
meat manufacture 169, 279–282
media
  crisis communication 270, 271, 274, 275
  mobile phone industry 105, 106, 107, 108
  selectivity 246
  smallpox 178, 179, 184
medical profession 105, 106, 107, 108
mental models
  health and safety 123
  knowledge 203
  smallpox vaccination 174, 175–178, 180, 185
  systems trust 249, 250
mere process feedback 251–256
mind reading 192, 197, 201
minority endorsement *see* consensus information
minority groups 63, 73, 79, 83, 84
mixed-methods approaches 119
mobile phones 102–111, 129
Montegrano, Italy 83
moralistic trust 73, 75, 76–77, 77–79, 82
  see also generalized trust
morality
  importance 35
  information 7, 8, 9, 21–22, 27, 28, 268
MSI *see* mass sociogenic illness
multiculturalism 64
multiple regression analysis 230–231, 234–235

natural food 169
negative events 2
negotiated exchange 16
neighbourhood safety 85–88
new information 197–198
New Jersey, US 219–236
non-dilemma studies 16–17
normative issues 120
nuclear power 18, 20–21, 25–26, 111, 113, 117–118

objectivity 119, 213, 214, 224, 233, 245, 264
obligatory interdependence 193
officials 211–236
online bookstores 247
openness 101–102, 118
   see also transparency
optimism 73, 77, 78, 79, 88
   see also moralistic trust
organizational level 117–138, 148–149
   see also institutional level
origin labels 169
outcome feedback 263
outcome uncertainty 152
out-groups 52, 55, 56, 62–64, 83, 192
   see also in-groups; minority groups
output feedback 249

paranoia 200–201
participation 136, 146–147, 150, 155
particularized trust 75–76, 77, 79–80
past performance *see* performance
Pay Pal 191
PCAs *see* principal components analyses
perceptions
   competence 119
   danger 85–88
   discrimination ability 105–111
   expert versus public 145–146
   knowledge 213, 214, 231
   TCC model 1–49
performance information
   confidence and trust 11, 12, 33–34
   crisis communication 268, 282
   IDT 99–100, 101, 102
   perceptions 7, 8, 9, 27, 28
   referent of judgement 21
   wetlands management 211, 222
periodic surveys 13, 14
peripheral processing 189
permeability 137–138
personal importance 113
pessimism 78, 79–80, 83
Pew survey 85–88
planning board officials 217, 219–232
Plan of Attack 197
policy level 96, 148
positive deviants 61

positive events 2
positive role relations 63–64
precautionary approaches 112
predictions 31
principal axis factoring 227, 233
principal components analyses (PCAs) 130
prisoner's dilemma 53
privacy issues 84–85
process feedback 241–264
processing, information 32–33, 188–190, 199–201
prototypicality 55, 56, 59, 60, 64
psychological processes 192–198
public goods dilemma 54
purchasing decisions 160

radar operation 98
radioactive waste 129
random process feedback 257, 259, 260, 261
ratings
   TCC model 20, 23, 24, 25
   wetlands management 212, 214–217, 218–219, 224–225, 226–228, 232–236
rationality 148
recategorization 63, 64
reciprocity 16, 77, 78
recommendations 247–248, 250–251, 256
reconciliation of ideas 127
referent of judgement 21–22, 23, 36–39
regulatory knowledge 224, 225
relational assurances 190–192, 201, 203–204
relational trust 162
relative-importance ratings 216, 233, 235–236
   see also importance
reliance 117–138, 128–134, 187–204
   see also dependency
resilient trust 11, 13
response bias 99, 100, 101, 102, 103–104, 105–113
responsibility 244, 283
retrospective action 126
role experience 224, 225, 232, 233

role relations 63–64
route planners 250, 251–264
rule-based processing 32
ruminative mode of social interaction 200
Russians 81–82

safety 85–88, 161, 191, 280, 281
  see also health and safety
salespeople 241
saliency 55–56
salient values similarity (SVS)
  context-specific characteristics 201–202
  forest management 195, 196
  group level 192–194
  new information 197
  reliance and scepticism 135, 138
  wetlands management 222
  see also shared values; similarity; social trust
salmon consumption 165, 166
sanctions 15–16, 30
Sara Lee Corporation 279–282, 283
satisfaction 247
scientific level
  credibility 151
  governance of 118
  mobile phones 105, 106, 107, 108
  smallpox 179–180
  societal trust 144–145, 147–148
SDT see signal detection theory
search engines 262
selectivity 246
self-assessed knowledge 198–199, 202–203, 217
self-categorization 56, 58, 59
self-confidence 244
self-construal 62, 64
self-disclosure 51
self-enhancement 57
self-extension 56
self-reference 200
self-reported knowledge 214, 224, 225–226, 231, 233, 235
self-selection 13, 14
sentiments 126–128
shared group membership 16
shared values

food crisis 163, 164–165, 168–169
forest management 195
health and safety 125, 135–136
lack of information 211
moralistic trust 75
TCC model 6–7, 11, 18, 34
wetlands management 217, 222
see also salient values similarity; similarity; social trust
Shell Oil 269–273, 283
signal detection theory (SDT) 98–103
similarity 58, 246, 247
  crisis communication 282
  social trust 272
  wetlands management 217, 222, 229, 231, 232–233
  see also salient values similarity; shared values; social trust
Sjöberg, L. 18, 19
Slovic, P. 1–2, 18–19
smallpox vaccination 173–185
social level
  amplification of risk 95, 106, 160
  attraction 57, 59, 60
  categorization 55–56, 58
  crisis communication 278
  dependency 51
  desirability 7, 215, 216
    see also morality
  differentiation 193
  dilemmas 15–16, 54
  emotions 188–189
  evaluation 57
  identity 15, 16, 17, 51–64, 192–193
  judgements 95–114
  mistrust 83
  reliance 187–204
  representations 192
  scientific independence 147–148
social psychological level 52–57, 192–198
social trust 4, 30, 268, 271–272
  see also inter-group trust; shared values; system trust
soft drinks 273–279
somatization of fear 277–278
specification 6
specific performance criterion 4

staked credits 253–254, 255, 256, 257, 259, 260, 263
strategic trust 73, 74–75, 76, 77–79, 80, 82, 85
  see also knowledge
structural trust 162
  see also system trust
suburban residents 86, 87
superordinate groups 63, 64
SVS *see* salient values similarity
system trust 162, 241–264
  see also social trust; structural trust

tangibility of targets 21, 23, 36–39
targets of judgement 20–21, 22–23, 23–24, 36–39
TCC *see* trust, confidence and cooperation model
technology
  acceptance 145–146, 147
  emerging 14, 151–152
  hazards management 18
  institutions survey 128–134
  internet 84–85
  mobile phones 102–111
tentative knowledge 122–123
terrorism 173, 176, 177, 181, 184
toxicological science 144–145
transparency 101–102, 146, 147, 149–159
true positives and negatives 98–99
trust, confidence and cooperation (TCC) model 1–49

UK *see* United Kingdom
unification 6
United Kingdom (UK) 118, 121–128, 212
United States of America (US)
  elections 194
  immigrants 83, 84
  neighbourhood violence 85–88

relational assurances 191, 204
smallpox vaccination 173–185
wetlands management 212, 219–236
universalization of assessments 145
US *see* United States of America
utility theory *see* game theory

vaccination 173–185
validity 33
value information *see* morality information
values
  absolute 273
  consensual 97, 119
  consistency 195–196
  crisis communication 271–272, 281–282
  hazard-related 26, 27
  similarity 58, 246, 247
  societal trust 151–152
  see also salient values similarity; shared values
variability 152–154, 155
verbal protocols 125, 174
  see also focus groups
vested interests 130
vicarious learning 248
vigilance 112–113
violence 86–88
voluntary activities 73, 89
vulnerability 16, 17, 30, 34, 51

waste disposal 129, 269–273
watchdog organizations 132, 133, 134
wetlands management 211–236
wildfire management 199
wine tasting 251
workplace safety 121–128
worldviews 77–78, 79–80

zoning board officials 219–232